Soil Analysis: Recent Trends and Applications

Amitava Rakshit • Subhadip Ghosh •
Somsubhra Chakraborty •
Varughese Philip • Avishek Datta
Editors

Soil Analysis: Recent Trends and Applications

Springer

Editors
Amitava Rakshit
Dept. of Soil Science & Agri. Chemistry
Institute of Agricultural Science, BHU
Varanasi, Uttar Pradesh, India

Subhadip Ghosh
National Parks Board
Centre for Urban Greenery and Ecology
Singapore, Singapur, Singapore

Somsubhra Chakraborty
Agricultural & Food Engineering Dept.
Indian Institute of Technology Kharagpur
Kharagpur, West Bengal, India

Varughese Philip
Soil Management Lab, Plant Science
and Health, Horticulture and Community
Gardening Division
National Parks Board
Singapore, Singapore

Avishek Datta
Asian Institute of Technology
Dept of Food, Agriculture & Bioresources
Klong Luang, Pathum Thani, Thailand

ISBN 978-981-15-2038-9 ISBN 978-981-15-2039-6 (eBook)
https://doi.org/10.1007/978-981-15-2039-6

This Springer imprint is published by the registered company Springer Nature Singapore Pte Ltd.
The registered company address is: 152 Beach Road, #21-01/04 Gateway East, Singapore 189721, Singapore

Gratefully and affectionately dedicated to our wives and our children who encouraged us to fly towards our dream

Preface

Soil science as profession can address the challenges of the recently adopted UN Sustainable Development Goals in the most befitting manner. The sustainability of human societies depends on the wise use of natural resources. Soils contribute to basic human needs like food, clean water, and clean air, and are a major carrier for biodiversity. It has been repeatedly shown that a proper soil analysis is important for several aspects: to optimize crop production, to guard the environment from contamination by runoff and leaching of excess fertilizers, to assist in the diagnosis of plant culture problems, to improve the nutritional balance of the growing media, and to save money and conserve energy. Updating the analysis is necessary for achieving more precision and accuracy in the estimated parameters. The book *Soil Analysis: Recent Trends and Applications* is going to provide the synopsis of the analytical procedures used for soil analysis. The book will encompass the common physical, chemical, and biological analytical methods used for agriculture and horticulture. The content will help a range of different users even with limited laboratory instrumentation facilities. It will assist students, teachers, soil scientists, and laboratory technicians to choose appropriate methods to imply for soil analysis. This compiled book will have experienced authors from various institutions and laboratories worldwide. This is going to be a pioneer book on the soil analysis as it is the first attempt to combine commonly used soil analytical procedures needed for agriculture and horticulture.

Varanasi, India	Amitava Rakshit
Singapore, Singapore	Subhadip Ghosh
Kharagpur, India	Somsubhra Chakraborty
Singapore, Singapore	Varughese Philip
Klong Luang, Thailand	Avishek Datta

Acknowledgements

An edited book of this outlay does not become a reality without inputs of several enthusiastic souls. Our heartfelt gratitude to all the contributors who helped us in our venture to bring this book to light. We are grateful to Prof. Panjab Singh, Prof. D. K. Das, and Prof. H. B. Singh for their steady encouragement and support. We hope that this book will be a help to soil scientists, agronomists, agricultural engineers, plant physiologists and other scientists, and especially to those in the field who are actively concerned with improving their knowledge on the wide array of tailor-made soil test methodologies and its latest advancements.

Finally, the production team members deserve special appreciation for guiding us through the process of publishing a new work. Last but not the least, we should thank our family, immediate and extended who always encouraged us to continue the massive task. In spite of the best efforts, it is possible that some errors may have crept into the compilation. We shall be highly obligated to receive constructive comments and suggestions from the readers for further improvement in the future editions.

Contents

1 **Soil Analysis: A Relook and Way Forward** 1
Dibyendu Mukhopadhyay

2 **Application of Statistical Techniques in Soil Research** 19
Dibyendu Deb, Shovik Deb, and Ashok Choudhury

3 **Monitoring and Impact Assessment of Climate Change on Agriculture Using Advanced Research Techniques** 33
Dibyendu Chatterjee, Saurav Saha, Barsita Swain, Debashis Chakraborty, Amaresh Kumar Nayak, Himanshu Pathak, and Madan Pal Singh

4 **Advancement in Soil Testing with New Age Sensors: Indian Perspective** 55
R. K. Swetha, Swagata Mukhopadhyay, and Somsubhra Chakraborty

5 **Isotopes and Tracer Techniques for Soil Analysis** 69
Manoj Shrivastava, P. C. Srivastava, and S. F. D'Souza

6 **Protocols for Determination and Evaluation of Organic Carbon Pools in Soils Developed Under Contrasting Pedogenic Processes and Subjected to Varying Management Situations** 87
Rakesh S, Deepranjan Sarkar, Shikha, Ardith Sankar, Abhas Kumar Sinha, Prabir Mukhopadhyay, and Amitava Rakshit

7 **Analytical Strategies for Arsenic Estimation** 107
Arghya Chattopadhyay, Amitava Rakshit, A. P. Singh, Deepranjan Sarkar, Nitesh Kumar Singh, and Aditi Chourasia

8 **Approach to Study Clay-Organic Complexes** 119
Nintu Mandal, Arnab Bhowmik, Pritam Ganguly, and Samar Chandra Datta

9 **Recent Trends in Soil Salinity Appraisal and Management** 143
Nirmalendu Basak, Arijit Barman, Parul Sundha, and Arvind Kumar Rai

10 Modern Sample Preparation Techniques for Pesticide Residues Analysis in Soil 163
Prithusayak Mondal

11 Characterization of Nanomaterials Using Different Techniques . . . 187
Ritika Joshi, Ashish Khandelwal, Manoj Shrivastava, and S. D. Singh

12 Soil Health Assessment 199
Kavya Krishnan, Robert Schindelbeck, K. S. M. Kurtz, and Harold van Es

13 Soil Health Indicators: Methods and Applications 221
M. Raghavendra, M. P. Sharma, A. Ramesh, A. Richa, S. D. Billore, and R. K. Verma

14 Indexing Methods of Soil Quality in Agro-Ecosystems: An Overview of Indian Soils and Beyond 255
Debarati Bhaduri, Tapan Jyoti Purakayastha, Pragati Pramanik, Ajoy Saha, Bibhash Chandra Verma, Mohammad Shahid, and Debasish Saha

15 Nanobiosensors: Recent Developments in Soil Health Assessment 285
Nintu Mandal, Samrat Adhikary, and Rajiv Rakshit

16 Forensic Pedology: From Soil Trace Evidence to Courtroom 305
Tan Wei Ling Stella, Sanjay Swarup, Mandi See Suet Ning, Nur Qasrina Bte Iskandar Lim, Samantha Phua Mun Lin, Terry Tan Boon Jay, and Subhadip Ghosh

17 Harnessing Soil Microbiomes for Creating Healthy and Functional Urban Landscapes 325
Aditya Bandla, Shruti Pavagadhi, and Sanjay Swarup

About the Editors

Amitava Rakshit an IIT, Kharagpur alumnus is presently the faculty member in the Department of Soil Science and Agricultural Chemistry at Institute of Agricultural Sciences, Banaras Hindu University (IAS, BHU). Dr. Rakshit worked in the Department of Agriculture, Government of West Bengal in administration and extension roles. He has visited Scandinavia, Europe and Africa pertaining to his research work and presentation. He was awarded with TWAS Nxt Fellow (Italy), Biovision Nxt Fellow (France), Darwin Now Bursary (British Council), Young achiever award and Best Teacher's Award at UG and PG level. He is serving as review college member of British Ecological Society, London since 2011. He is an active member of Global Forum on Food Security and Nutrition of FAO, Rome and Commission on Ecosystem Management of International Union for Conservation of Nature. He has published 70 research papers, 35 book chapters, 28 popular articles, 1 manual and authored 20 books.

Subhadip Ghosh is a Senior Researcher (Urban Soils) with Centre for Urban Greenery and Ecology (Research), National Parks Board, Singapore. He holds his Bachelor of Science in Agriculture with Honours from BCKV, India and Master's in Soil Science from PAU, Punjab, India with national scholarship. He completed his Ph.D. from the University of New England, Australia and postdoctoral research with Primary Industries and Innovation Centre, New South Wales (NSW), Australia. Dr. Ghosh has published more than 60 scientific articles. He is an Adjunct Lecturer at the University of New England, NSW, Australia. His major fields of research interest are soil quality,

GHG emission, carbon sequestration and land-use change, biochar studies, environmental soil science and climate change, nutrient dynamics, and urban and peri-urban soils.

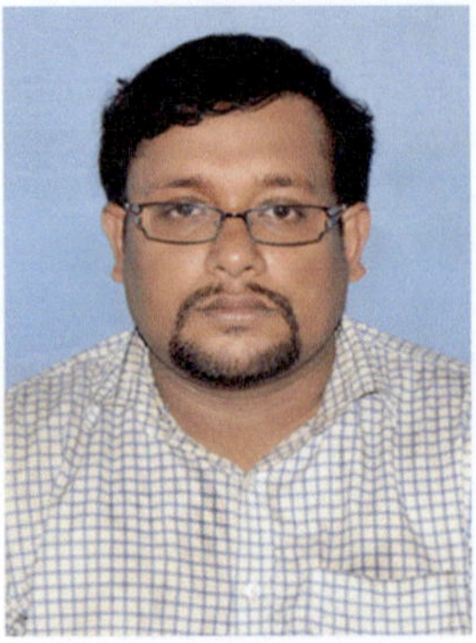

Somsubhra Chakraborty a Louisiana State University (USA) alumnus is presently the faculty member in the Agricultural and Food Engineering Department at Indian Institute of Technology Kharagpur, India. Dr. Chakraborty has formerly served as the faculty member of Vivekananda University and Uttar Banga Krishi Viswavidyalaya, West Bengal. He visited USA, Kenya, Australia, China, Thailand during his academic career. He got the Australia Awards Fellowship (2016). He is currently serving as the editorial board member of Geoderma (Elsevier). He has published more than 60 research papers, several book chapters and 1 book.

Varughese Philip the Deputy Director (Soil Management) of the Agri-Food & Veterinary Authority of Singapore. A trained agronomist, he holds a PhD from the Indian Institute of Technology, Kharagpur. He is involved in laboratory diagnostics, advisories on soils and standards setting. He had previously worked on various aspects of agronomy and soils of natural rubber (*Hevea brasiliensis*). As an adjunct lecturer at the Ngee Ann Polytechnic Singapore, he taught soil science at the Horticulture and Landscape Management Diploma programme. He has authored several peer reviewed publications and chapters in books. He co-chaired preparation of the ASEAN guidelines on Soil and Nutrient Management (2017) and is currently a member of the Technical committee on Solid Waste, SPRING Singapore.

Avishek Datta is an Associate Professor of Agricultural Systems at the Department of Food, Agriculture and Bioresources, Asian Institute of Technology, Thailand. Dr. Datta has more than 11 years of international experience in research, consultancy, and capacity building in the areas of sustainable crop production, crop water management, climate change and food security, climate-smart agriculture, crop stress tolerance, conservation agriculture, organic farming, and integrated nutrient and pest management. A total of

7 Doctoral and 45 Master students have been graduated under his direct supervision. He has published 8 book chapters and more than 90 papers in international peer-reviewed journals. He has implemented several research, development, and outreach projects in various countries including Australia, Bangladesh, Cambodia, Laos, Myanmar, Nepal, Thailand, Timor-Leste, USA, and Vietnam funded by many international donor agencies.

Soil Analysis: A Relook and Way Forward

1

Dibyendu Mukhopadhyay

Abstract

It requires to get familiar and accustomed with the changing scenario of the scientific tools and technology in the ecosystems. The cutting-edge technology in agricultural and allied sectors extends its fecundity to meet demands for food, energy and water. The natural resource conservation and its management through the minimum external input has become a challenge to the scientific community. In order to face the troubleshoot, a realistic mechanism of transport, uptake and accumulation of nutrients and toxicants in soil–water–plant continuum through the improved analytical tools such as high-resolution mass spectrometry (HRMS) with liquid chromatography (LC) and LC/MS/MS needs to be understood. In order to create environment in the laboratory same as that in the field, advance simulation will be required to facilitate a wider range of temperature, water potential and light. The temperature fluctuation, humid and sub-humid environment and or near freezing condition when the biological activity tends to minimum can be created through advanced technological tools in the laboratory. Accurate and timely analysis of samples (soil/plant/water) will help judicious recommendation of fertilizers to crops grown at different agroclimatic region, where accuracy and precision of the produced analytical data will signify the validation of the experiment, thus minimizing the error. In this regard, an accurate soil sampling is required to establish a correct database, which is possible through repeated internal checks in the analysis. Proper identification of the sources (point/diffuse) is required before interpretation of the data set. However, soil being the major source of nutrients for crops can also provide support to the plant growth. Hence, soil health and its maintenance are the key issues to sustain crop productivity, which is assessed by the quality indicators and sustenance of the crops grown on them. However, the policy may be framed on the platform based

D. Mukhopadhyay (✉)
Department of Soil Science and Agricultural Chemistry, Uttar Banga Krishi Viswavidyalaya, Pundibari, Cooch Behar, West Bengal, India

A. Rakshit et al. (eds.), *Soil Analysis: Recent Trends and Applications*,
https://doi.org/10.1007/978-981-15-2039-6_1

on "Strategic and fundamental research" for developing innovative models in agricultural systems. The chapter deals mainly with the conventional methods *vis-a-vis* the modern approach of soil analysis, so as to maintain the balance of the data set for interpretation and validation of the results.

Keywords
Soil analysis · Available nutrients · Fertility map · Analytical instruments · Labile pools · Soil reaction

1.1 Introduction

The sustainable agriculture is facing a problem for not having the fertilizer and water to access at ease under the stress-based ecosystems. It requires proper management to get sufficient food for the ever-growing population from the available soil and water resources by employing technological innovations in agricultural science (Augustine and Lane 2014). It is indeed the basic and strategic research which can play a pivotal role in propelling agricultural productions. Hence, to meet demands for food, energy and water, the focus will be to utilize cutting-edge molecular and microscopic tools in agricultural and allied sectors. Among the external inputs, biosolids, wastewater effluents, manures and fertilizers are applied to lands for improving crop production. Therefore, it requires to find out the mechanism of transport, uptake and accumulation of nutrients and toxicants in soil–water–plant continuum through the improved analytical tools such as high-resolution mass spectrometry (HRMS) with liquid chromatography (LC) and LC/MS/MS (Richardson and Ternes 2014). Besides, the nature of the colloidal contaminants with its own reaction mechanism can govern transport and bioavailability in the environment (Praetorius et al. 2014). The engineered nanoparticles, pathogenic microorganisms (virus, bacteria and protozoa) and colloid-associated phosphorus, radionuclides, heavy metals or organic toxins are some of the colloidal contaminants, and the analysis of which can be done by the advanced and accurate analytical tools such as single particle inductively coupled plasma-mass spectrometry (SP-ICP-MS) (Mitrano et al. 2012), hyperspectral spectroscopy, microbial pathogen sensor and droplet digital PCR.

In order to create environment in the laboratory same as that in the field, advance simulation will be required to facilitate a wider range of temperature regime, water potential and light. The temperature variation, humid and sub-humid environment condition or sometimes like near freezing condition when the biological activity tends to minimum can be created through advanced technological tools in the laboratory. Accurate and timely analysis of plant nutrients will help understanding their supplying capacity from various sources. In this regard, the chemical analysis of soils, plants, water and fertilizers become specific and specialized in terms of facilities and chemicals required for the estimations. In view of this, a common analytical facility can be developed for the soil–plant–water–fertilizer–biofertilizer testing under one roof and thus minimizing the cost of analysis per sample.

1.2 Soil Sampling

The purpose of sampling is important to consider the appropriate methods and procedure for obtaining soil samples required for engineering and agricultural purposes. Besides, fruitful soil sampling is essential for soil fertility evaluation and fertilizer recommendations for crops. Hence, the efficiency of a soil testing depends on the care and skill with which soil samples are collected and prepared to avert the sampling error. It is in the field where the utmost care is to be taken during the sampling before being analysed in the laboratory, where 1–2 g sample will be representative of the 0–20 cm depth of soil for evaluating soil fertility. It requires to have some basic knowledge and understanding before going into any physico-chemical or biological analysis of the samples. Some of the information pertaining to frequently used chemicals (Table 1.1) is required for the analytical purpose of the samples.

1.2.1 Need for Soil Analysis

An accurate soil sampling is required to establish a correct database, which is possible through repeated internal checks in the analysis along with the periodical monitoring. Proper identification of the sources (point/diffuse) is to be made before reporting the data. Specific standard methodology is to be followed to validate the programme through series of trials. The site specific suitability of the method can be established by sample identification, standards, chemicals, pipettes, dispensers, glassware, calibration procedure and equipment.

1.2.2 The Future of Soil Chemistry

The concept of soil chemistry (James 1993) is primarily based on (Fig. 1.1) the land use pattern of an area (domesticated, wild and engineered soils) and their

Table 1.1 Strength of commonly used acids and alkali

Reagent/Chemical	Normality (approx.)	Molarity (approx)	Formula weight	Specific gravity (approx.)
Nitric acid	16.0	16.0	63.0	1.42
Sulphuric acid	35.0	17.5	98.0	1.84
Hydrochloric acid	11.6	11.6	36.5	1.19
Phosphoric acid	45.0	15.0	98.0	1.71
Perchloric acid	10.5	10.5	100.5	1.60
Ammonium hydroxide	15.0	15.0	35.0	0.90

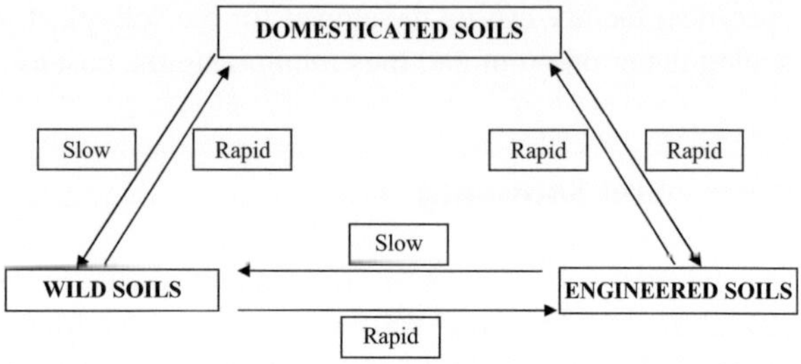

Fig. 1.1 Rate of transformation of wild, domesticated and engineered soils (James 1993)

interconversion rate of change of the state. It is the chemical processes within the soil matrix which governs the release of ions in available pools from rocks and minerals. The future of soil chemistry stands on understanding the mechanisms of the changing reaction kinetics of the substances into products. An acre of soil may be used for cultivation or for sanctuary or for building construction depending upon the requirement of the society. However, the future of soil chemistry relies more on basic and fundamental research work on sorption–desorption characteristics of plant nutrients, the intricacy of the metal-humus complexation and its stability (Mukhopadhyay and Sanyal 2004; Ray et al. 2018), the ligand exchange mechanisms of the adsorbed ions from the soil and the surface charge characteristics of the clay colloids governing the nutrient mobility. In this regard, advanced and innovative technological tools are to be employed in the field of soil chemistry for understanding the role of individual ionic species in sustainable crop production.

1.3 Soil Analysis

Soil being the major source of nutrients for crops can also provide support to the plant growth. Soil health and its maintenance are the key issues to sustain crop productivity. The health of soil is assessed by the quality indicators and stand of the crops grown on them. A scientific assessment is possible through physical, chemical and biological analysis of the soils, such as N, P, K, Ca, Mg and S the macronutrients, while Fe, Zn, Cu, Mo, Mn, B and Cl the micronutrients. It is necessary to assess the capacity of a soil to supply nutrients in order to replenish the remaining amounts of needed plant nutrients (total crop requirement—soil supply). Soil testing is also used in managing environmental issues to reduce non-point source of pollution from agriculture. For example, specific soil test value for the nutrients may be considered as nutrient index. Thus, soil-testing laboratories can play important roles to maximize production potential of crops.

1.3.1 Soil Analysis Processes

A standard procedure has been developed for the determination of available nutrients required for plant growth and nutrition. However, there are modifications on the procedure for precise validation of the methods. It is because of the variability in soils that demands for the closer look on the reproducibility of the data set in the field of mineralogy, weathering, soil formation, soil classification, soil chemistry and fertility or allied sectors. But, recent approach is focussed on to sustain the food security of a country which has been achieved due to advancement in the field such as chemistry, physics, mineralogy, genesis, conservation and plant nutrition. Besides, constant research is going on understanding the global carbon stock and ways to reduce carbon dioxide levels in the atmosphere. In this regard, the carbon sequestration in soils and minimizing the use of fossil fuel are to be framed. In order to have a sustained research in soil science, it requires to have a closer look on the modern research approach to meet the demand for food for the increasing population on earth. The platform may be based on "strategic research", "research for development", private-partnership model and balancing acts.

1.3.2 Analytical Methods

The following parameters are generally determined in a soil-testing laboratory:

1. Soil texture
2. Soil structure
3. Cation exchange capacity (CEC)
4. Soil moisture
5. Water holding capacity (WHC)
6. pH
7. Electrical conductivity (EC)
8. Lime requirement (LR)
9. Gypsum requirement (GR)
10. Oxidizable organic carbon
11. Total N
12. Mineralizable N
13. Inorganic N
14. Available P
15. Available K
16. Available S
17. Calcium
18. Calcium plus magnesium
19. Micronutrients—available Zn, Cu, Fe, Mn, B and Mo
20. Heavy metals (Pb, Cd, Cr, Ni, As, Hg, etc.)

1.3.2.1 Soil Texture

Soil texture (or particle size distribution) is the basic and a stable soil property that influences the physical and chemical characteristics of the soil.

Table 1.2 Classification of soil separates according to the size

USDA

Clay (mm)	Silt					Sand (mm)	Gravel (mm)
	Very fine (mm)	Fine (mm)	Medium (mm)	Coarse (mm)	Very coarse (mm)		
	0.002	**0.05**	**0.1**	**0.25**	**0.50**	**1.00**	**2.00**

ISSS/IUSS

Clay	Silt	Sand		Gravel
		Fine	Coarse	
	0.002 mm	**0.02 mm**	**0.2 mm**	**2.0 mm**

The two methods in general use for estimating particle size or soil texture are:

1. International pipette method;
2. Bouyoucos hydrometer method.

The soil scientists and agronomists mostly preferred two systems of classification, namely U.S. Department of Agriculture (USDA) and the International Society of Soil Science (ISSS) which has been renamed as International Union of Soil Science (IUSS) (Table 1.2).

1.3.2.2 Soil Structure

Soil structure is defined as the relative arrangement of the soil particles (sand, silt and clay). With regard to structure, the primary and secondary (aggregate) particles are the active fractions to make-up the stability of the soil. The aggregation requires cementation or binding together the floccules with different forces. Aggregation = Flocculation + Cementation. The size, shape and character of the soil structure may vary (e.g. cube like, prism like, etc.). On the basis of size, the soil structure is classified as:

1. Very coarse: >10 mm
2. Coarse: 5–10 mm
3. Medium: 2–5 mm
4. Fine: 1–2 mm
5. Very fine: <1 mm

Depending on the stability of the aggregate and the ease of separation, the structure is characterized as:

1. Poorly developed
2. Weakly developed
3. Moderately developed
4. Well developed
5. Highly developed

Dry Aggregate Analysis
The size distribution of dry clods is measured by dry sieving analysis performed on an air-dried bulk soil sample, either manually or with the help of a rotary sieve shaker (Gupta and GhilDyal 1998).

Wet Aggregate Analysis
For wet aggregate analysis, the wet sieving technique is followed to determine the size of the water stable aggregates in arable soil which was conceived by Yoder (1936) and subsequently by van Bavel (1953).

$$\mathrm{MWD} = \sum\nolimits_{i=1}^{n} \mathrm{XiWi}$$

$$\mathrm{GMD} = \exp\left[\sum\nolimits_{i=1}^{n} \log\mathrm{XiWi} / \sum\nolimits_{i=1}^{n} \mathrm{Wi}\right]$$

where MWD = mean weight diameter, GMD = geometric mean diameter, Xi = mean diameter of each size fraction and Wi = proportion of the total sample weight in the corresponding size fractions.

1.3.2.3 Cation Exchange Capacity (CEC)

The total number of exchangeable cations a soil can hold is called its cation exchange capacity (CEC). The higher the CEC, the more cations it can retain. The CEC can be expressed in terms of milli-equivalents per 100 g of soil (me/100 g) or in centimoles of positive charge per kilogram of soil [$\mathrm{cmol(p^+)kg^{-1}}$], which is numerically equal to me/100 g. The CEC of the soil depends on the kind and amount of clay and organic matter present. The CEC is determined using a solution buffered to maintain a certain pH (pH 7.0 or 8.2) or sometimes at unbuffered solution at the actual soil pH. The unbuffered method measures only the effective cation exchange capacity of the soil. Effective cation exchange capacity (ECEC) is another method of measuring the CEC by considering extractable Al.

$$\mathrm{ECEC} = \Sigma 1\mathrm{M\ NH_4} - \text{acetate extractable bases} + 1\mathrm{M\ KCl} - \text{extractable Al}$$

1.3.2.4 Soil Moisture

The gravimetric method of moisture estimation is most widely used where the soil sample is placed in an oven at 105 °C for 24 h and dried to a constant weight. The difference in weight is considered to be the water present in the soil sample.

$$\text{Water content} = (\mathrm{W_1} - \mathrm{W_2})/\mathrm{W_2} \times 100$$

Where, W_1 = initial weight of soil and W_2 = oven dry weight of soil.

Water Holding Capacity (WHC)

The WHC of soil is the amount of water held in the soil after the excess gravitational water is drained away and after the rate of downward movement of water has practically ceased (Veihmeyer and Hendrickson 1931). It is observed that the stage of field capacity is attained in the field after 48–72 h of saturation which is considered as the upper limit of plant-available soil moisture.

1.3.2.5 Soil pH

The soil pH is the negative logarithm of the active hydrogen ion (H^+) concentration (gmol/L) in the soil solution, where, 10 g soil is weighed in a 100 ml beaker and 25 ml of distilled water is added to maintain the ratio (soil: water :: 1:2.5). The pH of the soil-water suspension is measured by a pH meter after stirring the suspension with a glass rod for 25 minutes. It is the measure of soil alkalinity, acidity or neutrality. They are the important parameters of soils as soil pH has a considerable influence on the availability of nutrients to crops. It also affects microbial population in soils. Most nutrient elements are available in the pH range of 5.5–6.5. Acid soils need to be limed before they can be put to normal agricultural production for which the determination of lime requirement (LR) is essential. In some cases, lime potential (LP) is also considered to measure the actual pH (0.01 M $CaCl_2$) of the given soil. From the Schofield ratio law at equilibrium

$$[H^+]/\left[\sqrt{}Ca^{+2}\right] = \text{constant}$$

$$LP = pH - 1/2\ pCa$$

Similarly, the alkali soils (pH 8.5) need to be reclaimed with gypsum in order to remove the Na^+ from the matrix.

1.3.2.6 Lime Requirement (LR)

Crop yields are normally high in soils with pH values ranging from 6.0 to 7.5. Lime is added to raise the pH of acid soils and the amount of lime required to raise the pH to an optimal level is called lime requirement. Various methods are available for determining the lime requirement. Liming materials such as $CaCO_3$, CaO, Ca, Mg $(CO_3)_2$ and $Ca(OH)_2$ are normally used as the liming materials for the reclamation of the acid soil.

$$\text{Colloid}]^{H+}{}_{Al}{}^{+3} + 2Ca^{+2} + HCO_3 \leftrightarrow \text{Colloid}]2Ca^{+2} + Al(OH)_3 + H_2O + CO_2$$

The most widely used method for determining lime requirement (LR) is the Shoemaker, Mclean and Pratt (SMP) single buffer method where a ready reckoner table is used.

1.3.2.7 Electrical Conductivity (ECs)

Electrical conductivity is a measure of the ionic transport in a solution between the anode and cathode. This means, EC is normally considered to be a measurement of the dissolved salts in a solution. The soil:water as 1:2.5, i.e., 10 g soil with 25 mL distilled water is used to determine the electrical conductivity (EC) of the soil. The EC is the index of salinity of the soil and often measured by a standard EC meter.

1.3.2.8 Gypsum Requirement (GR)

Sodium hazards in soil sometimes become a big issue for agriculture. Hence, the replacement of sodium from the exchange complex by the addition of estimated gypsum ($CaSO_4$, $2H_2O$) is important (Schoonover 1952). The release of sodium from the exchanged complex is replaced by the Ca^{+2} by leaching of soils. The quantity of Ca^{+2} reduced is equivalent to the Ca^{+2} exchanged with Na^{+1}. It is equivalent to the gypsum requirement of the soil when "Ca" is expressed as $CaSO_4$.

$$\text{Clay}]\ 2\text{Na}^{+1} + \text{CaSO}_4 \leftrightarrow \text{clay}]\ \text{Ca}^{+2} + \text{Na}_2\text{SO}_4 \downarrow$$

1.3.2.9 Organic Carbon/Organic Matter

Among the various methods of estimating organic matter (OM) in soil, loss of weight on ignition can be used as a direct measure of the OM content in the soil. It can also be expressed as the content of organic C in the soil. It is generally assumed that OM contains about 58% organic C. The organic matter/organic C can also be estimated by volumetric and colorimetric methods. Potassium dichromate ($K_2Cr_2O_7$) is involved as an oxidizing agent in the estimation of oxidizable organic carbon. Soil organic matter (SOM) content in a soil can be used as an index of N availability (potential of a soil to supply N to plants). Different methods are:

1. Loss of weight on ignition
2. Volumetric method (Walkley and Black 1934)
3. Colorimetric method (Datta et al. 1962)

1.3.2.10 Total Nitrogen

In total N, the sum of the forms of inorganic N, such as NH_4^{+1}, NO_3^- and NH_2 (urea) and the organic N compounds such as proteins, amino acids and other derivatives is considered. Based on the available form of N present in a sample, a particular method is adopted for determining the total N value. For the conversion of organic N materials into simple inorganic ammoniacal form, sulphuric acid is used for reducing nitrates into ammoniacal form, the modified Kjeldahl method is adopted with the use of salicylic acid or Devarda's alloy. On completion of the digestion, all organic and inorganic salts are converted into ammonium form, which is distilled and estimated using standard acid.

1.3.2.11 Mineralizable Nitrogen

The mineralizable N in soil is measured as an index of plant-available N content (Subbiah and Asija 1956). The mineralizable N is estimated using alkaline $KMnO_4$, when organic matter is oxidized and hydrolysed to liberate ammonia which is condensed and absorbed in boric acid and is titrated against standard acid. A uniform time and heating temperature is maintained for the best results. The use of glass beads checks bumping, while liquid paraffin checks frothing during heating (as recommended in total N estimation by the Kjeldahl method).

1.3.2.12 Inorganic N–NO_3^- and NH_4^+

Inorganic N in soil is present as NO_3^-–N and NH_4^+–N. Because of the less detectable amounts of nitrite in soil, its determination is normally restricted except in neutral to alkali soils following the application of NH_4 or NH_4-forming fertilizers. There are various methods for the determination of NO_3–N and NH_4^+–N than the methods of extraction (Keeney and Nelson 1982) ranging from specific ion electrode to colorimetric techniques, micro-diffusion, steam distillation and flow injection analysis. Steam distillation is often preferred while using ^{15}N, although, for routine analysis, the phenoldisulphonic acid method for NO_3 and the indophenol blue method for NH_4 estimation are recommended.

1.3.2.13 Available Phosphorus

The two methods most commonly used for determining the available P in soils are:

(1) Bray's method no. 1 for acid soils and (2) Olsen's method for neutral and alkali soils. In these methods, specific coloured compounds are formed with the addition of appropriate reagents in the solution, the intensity of which is proportionate to the concentration of the element being estimated. The colour intensity is measured spectrophotometrically. In spectrophotometric analysis, light of definite wavelength (not exceeding 0.1–1.0 nm in bandwidth) extending to the ultraviolet region of the spectrum constitutes the light source. The photoelectric cells in the spectrophotometer measure the light transmitted by the solution. The plot of absorbance against concentration of the coloured solution will produce the standard curve (Fig. 1.2) from which the concentration of the unknown solution can be measured.

$$A = 2 - \log T;$$

where A = absorbance and T = % transmittance.

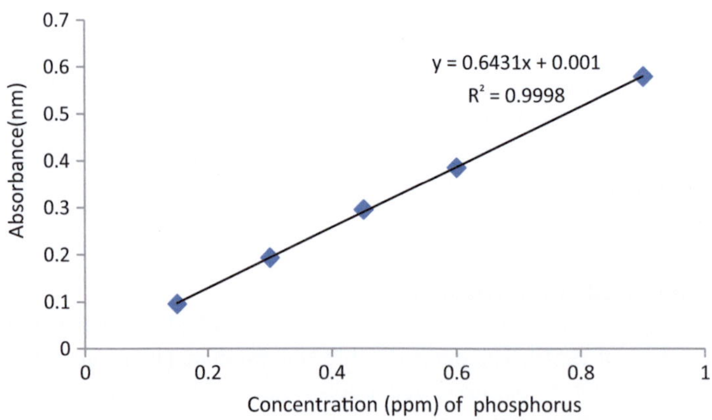

Fig. 1.2 Standard curve of the solution

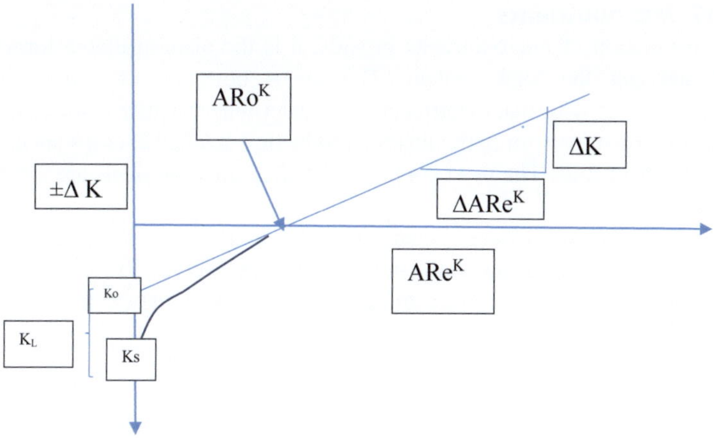

Fig. 1.3 *Q/I* curve for potassium estimation. $PBC^K = \Delta K/ARe^K$, $\pm\Delta K$ = gain or loss of exchangeable K, PBC^K = potential buffering capacity of potassium, K_L = labile or exchangeable K, K_s = the number of specific sites for K, K_o = non-specific sites, ARe^K = activity ratio of K at equilibrium solution = ${}^{a}{}_{K}/({}^{a}{}_{Ca} + {}^{a}{}_{Mg})^{1/2}$

1.3.2.14 Available Potassium

Potassium present in the soil is extracted with neutral (pH 7.0) normal (1N) ammonium acetate solution. This is considered as plant-available K in the soils. It is estimated with the help of a flame photometer (Toth and Prince 1949). The quantity (*Q*)–intensity (*I*) relationship is often used to predict the availability of K to plants. The Q/I relation is represented by the curve in which the quantity factor (ΔK) is obtained as the difference between the initial and equilibrated solution K which is plotted in *Y* axis and the corresponding intensity factor in the *X* axis (Fig. 1.3).

1.3.2.15 Available Sulphur

The available S in soils occurs mainly as adsorbed SO_4 ions which may be replaced by phosphate ions. The extraction is carried out using $CaCl_2$ solution, although the former is considered better for efficient replacement of SO_4 ions. The SO_4 in the extract can be estimated turbidimetrically using a spectrophotometer. The barium sulphate precipitation method of estimating soil-available S (Singh et al. 1999) is also used for analysis.

1.3.2.16 Exchangeable Calcium and Magnesium

Exchangeable cations are usually determined by the neutral normal ammonium acetate solution of soil. Extraction is carried out by shaking the soil–extractant mixture, followed by filtration or centrifugation. Calcium and Mg are determined either by the EDTA-titration method or using an atomic absorption spectrophotometer (AAS) after the removal of ammonium acetate and organic substances. The EDTA-titration method developed by Cheng and Bray (1951) is preferred on account of its accuracy, simplicity and speed.

1.3.2.17 Micronutrients

For the estimation of micronutrients in soils, it is the plant-available form that is critical and not the total content. The major objective of soil testing for micronutrients, as with macronutrients, is to determine whether a soil can supply adequate micronutrients for optimum crop production or whether crops are grown on nutrient deficient soils. The most commonly studied micronutrients are Zn, Cu, Fe, Mn, B and Mo.

Diethylenetriaminepentaacetic acid (DTPA) is a common (universal) extractant which is widely used for the extraction of elements such as Zn, Cu, Fe and Mn (Lindsay and Norvell 1978). The estimation of elements in the extract is done with the help of an AAS.

Available Boron

The available boron (B) is determined by the hot water extraction of soil as developed by Berger and Truog (1939). Boron can be analysed by colorimetric methods by carmine, azomethine-H and most recently by inductively coupled plasma (ICP) and atomic emission spectrometry.

Available Molybdenum

Molybdenum is present in very small amounts in igneous and sedimentary rocks. The Mo is extracted by ammonium acetate and/or ammonium oxalate and can be estimated by both the AAS and colorimetric methods, with preference for the latter owing to the formation of oxide in the flame in the case of estimation by AAS. Sometimes, ammonium oxalate is used as a better extractant for Mo.

Taking into consideration all the nutritional approach for soil testing, the objective is to advocate actual fertilizer dose to the crop to get maximum yield (95–100%), where critical soil test level is obtained (Pal 2016).

1.3.2.18 Heavy Metals

Preparation of the standard curve after the calibration of the instrument using calibration blank is used for determining the concentrations of heavy metals (Cd, Cr, Pb, As, Hg, Co, Ni, etc.) in atomic absorption spectrophotometer. For the determination of arsenic and mercury, the hydride generator unit is required as an attachment to the AAS. Final concentrations of the metals in the soil samples are calculated using the following formula:

$$\text{Concentration (mg/kg)} = [\text{Concentration (mg/L)} \times V]/W$$

where $V=$ final volume (50 mL) of solution and $M=$ initial weight (0.5 g) of sample measured. Sometimes, for the multinutrient analysis, the auto-analyser is used in the laboratory. Besides, for separation of compounds from a mixture and to identify, quantify or purify the individual components the instrument like HPLC (High Performance Liquid Chromatography) becomes useful to the end user. Again, for pesticide residual action in a compound or to detect and measure contaminants from admixtures, the GC-MS is extensively used for the analysis.

1.3.2.19 Proximal/Hyperspectral Methods of Soil Analysis

Hyperspectral imaging (HIS) has become an effective tool in civil, environmental and military applications over the last three decades or so. The sensor-based technologies are spreading over the surfaces of earth in respect to the spatial, spectral and temporal resolutions. The hyperspectral imaging is now being used in remote sensing applications for identification and differentiation of varying spectral signatures. The ground-based hyperspectral imaging has improved its momentum in the research on electronic imaging for food inspection, forensic science, medical surgery and diagnosis and military applications. The remote sensing and hyperspectral applications on precision agriculture and water resource management have been fruitful in the research field.

For identification and diagnosing disease infestation, water stress, nutrient content and insect attack of the crops, manual visual inspection from the ground is required when the symptoms appear sometimes late at the growth period of the plants which sometimes becomes difficult to restore the plant health. But, the airborne and ground-based HSI methods have made possible the evaluation of crop stresses, analysing soil and vegetation characteristics quite ahead over that of the traditional scouting methods.

1.3.2.20 Modern Approach

Recently, the robotic system is being introduced to monitor the entire laboratory operation (sample preparation to calculating and reporting of the analysis) through the continuous innovative approach. The type of robot that may work best in a soil-testing laboratory is one that travels on a track, the length of which can be essentially unlimited (Munter 1990). Through the use of a laboratory information management system (LIMS), the automatic control system, computations and data management can be carried out efficiently. Some applications are given below:

The Presidedress Soil Nitrate Test (PSNT)

In most of the areas, the PSNT is primarily recommended for use on fields where there are significant organic N sources such as manure, biosolids applications or leguminous crops in rotation.

- Chlorophyll meter N test

  An alternative to the PSNT soil test used in some zones is the chlorophyll meter test. A chlorophyll meter (SPAD meter) is used to estimate the N status on field crops.

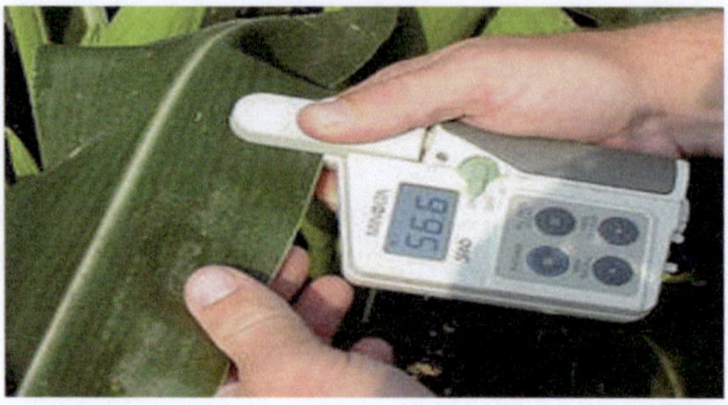

SPAD meter

Besides, global positioning system (GPS) instrument is being used to specify the position of the sample site based on the latitude and longitude of a particular location and geo-referencing the data for image analysis for mapping interpretation.

Android App

Online data entry: To save the time and cost, online data from the field is transferred to the laboratory for analysis and its subsequent online feedback is received by the users.

Communications of Recommendation (SMS, Internet, Health Card)

The report of the analysed sample (soil/plant/water) is sent through SMS/e-mail/soil health card to the farmers with recommendations.

Radioisotopes

The amount of available soil nutrient can be calculated using the formula $A = B(1 - Y)/Y$, where A = amount of available soil nutrient, B = amount of fertilizer nutrient applied and Y = fraction of nutrient in plant obtained from the fertilizer by tracer technique.

Radiotracer techniques are used in the fields of plant physiology, soil chemistry and plant biochemistry through which small amount of labelled nutrient can be measured in the bulk of nutrients. Besides, isotope dilution analysis (IDA) is used in chemical analysis which is based on the mixing (or dilution) of a radioisotope or a separated stable isotope with its natural isotopes in the sample. The total quantity of the substance in the original mixture can be determined from the fraction of the labelled isotope present in the sample.

If the known amount (m_1) of isotopic tracer having activity A_1 is mixed with M_1 weight of the unlabelled compound, then the activity of an aliquot Z_1 from the mixture is A_2.

Hence, the specific activity of the tracer, $S_1 = A_1/m_1$ and the specific activity (S_2) of the mixture ($M_1 + m_1$) will be $S_2 = A_2/Z_1$.

If the total radioactivity is unchanged,

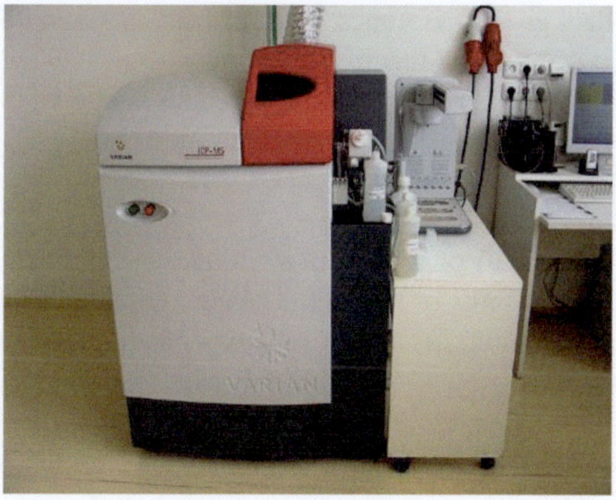

Inductively coupled plasma mass spectrometry (ICP-MS)

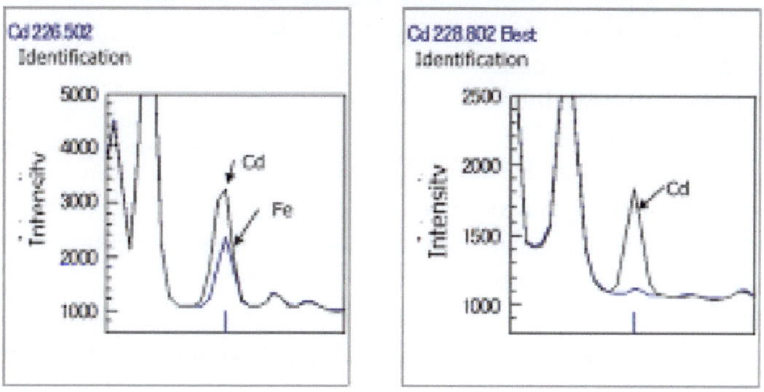

Fig. 1.4 Spectrum of sample from ICP-MS

$$S_1 m_1 = S_2\,(M_1 + m_1)$$

$$M_1 = m_1\,(S_1/S_2 - 1) = Z_1\,(A_1/A_2) - m_1 \tag{1.1}$$

As $m_1 << M_1$, Eq. (1.1) can be written as

$$M_1 = m_1 \cdot S_1/S_2 = Z_1 A_1/A_2$$

where A_1/A_2 is the fractional recovery of the isotopic tracer in the weight Z_1 of the sample.

ICP-MS

Inductively coupled plasma mass spectrometry (*ICP-MS*) is a type of mass spectrometry which is capable of detecting metals and several non-metals at concentrations as low as one part in 10^{15} (part per quadrillion, ppq) on non-interfered low-background isotopes. The instrument is suitable to study the fraction of the ionic species of the metal or non-metal substances in a sample (i.e. Fe^{+3}/Fe^{+2} or Mn^{+4}/Mn^{+2}) from which the dominance of a particular ionic species in a system (soil–plant–water) can be identified (Fig. 1.4).

Village Fertility Map

The fertility status of the soil of a particular village will delineate the area under low/medium/high fertility status of the soil. The change in status over time may be ascertained (temporal variation) for further recommendation. The soil test-based and site-specific nutrient management would help achieving the desired yield of a particular crop.

The nutrient index value (NIV) is used to prepare soil fertility map

$$\mathrm{NIV} = (N_l + 2N_m + 3N_h)/(N_l + N_m + N_h)$$

where N_l, N_m, N_h are the number of soil samples falling under the category of low, medium and high, respectively, of the nutrient status.

If, NIV < 1.50—low category
NIV $= 1.50$–2.50—Medium
NIV > 2.50—High

Validation of Analytical Procedures

The performance characteristics of a method/procedure are determined through validation which is a prerequisite for assessing the acceptance of the generated data for the intended purpose. The process of validation is site specific. Accuracy and precision of the produced analytical data will signify the validation of the experiment. Multilocational trials/experiments will bring closer the reproducibility of the results and hasten the validation and thus minimize the error.

Interpretation

Interpretation of the data is done normally by response curve or by ground truth observations. This will develop the fertility index of a particular zone, based on that the interpretation and recommendation for the fertilizer application can be ascertained. Soil being the dynamic natural ecosystem is the most precious materials on earth which should be in good health to sustain the supplying capacity of nutrients to the crops. Whatever the data being generated from soil through classical or modern tools should best be utilized for up-scaling the demand scenarios on global to local scales on assessing the risk hazards. An approach should be there from all specific corners to develop a comprehensive database management system (DBMS) so as to retrieve the data set for the betterment/improving the yield of the crop, whenever there is a need to do so.

Some Images of Sophisticated Instruments

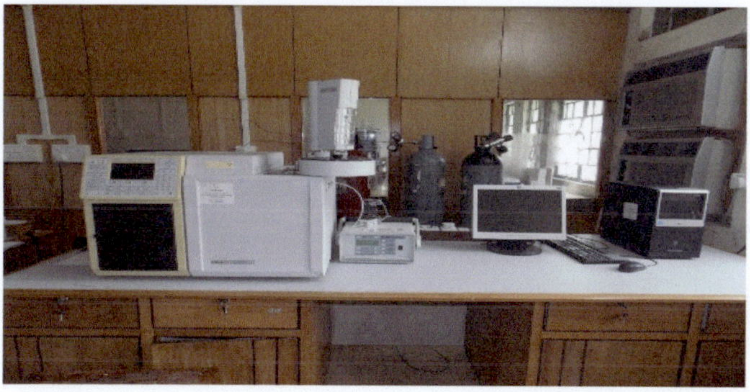

GCMS

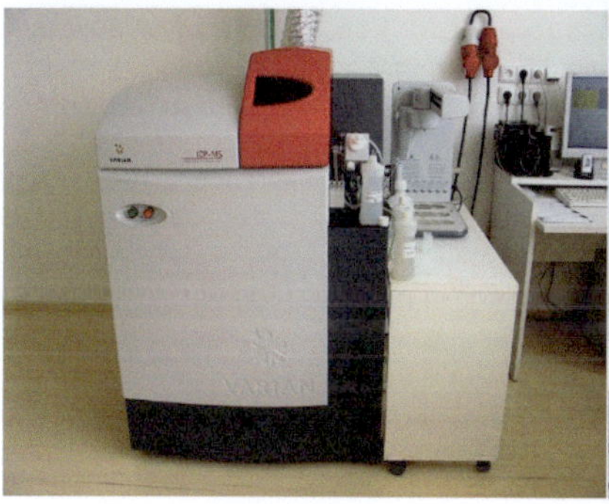

ICPMS

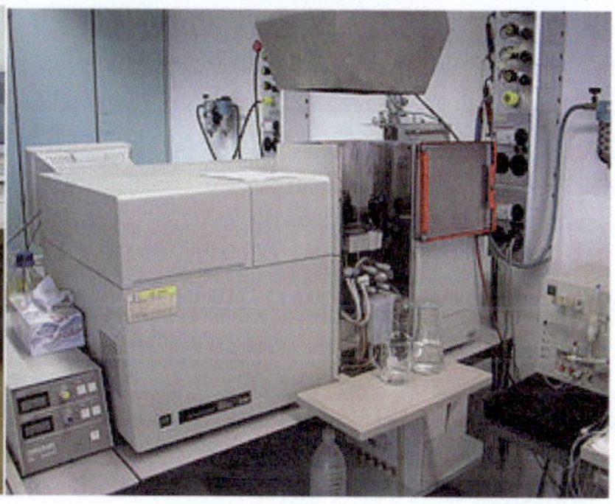

AAS

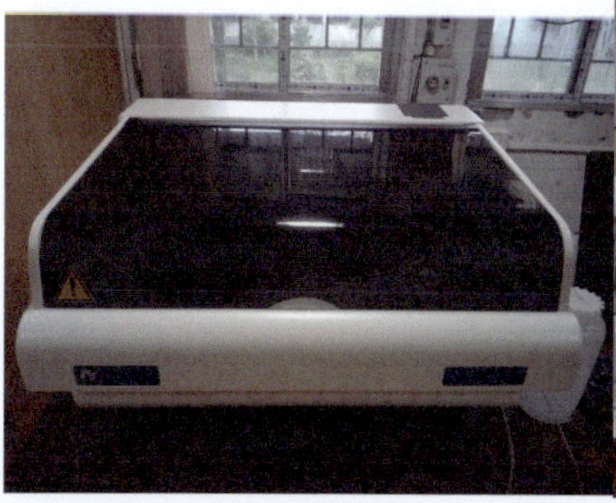

AUTOANALYSER

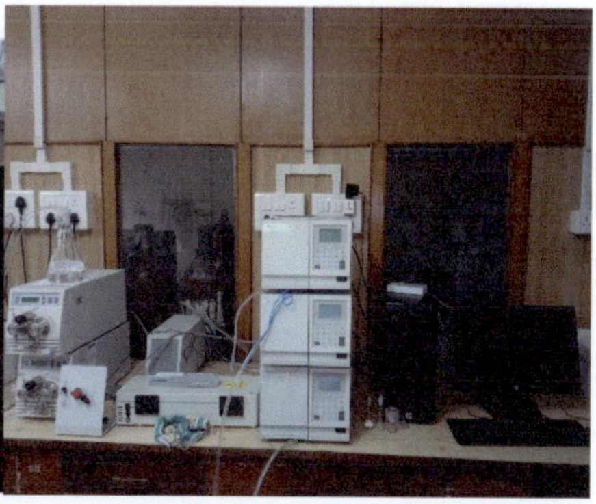

HPLC

References

Augustine NR, Lane NF (2014) Restoring the foundation: the vital role of research in preserving the American dream. American Academy of Arts and Sciences, Cambridge

Berger KC, Truog E (1939) Boron determination in soils and plants. Ind Eng Chem Anal Ed 11:540–545

Cheng KL, Bray RH (1951) Determination of calcium and magnesium in soil and plant material. Soil Sci 72:449–458

Datta NP, Khera MS, Saini TR (1962) A rapid colorimetric procedure for the determination of the organic carbon in soils. J Ind Soc Soil Sci 10:67–74

Gupta RP, GhilDyal BP (1998) Theory and practices in agrophysics measurements. Allied Publishers, New Delhi

James BR (1993) The future of soil chemistry in the north east; lessons from colonial new England. In: Agricultural research in the north-eastern United States: critical review and future perspectives. ASA, Madison

Keeney DR, Nelson DW (1982) Nitrogen inorganic forms. In: Methods of soil analysis, part 2. ASA, Madison, pp 643–668

Lindsay WL, Norvell WA (1978) Development of a DTPA soil test for zinc, iron, manganese, and copper. Soil Sci Soc Am J 42:421–448

Mitrano DM, Barber A, Bednar A, Westerhoff P, Higgins CP, Ranville JF (2012) Silver nanoparticle characterization using single particle ICP-MS (SP-ICP-MS) and asymmetrical flow field flow fractionation ICP-MS (AF4-ICP-MS). J Anal At Spectrom 27(7):1131–1142

Mukhopadhyay D, Sanyal SK (2004) Complexation and release isotherm of arsenic in arsenic-humic/fulvic equilibrium study. Aust J Soil Res 42:815–824

Munter RC (1990) Advances in soil testing and plant analysis analytical technology. Commun Soil Sci Plant Anal 21(13–16):1831–1841

Pal SK (2016) Textbook of soil science. Oxford and IBH Publishing Company, New Delhi

Praetorius A, Tufenkji N, Goss KU, Scheringer M, von der Kammer F, Elimelech M (2014) The road to nowhere: equilibrium partition coefficients for nanoparticles. Environ Sci Nano 1 (4):317–323

Ray SK, Barman KK, Mukhopadhyay D (2018) Stability of organo-zinc complex in application of inorganic and organic nutrients to rice (Oryza sativa) growing soils of West Bengal (India). Int J Chem Stud 6(2):2157–2164

Richardson SD, Ternes TA (2014) Water analysis: emerging contaminants and current issues. Anal Chem 86(6):2813–2848

Schoonover WR (1952) Examination of soils for alkali. University of California Extension Services, Berkeley

Singh D, Chhonkar PK, Pandey RN (1999) Soil plant water analysis – a methods manual. IARI, New Delhi

Subbiah BV, Asija GL (1956) A rapid procedure for the determination of available nitrogen in soils. Curr Sci 25:259–260

Toth SJ, Prince AL (1949) Estimation of cation exchange capacity and exchangeable Ca, K and Na contents of soils by flame photometric techniques. Soil Sci 67:439–445

Van Bavel CHM (1953) Report of the committee on physical analysis 1951-1952. Soil Sci Soc Am Proc 17:416–418

Veihmeyer FJ, Hendrickson AH (1931) The moisture equivalent as a measure of the field capacity of soils. Soil Sci 32:181–194

Walkley AJ, Black CA (1934) Estimation of soil organic carbon by the chromic acid titration method. Soil Sci 37:29–38

Yoder RE (1936) A direct method of aggregate analysis of soil and a study of the physical nature of erosion losses. J Am Soc Agron 28:337–351

Application of Statistical Techniques in Soil Research

2

Dibyendu Deb, Shovik Deb, and Ashok Choudhury

Abstract

A good research on soil science needs some basic ideas about different statistical tools. May it be pedology or edaphological study, statistics confirms the robustness of test as well as helps to comprehend the outcome. This chapter tries to explain some basic statistical tools with few examples of their possible applications in soil research. Within limited scope, coverage of topics has been restricted from central tendency and dispersion to different test of significance only. The importance and differences among tests of significance (t, Z, F, χ^2) and analysis of variance techniques (ANOVA, MANOVA, ANCOVA) have been discussed in terms of their practical applicability. In addition, the initiation of any problem with the construction of hypothesis and solving that problem with approval/disapproval of hypothesis has been detailed.

Keywords

Soil research · Statistical tools · Test of significance · Analysis of variance

2.1 Introduction

Research in soil science has seen remarkable advances in last few decades and application of different statistical tools helped a lot to achieve this feat. This is a fact that when on the one hand researchers are making substantial progresses in using advance concepts like geostatistics, mathematical morphology, fractal geometry, machine learning, etc. for studying soils, there are people working in laboratories

D. Deb
Indian Grassland and Fodder Research Institute, Jhansi, India

S. Deb (✉) · A. Choudhury
Department of Soil Science and Agricultural Chemistry, Uttar Banga Krishi Viswavidyalaya, Cooch Behar, India

A. Rakshit et al. (eds.), *Soil Analysis: Recent Trends and Applications*,
https://doi.org/10.1007/978-981-15-2039-6_2

and getting no help of statistical tools. Things get worse when without proper statistical concept different statistical software packages are used. The situation leads to inconclusive understanding as the basic concept is violated. Referring conventional statistical textbooks prior to analyzing experimental data generally does not appear to be a preferred solution for persons with non-mathematical background. In this context, the focus of this chapter is to provide a quick idea about some basic statistical tools, which can be very important for soil science research. These predominantly involve studies conducted in the field as well as in the laboratory and need statistics to validate the results of the investigations.

Alike many other disciplines, any kind of soil research involves a particular flow of processes to reach a valid conclusion. Procedural guideline of conducting research determines the order of the steps which are as follows: identifying research problem, extensive literature survey, formulation of hypothesis, determining design of experiment, collecting the data, project execution, data analysis, testing of hypothesis, generalization–interpretation, and preparation of the report and presentation of the results. Identification of research problem as the first step of the process is quite obvious as any research activity addresses theoretical or practical problems based on the subject and background knowledge. Formulation of hypothesis and its testing are the two steps, which determine validity of the experiment conducted and guide the researcher towards drawing conclusive inferences. Data analysis gives meaning to raw data, which otherwise would be meaningless. Design of experiment is laying out the strategy for application of scientific method to study the problem. It is a very crucial step as if it gets faulty, no amount of statistical manipulation can lead to draw a valid inference.

It has also been established that many apparently deterministic processes are basically inherently stochastic in nature. And to deal with these processes, specialized statistical tools are there, which can improve the ability of researchers to analyze and interpret voluminous data (Jayade et al. 2015). The statistical tools for soil research may be very simple or complicated, and few of them may be very specific for certain purposes (Panse and Sukhatme 2000; Das 2008). Research activities in classical agricultural disciplines like crop production studies or crop improvement program generally have research designs and more structured statistical guidance, which is not the case for research in soil science. Field research studies have usually been evaluated critically in case of soil science. Furthermore, pedometricians have started establishing ranges of field sampling designs for geostatistical and related spatial statistical problems for the last few years (Pennock 2004; Deb and Chakraborty 2018). The following section focuses on statistical methods pertaining in soil research including very basic statistical key jargons, classical concepts, and inferential statistics.

2.2 Statistical Tools

Statistical tools are important to reveal the nature of data, its distribution pattern, as well as to manage, organize, and present the data in a manner so that some meaning and conclusion can be drawn from it. Therefore, these basic tools are very important for making the raw data amenable for further advance statistical analysis. Before discussing about them, first a basic idea of data and variable has been presented below.

2.2.1 Types of Variables and Data

Statistics is all about variables. In soil science, analyzing different soil parameters/characters actually means dealing with several variables. Variables in statistics are data items and statistical operations work upon them. The characteristic of individual member of a population is called variable. Variables are of two types, qualitative and quantitative (Fig. 2.1). Some information can be measured in some scale and thus they are called as quantitative variables. Any variable is called qualitative when it can be categorized into distinct groups depending upon some characteristics and cannot be measured quantitatively. For example, soil sand, silt, and clay % are quantitative variables, while soil texture (suppose sandy clay loam) is a qualitative variable. Quantitative variable again can be divided into discrete and continuous variables. If the variable assumes only whole numbers or integers (like 1, 2, 3,. . .) then it is discrete, and in case of continuous variable, it can have any value (like 1.234, 2.016,. . .).

Example of discrete variable is number of soil samples collected for a specific study while soil CEC is a continuous variable (e.g., 10.23 cmol (p+) kg^{-1}). Besides this, a soil data can be nominal (unordered) (for example, soil texture: clay soil, sandy soil) or ordinal (ordered or ranked) (for example, soil available N status: low, medium, high), interval (for example, soil available N status: <272 kg ha^{-1} means low, 272–544 kg ha^{-1} means medium, >544 kg ha^{-1} means soil high in N) or ratio (for example, C:N ratio).

Fig. 2.1 Types of variable

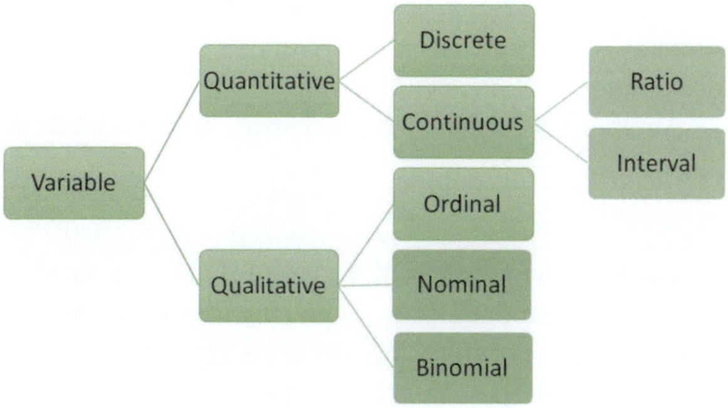

2.2.2 Measure of Central Tendency and Dispersion

To start with statistical analysis, some jargons will be discussed first in a simple manner. These terms are frequently used in research activities and very critical as they reveal the very basic nature of data through measuring central tendency or dispersion. One of such term is "mean" or average which is sum of the observations divided by number of observations and it is a measure of central tendency as it lies within the range of data. The mean (arithmetic mean) of a set of n observations (x_1, $x_2, \ldots, x_n$) is $\overline{x} = \sum \frac{x_i}{n}$. Mean is the easiest to calculate, rigidly defined, less susceptible to sampling fluctuations, and provides a good comparison between two series. Except mean, two other very popular measures of central tendency are "median" and "mode." Like mean, these two measures also summarize the data with a single number representing a critical data point in the data set. If distribution is asymmetric, median provides most preferable measure of location by picking the middle one (or taking the mean if there are two middle numbers) after arranging the data set in order. The beauty of median is as it is a positional average, it does not get affected by extreme values and can be calculated even in case of qualitative factors. Mode is the most frequent number, i.e., mode of a series is the particular value which appears highest number of times. Mode also can be more than one for those distributions, which have equal peaks. As an example, if few soils have pH values 5.0, 5.5, 6.0, 6.5, 7.0, 8.0, 8.0, 8.5, and 9.0, the mean is (5.0 + 5.5 + 6.0 + 6.5 + 7.0 + 8.0 + 8.0 + 8.5 + 9.0)/9 = 7.05, median is 7.0 (as the middle number), and mode is 8.0 (highest frequency).

When it comes to measure of dispersion for observations in hand, "variance" is the most important technical term in soil science. It is not possible to imagine a stretch of soil without any kind of heterogeneity as it is a natural body. Farmers are always contributing towards this variation with their different types of cultivation practices. So, what is this variance? The variance is a set of n observations (x_1, $x_2, \ldots, x_n$), which can be estimated using the following expression:

$$\sigma^2 = \frac{1}{n} \sum_{i=1}^{n} (x_i - \overline{x})^2$$

where $\overline{x}$ is the mean of the data. It is simply the sum of the squared differences of individual observations from their overall mean. The square root of variance is called standard deviation and more frequently used due to having the same unit as the observations.

Another key term is "standard error" which tells us the deviation of sample means ($\overline{x}$) from the true population mean (μ) and expressed as S.E. $= \sqrt{\frac{s^2}{n}}$ where $s^2 = \frac{1}{n-1} \sum_{i=1}^{n} (x_i - \overline{x})^2$.

The term "coefficient of variation (C.V.)" is another way of expressing existing variation in pure number nullifying the effect of different measurements. It is calculated by dividing the standard deviation by mean and then multiplied by 100 to get the percentage variation $\left(\text{C.V.} = \frac{StandardDaviation}{Mean} \times 100\right)$. This makes

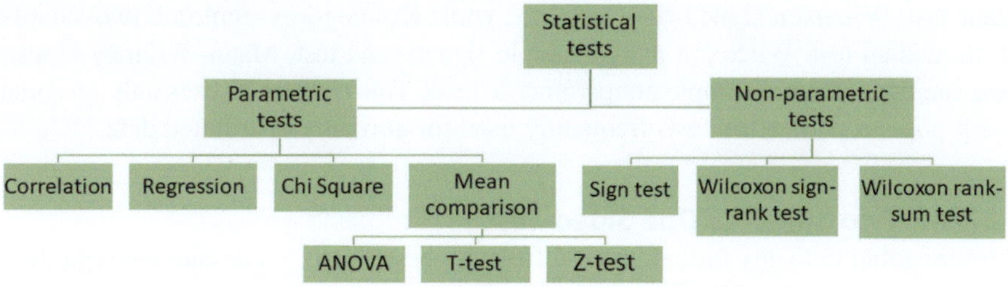

Fig. 2.2 Different statistical tests

things easy for researchers as they can easily compare the variation between different experiments or regions in one go. For example, ten soil samples have organic carbon content 5.6, 5.8, 6.9, 4.7, 7.1, 7.8, 6.2, 5.9, 6.4, and 7.8%. Therefore, the variance, standard error, and coefficient of variation are 0.88, 0.30, and 14.64, respectively.

2.2.3 Statistical Tests

Research problems very often involve comparison of several processes like performance of different varieties, effect of different fertilizer treatments, comparison of crop responses to different dose of nutrients, etc. Different statistical tools like F test, t test, regression analysis etc. make these tasks easier and help the scientist to establish superiority of any of the methods over others quantitatively. F test and t test are used to compare whether two sets of normally distributed data are similar or dissimilar, i.e., if they belong to the same population. The following figure (Fig. 2.2) details the different statistical tests.

2.2.3.1 Parametric and Nonparametric Tests

As the title of this section suggests, there are two types of tests, namely parametric and nonparametric tests available to choose for comparing measurements. Parametric test method is based on stringent assumptions (which may or may not be fully valid) about the population from where the sample is drawn. Generally, these assumptions are about the form of probability distribution, accuracy of observations, etc. and applied on data, which are primarily measured in interval or ratio scale. Very frequently used tests like t tests (one-sample t test, paired t test, unpaired t test), simple linear regression, multiple linear regression, and one-way ANOVA are examples of parametric tests. The theory of parametric test was first given by J. Neyman in 1928 and Karl Pearson in 1933 (Agarwal 2007).

Often situations arise, where the assumptions do not appear valid and the experimenter has no idea about the parameters or if they are normally distributed. To handle such situations, the experimenter has to take resort to nonparametric test which unlike parametric test entails very mild assumptions about the distributions like its continuity, symmetry, randomness, and independence. Some examples of one-sample nonparametric tests are Kolmogorov–Smirnov test, ordinary sign test, runs test, Wilcoxon signed-rank test, etc., while Kolmogorov–Smirnov two-sample test, median test, Wilcoxon paired sample signed-rank test, Mann–Whitney U-test, etc. are two- or more-sample nonparametric tests. This chapter focuses only on some very popular parametric tests frequently used for normally distributed data.

2.2.3.2 Two-Sided Vs One Sided Test

Before going in to any further about different types of tests, an orientation regarding this basic concept of statistical significance test is required. Knowledge about alternative ways of testing the significance of any parameter helps to set the hypotheses. To conclude about a test result, the *p* value provided by the statistical tool is very important. Based on this p value, only the significance of any test gets analyzed, be it an ANOVA or a regression or any other kind. The first and most important step for conducting a test is the formulation of hypotheses, i.e., making a statement or supposition, which has to be proved or disproved. A "null hypothesis" (H_0) is one, which states that there is no significant difference between specified population. If there is symmetrical distribution for the test statistic, then there should be three alternative hypotheses (H_1), viz. 2 one-tailed tests and 1 two-tailed test. Rejection or acceptance of hypothesis depends on the value of the tests statistics. If the value of the difference is so small that it does not exceed a tabulated critical value of F or t, then it comes within 95% confidence about no significant difference, i.e., the null hypothesis is true.

Otherwise there should be two situations concerning both the comparison of standard deviation with F test and means $\bar{x}_1$ and $\bar{x}_2$ with t test: (1) whether A and B are different? (two-sided test) and (2) if they are different, then A is higher (or lower) than B? (one-sided test). A test of any statistical hypothesis, where the alternative hypothesis is one tailed (right/left tailed) is called one-tailed test (Fig. 2.3). As an example, if the mean of two populations gets tested, then the null and alternative hypothesis are

- H_0: $\mu = \mu_0$ against the alternative hypothesis
- $H_{1:}$ $\mu > \mu_0$ (right tailed test) or H_1: $\mu < \mu_0$ (left tailed test)

In case of two-tailed test of any statistical hypothesis, the null and alternative hypothesis are

- H_0: $\mu = \mu_0$ against the alternative hypothesis
- $H_{1:}$ $\mu \neq \mu_0$ (two-tailed test)

2.2.3.3 Correlation and Regression

Correlation is simply the relation between two or more variables. This is a very important statistic, particularly when dealing with a bivariate population and the variation of one variable in relation to the other is to be studied. For example, if the variables X and Y represent the soil clay content (X) and surface area (Y),

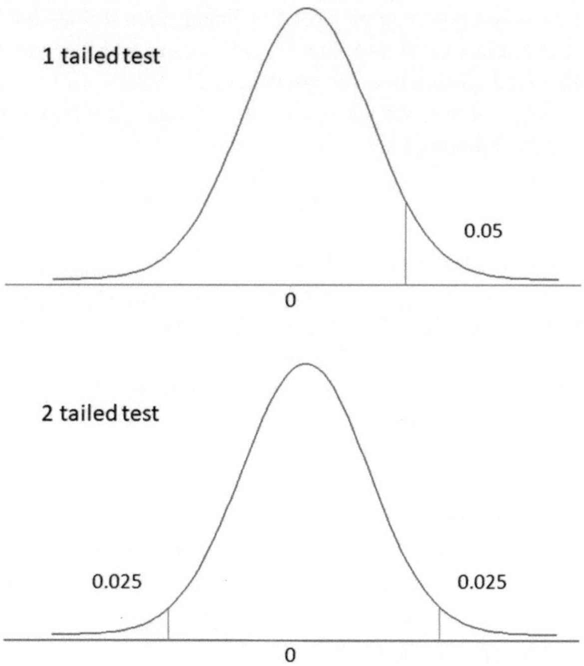

Fig. 2.3 One- and two-tailed test depicted in normal distribution

respectively, then the variation of X and Y are $E(X-\overline{X})^2$ and $E(Y-\overline{Y})^2$, respectively, and the covariance is $E(X-\overline{X})(Y-\overline{Y})$. So, according to the Karl Pearson formula of correlation (postulated in 1890), the correlation coefficient/product moment correlation coefficient (ρ_{xy}) between clay and surface area is as below:

$$\rho_{xy}=\frac{E(X-\overline{X})(Y-\overline{Y})}{\sqrt{E(X-\overline{X})^2E(Y-\overline{Y})^2}}=\frac{\mathrm{Cov}(X,Y)}{\sqrt{\mathrm{Var}(X)\mathrm{Var}(Y)}}=\frac{\mu_{12}}{\sqrt{\mu_{11}\mu_{22}}}$$

The range for this coefficient is from -1 to 1, i.e., $-1\leq\rho\leq 1$. When $\rho=1$ it indicates perfect positive linear association and implies that a unit increase in X causes a unit increase in Y. In contrast when $\rho=-1$ then it is perfect negative association and Y increases with decrease of X. When the variables have no association between them, the correlation coefficient is 0 and its manifestation is a completely scattered graph.

Regression analysis reveals the best functional relationship between a dependent variable (response variable) Y with one or more independent/explanatory variable (s) X. For example, the nitrogen application in soil (X) and increase in plant greenness (Y). The function relationship between these two types of variables is known as regression equation. The term "regression" for a linear relation between two variables was first coined by English scientist Sir Francis Galton (Bland and Altman 1994). If a regression equation is a linear one between the independent and dependent variables, then it is called simple linear regression equation. If the regression equation of Y on X is linear, then it does not necessarily suggest that the regression equation of X on Y is also linear and vice versa. The dependent variable is modeled as function of independent variables with respective constant parameters along with an error term. This error is again a random variable. A regression model can be expressed as:

$$Y = f(X_1, X_2 + \ldots X_n | \beta_0, \beta_1, \beta_2, \ldots, \beta_n) + \epsilon$$

where Y is the dependent variable, $X_1, X_2, \ldots X_n$ are independent variables, $\beta_0, \beta_1, \beta_2, \ldots, \beta_n$ are the regression coefficients, and ϵ is the error term (normally distributed with mean 0 and variance σ^2). This type of regression model is also known as a deterministic model. The regression coefficient β is a measure of change in the dependent variable Y for a unit change in independent variable X. The range of regression coefficient is $-\infty$ to ∞. For validity of the regression model, some assumptions are made like:

- For each dependent variable, the independent variables must be linearly independent, i.e., any of the predictors cannot be expressed as linear combination of the other predictors and they are normally distributed about the mean $\mu_{Y\,|\,X} = \eta = \beta_0 + \beta_1 X$ and variance $\sigma^2_{Y|X} = \sigma^2_{Y \cdot X}$
- Homoscedasticity: The error variance is constant.
- The errors are uncorrelated and follow a normal distribution.
- The mean (μ) of the population of dependent variable satisfies the regression equation, i.e., $\mu = \beta_0 + \beta_1 (X - \overline{X})$

Along with these basic statistical tools, statistical tests are also important for standard soil research. In soil science studies, experimenters are always interested to compare some characteristic like mean, variation, etc. of a group for a specified value or comparison between two or more groups. Here comes statistical test of significance to help. Although statistical inferences being popularly used dates back almost 300 years, statistical tests were lately systemized by famous statisticians like F. J. Neyman (in 1928) and Karl Pearson (in 1933) (Batanero 2000).

2.2.3.4 F Test

This test uses the F distribution to check if the variances of two normal populations are equal. The null hypothesis suggests that the population variances are equal. So, here the test checks

$$H_0 : \sigma_1^2 = \sigma_2^2 \text{ against } H_1 : \sigma_1^2 \neq \sigma_2^2;$$

If the estimates of σ_1^2 and σ_2^2 are s_1^2 and s_2^2, respectively, with sample sizes n_1 and n_2, then the test statistic for this is the ratio of the two variances:

$$F = \frac{s_1^2}{s_2^2}$$

As the statistic is calculated (F_{cal}), it is compared with the table value of $F_{1-\frac{\alpha}{2}, \{(n_1-1)(n_2-1)\}}$ (F_{tab}, critical value). Now if $F_{cal} < F_{tab}$ means with 95% confidence it can be said that the precision is almost same, i.e., the null hypothesis $H_0 : \sigma_1^2 = \sigma_2^2$ is accepted. In this specific case, there is 5% chance that the conclusion is wrongly drawn. For more assurance, 99% confidence level should be used. F test is also very useful while comparing several means in the analysis of variance (ANOVA) (discussed later). In farming system research in particular, the performance between different level of fertilizers or different type of crop verities is needed to be compared and F test through ANOVA is the best option.

2.2.3.5 t Test

Depending on the nature of the two sets of data to be compared, there are variants of t test. The most common types are as follows: (1) t test for testing significance of the correlation coefficient, (2) Student t test, (3) Cochran t test, and (4) Paired t test.

1. *t test for testing significance of the correlation coefficient*: Here, the proposed null hypothesis infers no correlation between the variables x_i, y_i $(i = 1, 2, \ldots, n)$, which follows a bivariate normal distribution with n variables, i.e., the correlation coefficient is zero. Thus, here the goal is to test the correlation coefficient ρ (both for Pearson product moment correlation and Spearman rank correlation coefficients) with the hypotheses $H_0 : \rho = 0$ against $H_1 : \rho \neq 0$. The test statistic is

$$t = \frac{r\sqrt{n-2}}{\sqrt{1-r^2}} \sim t_{n-2}$$

The H_0 is rejected if $|t| \geq t_{n-2, \alpha/2}$, otherwise it is accepted at specified level of significance α.

2. *Student's t test*: In 1908, W.S. Gossset (who was popular as Student) derived a new distribution and test statistic known as *t*. The value of *t* depends on the sample size "*n*." This test is used to compare two sets of data with very similar standard deviation. Thus, $H_0 : \mu_1 \neq \mu_2$ or $\mu_1 > \mu_2$ or $\mu_1 < \mu_2$. Here, the sample size is smaller, i.e., $n_1, n_2 < 30$ from a normal population $N(\mu, \sigma^2)$ with σ^2 unknown. In this case, the test is a two-tailed test and the statistic is as follows:

$$t_{cal} = \frac{|\overline{x_1} - \overline{x_2}|}{s_p} \sqrt{\frac{n_1 . n_2}{n_1 + n_2}}$$

where $\overline{x_1}$ = mean of data set 1, $\overline{x_2}$ = mean of data set 2, s_p = "pooled" standard deviation of the sets, n_1 = number of data in set 1, and n_2 = number of data in set 2.

The pooled standard deviations s_p are calculated using the formula:

$$s_p = \sqrt{\frac{(n_1 - 1)s_1^2 + (n_2 - 1)s_2^2}{n_1 + n_2 - 2}}$$

where s_1 = standard deviation of data set 1, s_2 = standard deviation of data set 2, n_1 = number of data in set 1, and n_2 = number of data in set 2. To compare the calculated value with the table value or critical value, the degree of freedom should be $n_1 + n_2 - 2$

3. *Cochran's approximate t test*: Cochran's variant of t test is required when the standard deviation of independent sets differs significantly. That is, it tests equality of means of two normal populations with different variances, i.e., to test $H_0 : \mu_1 = \mu_2$ against $H_0 : \mu_1 \neq \mu_2$ or $\mu_1 > \mu_2$ or $\mu_1 < \mu_2$ when $\sigma_1^2 \neq \sigma_2^2$. This test is applicable particularly in case of small sample data sets ($n_1, n_2 < 30$). The t value is calculated with the following statistic:

$$t_{\text{cal}} = \frac{|\overline{x_1} - \overline{x_2}|}{\sqrt{\frac{s_1^2}{n_1} + \frac{s_2^2}{n_2}}}$$

The calculated t value (t_{cal}) with the critical value can be obtained using the following equation:

$$t_{\text{critical}} = \frac{t_{\alpha,n_1-1}\frac{s_1^2}{n_1} + t_{\alpha,n_2-1}\frac{s_2^2}{n_2}}{\frac{s_1^2}{n_1} + \frac{s_2^2}{n_2}}$$

Now, if the calculated value is less than the critical value ($t_{\text{cal}} < t_{\text{critical}}$), then the null hypothesis (the means are not significantly different) is accepted. It is noteworthy that for $n > 30$, the Student and Cochran tests are almost same.

4. *Paired t test*: If normally distributed two variables X_1 and X_2 have a strong correlation ρ between them and the difference between the paired values is distributed with mean μ_d, then in paired t test, the null hypothesis becomes $H_0 : \mu_d = 0$ against alternative hypothesis $H_1 : \mu_d \neq 0$ with the statistic $t = \frac{\overline{d}}{\left(\frac{s_d}{\sqrt{n}}\right)}$ based on n paired values. Here, the degrees of freedom of t is $(n - 1)$, $\overline{d} = \frac{1}{n}\sum_{i=1}^{n} d_i$ and $s_d^2 = \frac{1}{n-1}\sum_{i=1}^{n} \left(d_i - \overline{d}\right)^2$.

   The test criterion is same as others, i.e., accept H_0 if calculated value of test statistic is smaller than table value of t at $(n - 1)$ degree of freedom at specified level of significance, otherwise reject H_0.

2.2.3.6 Z Test

The purpose of Z test is similar to t test. The only difference is large sample, i.e., $n > 30$ and hence, it is assumed that the population variance is known. If $x_1, x_2, \ldots, x_n$ is a random sample of size n ($n > 30$) from a normal population $N(\mu, \sigma^2)$, then the sample mean also follows a normal distribution with mean μ and variance $\frac{\sigma^2}{n}$, i.e.,

$\bar{x}\tilde{N}\left(\mu, \frac{\sigma^2}{n}\right)$. Now, if the hypotheses are: $H_0 : \mu_1 = \mu_2$ and $H_1 : \mu_1 \neq \mu_2$, then the test statistic is $Z = \frac{\bar{x} - \mu_2}{\frac{\sigma}{\sqrt{n}}}$ and it follows a normal distribution with mean 0 and variance 1, i.e., $Z \sim N\,(0, 1)$.

Suppose, the difference of means of two samples of size n_1, n_2, with sample means $\bar{x}_1, \bar{x}_2$ (following normal distribution), drawn from two populations with means μ_1, μ_2 and variances σ_1^2, σ_2^2, is need to be tested. In that case, test hypotheses should be: $H_0 : \mu_1 = \mu_2$ and $H_1 : \mu_1 \neq \mu_2$ Here, the test statistic is

$$z = \frac{\bar{x}_1 - \bar{x}_2 - \mu_1 - \mu_2}{\sqrt{\frac{\sigma_1^2}{n_1} + \frac{\sigma_2^2}{n_2}}}$$

Example: Let us assume a case in soil research, where 900 soil samples have been collected from a particular research field and the mean organic carbon content of these samples is 0.56% and standard deviation 0.16. Based on these samples, can it be concluded that the mean organic matter content of the whole field is 0.5%?

To test this particular problem, in case of Z test the hypotheses will be:

$$H_0 : \mu = 0.5 \text{ vs } H_1 : \mu \neq 0.5 \text{ (two – tailed test)}$$

If there are another set of soil samples of size 600 with sample mean 0.65 and the interest is to test whether both the samples are from the same fields, then Z test should be applied with the hypotheses: $H_0 : \mu_1 = \mu_2$ vs $H_1 : \mu_1 \neq \mu_2$ (two-tailed test) where μ_1, μ_2 are the means of the populations the samples are drawn from and variances are σ_1^2, σ_2^2, respectively. The same problem can be tested with t test if the sample size is <30 and consequently variances are unknown.

Suppose the two samples are such that they can be paired, like if the second sample is obtained after any intervention on the field, which is expected to improve the organic carbon content of soil. In that case, the equality of population means will be tested with paired t test with hypotheses and statistic as suggested earlier.

2.2.3.7 χ^2 Test (Test for the Variance for a Specified Value)

This test is used when the aim is to check if a specified value is the variance of a normal distribution. Here, the hypotheses are: $H_0 : \sigma^2 = \sigma_0^2$ and $H_1 : \sigma^2 \neq \sigma_0^2$.

So, the test statistic here is

$$\chi^2 = \sum_{i=1}^{n} \frac{(x_i - \bar{x})^2}{\sigma_0^2} \sim \chi_n^2$$

where μ is not known. Here $x_1, x_2, \ldots, x_n$ represent random sample of size n ($n < 30$) from a normal population $N(\mu, \sigma^2)$. Now if $\chi^2 \geq \chi^2_{\alpha/2,(n-1)}$ or $\chi^2 < \chi_{\left(1-\frac{\alpha}{2}\right),(n-1)}$ (in case of two-tailed test) and $\chi^2 \geq \chi_{(1\ -\ \alpha),\ (n\ -\ 1)}$ (in case of one-tailed test), then H_0 gets rejected. Otherwise, H_0 gets accepted.

Goodness of Fit Test

This is a very popular test to check the discrepancy between an observed frequency to that of theoretical frequency (Jaggi 2008). Therefore, here the null hypothesis is: H_0 : *The distribution is a good fit to the observed data*, against the alternative hypothesis H_1 : *not a good fit*. And the test statistic is

$$\chi^2 = \sum_{i=1}^{n} \frac{(O_i - E_i)^2}{E_i} \sim \chi^2_{n-r-1}$$

where O_i *and* E_i are the observed and expected frequencies of *i*th class ($i = 1,2,\ldots, n$) and r is the number of parameters estimated. Now H_0 is rejected if $\chi^2 \geq \chi^2_{n-r-1}$ at the specified level of significance α.

2.2.3.8 Analysis of Variance (ANOVA)

When any experiment on soil science has more than one influencing factors for comparison and they are distinguished by random effects, then analysis of variance is a most helpful statistical tool. It was developed by R.A. Fisher and is a collection of statistical models to analyze the differences between group means and their associated procedures (such as "variation" among and between groups) (Panse and Sukhatme 2000). In reality, ANOVA partitions the observed variance of a particular sample into components attributable to different source of variations. Thus, it provides a statistical test of whether or not the means of several groups are equal, and therefore generalizes the t test to more than two groups. As conducting multiple two-sample t tests may result in other statistical errors (type-I), ANOVA is useful in comparing three or more groups mean for statistical significance. The fundamental technique for ANOVA is to partition the total sum of square ($\sum (y_i - \overline{y})^2$) into the components related to the effects used in the model. The expression of a simplified ANOVA with one treatment at different levels is

$$\mathrm{SS}_{\mathrm{total}} = \mathrm{SS}_{\mathrm{treatment}} + \mathrm{SS}_{\mathrm{error}}$$

Finally, F test is performed in ANOVA to compare the factors of total deviation. As an example, in a single factor one-way ANOVA, statistical significance is tested by the following test statistic:

$$F = \frac{\text{Variance between treatments}}{\text{Variance within treatments}}$$

or

$$F = \frac{\mathrm{MS}_{\mathrm{treatments}}}{\mathrm{MS}_{\mathrm{Error}}}$$

If it is found that the calculated $F > F_{\mathrm{critical}}$, then the null hypothesis is rejected.

2.2.3.9 Analysis of Covariance (ANCOVA)

The ANCOVA is a linear model with one continuous dependent variable and one or more concomitant variable(s). As per assumption, a linear relationship exists between these variables. It is a hybrid process involving ANOVA and regression technique (Montgomery 2012). The model for this test is as below:

$$y_{ij} = \mu + \tau_i + \beta(x_{ij} - \bar{x}) + \epsilon_{ij}$$

where the dependent variable y_{ij} and concomitant variable x_{ij} present the jth observation of the ith categorical covariate group. μ is overall mean, $\bar{x}$ is global mean for covariate x, τ_i is independent variable, β is slope of the line, and ϵ_{ij} is the error which is independently and normally distributed with mean 0 and variance σ^2. The assumptions here are same as the linear regression model mentioned earlier in Sect. 2.3.3. ANCOVA tests the effect of independent variable on the dependent variable removing the variances, contributed by the covariates.

2.2.3.10 Multivariate Analysis of Variance (MANOVA)

It is a technique to compare multivariate sample means involving more than one dependent variables. While ANOVA compares the means of different groups, MANOVA compares different vector of means. MANOVA tests the statistical significance of mean differences on the basis of existing covariance between outcome variables. Unlike ANOVA, where univariate F value is used as the test statistic, a multivariate F value or Wilks' λ is calculated in MANOVA. The assumptions for MANOVA are not very different than ANOVA; it is assumed that (1) the dependent variables are normally distributed within groups, (2) the dependent variables are linearly related within themselves, and (3) the variances and covariances of dependent variables are homogenous (French et al. 2010; Sthle and Wold 1990).

Example: Suppose there is a research condition, where we have applied four different soil amendment interventions using any particular experimental design to improve the soil organic matter content. Now, ANOVA can be used to compare between the means of these four treatments or interventions. F test can be used to analyze the variance as it is capable to compare multiple group means.

In another research, suppose there are two independent variables (variety and fertilizers) and two dependent variables (yield and plant height). To deal with these dependent variables, MANOVA will be used as it is capable to analyze the variance between multiple vector means. If there is another explanatory variable like number of plants, which also may affect yield (or any other dependent variable) indirectly, then it can be used as a concomitant variable. ANCOVA should be used here to test precisely the effect of independent variables on the dependent variables. It will be beneficial over ANOVA as ANCOVA reduces total experimental error, eliminating the covariate contributed variances.

2.3 Conclusion

Studies about soil system and management need holistic understanding targeting better soil quality and sustainable agriculture. Along with the adequate knowledge in basic pedology and edaphology, it involves proper understanding of data structure and data analysis. To comprehend the significance of any trend and result, for unbiased estimates, conclusions, and towards appropriate interpretations, statistical tools are important. The statistical concepts discussed here within a limited scope are not exhaustive. However, proper knowledge of these tools can equip a soil scientist to conduct basic research and draw valid conclusion thereafter.

References

Agarwal BL (2007) Basic statistics. New Age International Publishers, New Delhi

Batanero C (2000) Controversies around the role of statistical tests in experimental research. Math Think Learn 2:75–97

Bland JM, Altman DG (1994) Regression towards the mean. BMJ 308:1499

Das NG (2008) Statistical methods, vol 2, 1st edn. McGraw Hill Education, New York

Deb S, Chakraborty S (2018) Digital soil science for identification of problem soil characteristics. In: Rakshit A, Sarkar B, Abhilash PC (eds) Soil amendments for sustainability: challenges and perspectives. CRC Press, Boca Raton, pp 51–56

French A, Macedo M, Poulsen J, Waterson T, Yu A (2010) Multivariate analysis of variance (MANOVA). http://online.sfsu.edu/~efc/classes/biol710/manova/manova.htm. Accessed Jan 2019

Jaggi S (2008) Tests of significance. In: IASRI E-Book. IASRI, New Delhi

Jayade KG, Deshmukh PD, Khot PG (2015) Statistical analysis software for agricultural research data analysis. Int J Adv Res Comput Sci Softw Eng 5:885–890

Montgomery DC (2012) Design and analysis of experiments, 8th edn. Wiley, Hoboken

Panse VG, Sukhatme PV (2000) Statistical methods for agricultural workers. ICAR Publications, New Delhi

Pennock DJ (2004) Designing field studies in soil science. Can J Soil Sci 84:1–10

Sthle L, Wold S (1990) Multivariate analysis of variance (MANOVA). Chemom Intell Lab Syst 9:127–141

3 Monitoring and Impact Assessment of Climate Change on Agriculture Using Advanced Research Techniques

Dibyendu Chatterjee, Saurav Saha, Barsita Swain, Debashis Chakraborty, Amaresh Kumar Nayak, Himanshu Pathak, and Madan Pal Singh

Abstract

Anthropogenic emission of greenhouse gas, destruction of vegetation, faulty agricultural practices, rampant use of fossil fuels bring a new challenge to the humanity in the form of climate change. In this chapter, we have tried to ensemble the information on modern techniques to monitor and measure the impact of climate change. The research facilities for assessing the impact of elevated CO_2 and temperature exposure experiment are broadly divided into closed and semi-open to open systems. Growth cabinets and phytotrons are the examples of closed system tools, while open top chambers (OTC), free air carbon dioxide enrichment (FACE), temperature gradient tunnels (TGTs), and free air temperature enrichment (FATE) technologies are of the semi-open to open type. Climate change monitoring on real time basis is measured using eddy covariance techniques. This system measures the fluxes of carbon dioxide, methane, water vapor, and heat. Monitoring of carbon dioxide and methane provide the idea of net carbon balance of an agricultural system, while monitoring energy balance is useful to understand the energy budgeting of the ecosystem.

Keywords

Climate change · Greenhouse gas · Open top chamber · Eddy covariance

D. Chatterjee (✉) · B. Swain · A. K. Nayak · H. Pathak
ICAR-National Rice Research Institute, Cuttack, Odisha, India

S. Saha
ICAR Research Complex for North Eastern Hill Region, Mizoram Centre, Kolasib, Mizoram, India

D. Chakraborty · M. P. Singh
ICAR-Indian Agricultural Research Institute, New Delhi, India

A. Rakshit et al. (eds.), *Soil Analysis: Recent Trends and Applications*,
https://doi.org/10.1007/978-981-15-2039-6_3

3.1 Introduction

Dominancy of progressive human civilization in the modern area accelerated the natural course of change in our climate through anthropogenic greenhouse gas (GHG) emission, destruction of natural vegetation/habitat, and genetic erosion of species diversity in our planet. Scientific evidences proved that the anthropogenic greenhouse gas (GHG) emission induced climate warming is unequivocal. The steady rise in atmospheric greenhouse gases (GHGs) concentration has reached to an unprecedented level over the past 800,000 years after the post-industrial revolution era (after 1750). The dominated anthropogenic sources of GHG emission since the mid-twentieth century already modified the natural composition of our earth atmosphere and its reaction chemistry with global biosphere reserve (IPCC 2013; WMO 2014). Among all the global anthropogenic GHGs, CO_2 emission from fossil fuel burning contributes the lion share of 56.6% (17.3% from biomass burning and 2.8% from cement production and natural gas flaring) followed by methane (14.3%), and nitrous oxide (7.9%). Synthetic hydrocarbons (HFCs and PFCs) also have the significant share in global radiative forcing (IPCC 2007). Carbon dioxide accounted the largest share of radiative forcing since 1990 (30% of total radiative forcing). The steady rate increase in the contribution of CO_2 is expected to grow in near future.

The response in crop eco-physiology towards this climate change phenomenon is obvious from individual leaf level to canopy level and from canopy level to agro-ecosystem level. However, the complex response pattern of individual crop species and the agro-ecosystems has advocated the need of long-term climate monitoring and quantification of the net changes in agro-ecosystem behavior over time. Development of modern advanced instrumentation facilities has enabled us to capture this complex response pattern of agro-ecosystems towards the ever-changing climate that also determines the direction of our future agricultural research progress. The engineered effort for the innovative research facility development has made some remarkable progress in the research arena of climate change monitoring and impact studies at ecosystem level. Since the early 1980s, climate change research has gained an accelerated momentum worldwide as a side co-branch from the mainstream air pollution research and gradually evolved as a distinct unfettered research field in the domain of environmental science research. The section includes real time monitoring of crop weather interaction phenomenon and assessment of the potential impact of climate change on the biophysical processes of indifferent natural and man-made agro-ecosystems. The research outcome often includes the successive alteration in species diversity under the diverse model generated projected GHGs emission scenarios in near future over variable projected time domain. In this section, our focus will be concentrated on the present-day mechanized research facilities available for environmental monitoring and impact of research in the field of environmental science.

3.2 Modern Research Facilities Involved in Climate Change Research

Quantitative investigation on the ongoing climate processes and its impact on terrestrial ecosystem is important to aid our better understanding of ecosystem health in the coming decades and for accurate prediction of the terrestrial feedback phenomenon on the global carbon cycle. The prospects of climate change impact studies include the inevitable effect of microclimate modification driven by elevated GHGs/other trace gases and air temperature on vegetation dynamics, viz. growth behavior, yield components, resource use efficiency, adaptive plant traits, etc. and below-ground processes in the rhizosphere soil environment. The increase in atmospheric CO_2 concentration from a base value of 280 ppm in the pre-industrial revolution era has reached to 400 ppm level in 2014 (Fig. 3.1a; Keeling Curve at Mauna Loa, Hawaii). During the last 12 years, the rate of increase has been measured as 2.242 ppm year^{-1} (Fig. 3.1b); it is predicted to reach around 570 ppm by the middle of this century and to 700 ppm by the end of twenty-first century if the present CO_2 injection rate is maintained. The subsequent rise in earth's average temperature from 2.0 to 4.5 °C is also obvious that might have the inevitable impact on our net agricultural productivity (IPCC 2007).

3.2.1 Climate Change Impact Evaluation Techniques on Crop Species

The available research facilities for elevated CO_2 and temperature exposure experiment are broadly two types (a) closed system and (b) semi-open to open systems. Closed systems experiments fail to simulate the natural environment variability. Those experimental facilties relies on the artificial light installation and air recirculation as evidenced in several growth cabinets and phytotron facilities. These facilities are precise and most suitable for studying phenomenological gene and expression studies under modified environment. The open field exposure systems are moderately more expensive but less precise than closed systems. At present all the newly developed experimental enrichment facilities are more focused to maintain the natural environment as far as possible to observe the net ecosystem response on real time basis. Several meta-analyses and experiments revealed that the response of elevated CO_2 exposure depends on the exposure facility used for a variety of crop species (Taub et al. 2008; Bunce 2016). A brief account of such facilities is as follows:

1. *Leaf Cuvettes (LCs)*: Leaf Cuvettes are designed for single leaf level experimentation (gas exchange measurements) under short-term impact of enriched CO_2 and air temperature for a variety of leaves and pods in different crop species (Fig. 3.2). The facility is equipped with quantum sensors, infra red (IR) sensors and LED (light emitting diode) based light controlling systems. However, these facilities are prone to possible diffusion leaks (Rodeghiero et al. 2007). The most commonly used open exhaust leaf cuvette system for climate change research

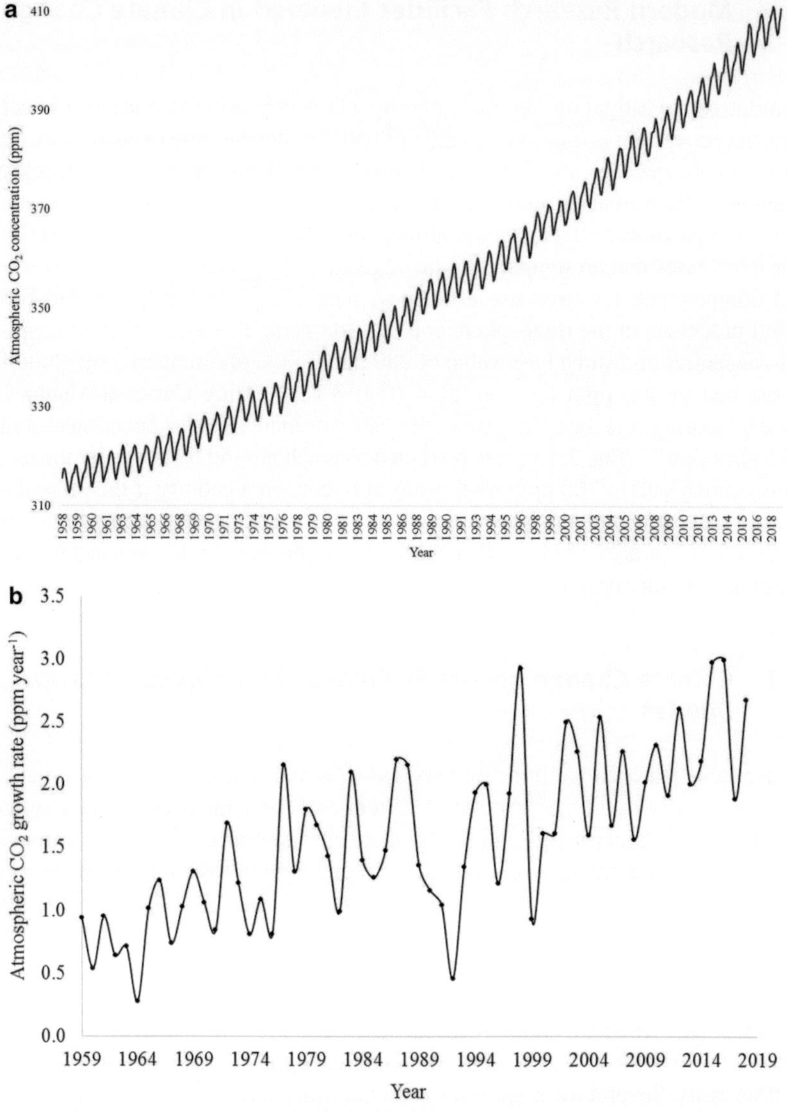

Fig. 3.1 (**a**) Keeling curve suggesting a steady state increase in atmospheric CO2 concentration (ppm) over past five decades (Source: Earth System Research Laboratory, accessed on 11th March 2019). (**b**) Rate of increase in atmospheric CO2 concentration (ppm/year) over past five decades (Source: Earth System Research Laboratory, NOAA, accessed on 11th March 2019)

experimentation is infrared gas analyzer (IRGA), to monitor leaf level gas exchange (including vapor exchange) and to measure leaf level photosynthesis and transpiration rates (Uprety et al. 2006). Leaf cuvettes are useful to study gas exchange phenomenon for small plant canopy grown under controlled chamber experiments to even perennial trees grown under open field condition.

Fig. 3.2 Automatic universal leaf cuvette (Source: PP Systems, Massachusetts, USA)

2. *Gas Exchange Canopy Chambers*: The gas exchange chamber facility is useful to capture canopy level response. There are two types of gas exchange chambers: (a) Steady state system: the gas entry and exit in the canopy chambers are maintained and a constant circulation of air is ensured to maintain the steady state condition (Fig. 3.3a, b). There is a serious problem of radiation quality (direct: diffuse ratio, UV fraction, etc.), the overpressure (partial pressure of target gas is different from natural air circulation) usually gave overestimated crop response towards modified microclimate inside the gas chambers. Inside RH (%) experienced huge fluctuation from canopy photosynthesis/transpiration that becomes crucial to maintain and impaired aerodynamic condition arises from the irregular stirring of air through fans. Moreover, the plastic sheets create artificial greenhouse effect that often modifies the internal thermal regime for crop growth. (b) Transient-state canopy chamber system: The system has movable upper part and transparent to photosynthetically active radiation-PAR (Pérez-Priego et al. 2010). The air circulation was maintained through vacuum pump with continuous monitoring of CO_2 and water vapor concentration in the airflow through IRGA is equipped over time (Müller et al. 2009).
3. *Soil-Plant-Atmosphere Research (SPAR)*: To overcome the problem of lower light quality in close artificial growth chamber facilities, USDA-SEA, Coastal Plains Soil and Water Conservation Research Centre (Florence) developed soil-plant-atmosphere research (SPAR) system (Phene et al. 1978). The system consists of a soil bin covered with an acrylic aerial chamber at top (Fig. 3.4), with a pressure-regulated water outlet, and automated irrigation system. The aerial enclosure has more transparency to solar radiation with its automated control on inside temperature and air circulation (Fig. 3.4). Ideally, the SPAR system allows >90% of the PAR and it is more convenient to monitor all

micrometeorological parameters and their interaction with crop canopy, viz. solar radiation, net radiation, air temperature, ultraviolet radiation, and other trace gases. It is now clear from the research outcome of SPAR climate research that the impact of atmospheric temperature rise has more harmful impact on plant growth than rise in atmospheric CO_2 levels (Allen et al. 2003; Prasad et al. 2003).

4. *Open top chambers*: The initial use of open top chamber was aimed to assess the impact of air pollution on plants. Rogers et al. (1983) started the use of OTCs for evaluation of enriched CO_2 impact on plants. At present, OTCs are mostly used to study the impact of ozone pollution on diverse plant community. The structure is mostly circular or square (Fig. 3.5a; Pal et al. 2004; Vanaja et al. 2008; Saha et al. 2011; Srinivasa Rao et al. 2018). The walls are usually made of polyvinyl chloride (PVC) sheets with >90% optical transmittance (<120 μ), installed on aluminum or galvanized iron frames with "frustum" (3-D portion of a cone or pyramid crossing between five parallel planes). For CO_2 circulation inside the chamber, a cylindrical double-walled plenum with inner side numerous perforations is installed around at base that is linked to the gas inlet pipe. Pure CO_2 gas is supplied inside the chamber with gas regulators and pressure gauze pipelines that are often mixed with air and injected inside the chamber. The OTCs

Fig. 3.3 (**a**) Controlled growth chamber facility at National Phytotron Facility at ICAR-IARI, New Delhi. (**b**) Controlled growth chamber facility at ICAR-CRIDA, Hyderabad (adapted from Srinivasa Rao et al. 2018)

Fig. 3.4 Soil Plant Atmosphere Research (SPAR) System consisting of 10 naturally-lit chambers at Mississippi State University, USA

a b

Fig. 3.5 Open top chamber facility at (**a**) ICAR-IARI, New Delhi. (**b**) ICAR-CRIDA, Hyderabad (adapted from Srinivasa Rao et al. 2018)

designed for multidisciplinary research in South East Asia are comprised of five sub-systems: (a) supply of pure and high concentration of CO_2, (b) system of valves regulators and flow meters, (c) CO_2 controlled chambers, (d) appropriate gas analyzer with feedback control, and (e) computed data acquisition and programming (Fig. 3.5; Uprety and Mahalaxmi 2000; Vanaja et al. 2008). The inside OTC temperature was maintained by air blower. However, the rise in inside temperature from 3 to 5 °C and 15 to 20% reduction in light intensity from the outside natural enviornment was inevitable under semi-arid environment of Delhi (Saha et al. 2012, 2015a). Relative humidity is also higher from the restricted mixing of air with plant transpired water (Saha et al. 2011, 2015b). The facility is simple and quite inexpensive. Additional cooling system in OTCs has the potential to remove the bias of overestimation in crop growth under elevated CO_2 (Bishop et al. 2014). The sidewalls of OTCs should be washed frequently with a gentle sprinkler of water for daily dust removal to maintain the optical transparency. The OTC system is also useful for studying rhizosphere C dynamics (Saha et al. 2016) or assessment of crop quality (Pal et al. 2003; Wang et al. 2012; Saha et al. 2015b, c).

5. *Free air carbon dioxide enrichment (FACE)*: To overcome shortcomings enclosed exposure systems field scale instrumentation systems having no containment were introduced. The free air CO_2 enrichment (FACE) was first developed in 1987 (Hendrey and Kimball 1990; Fig. 3.6). The basic design of this structure is still most popular and widely adopted by the researcher to study natural ecosystem functioning around the globe (Jones et al. 2014). FACE system maintains the identical condition of plant growth with minimum disturbance in crop microclimate, thereby most preferred for elevated CO_2 impact research in different field crops. The system consists of a series of vertical vent pipes installed in circular or hexagonal pattern around the plot and thereby injecting CO_2 towards the center portion (Fig. 3.6a, b). The sensor system is installed at the center for continuous monitoring of atmospheric CO_2 concentration, wind velocity, and wind direction, and the observations on daily weather were collected through computer-controlled systems. The main component of FACE facility is as follows: (a) CO_2 storage and distribution system, (b) FACE ring (Plenum), (c) sensors and actuators for environmental monitoring and controlling daily operations, and (d) automated electronic system for CO_2 monitoring using IRGA. Some of the well-recognized FACE systems are summarized in Table 3.1.

Fig. 3.6 FACE system installed in India (**a**) at ICAR-IARI, New Delhi, (**b**) ICAR-CRIDA, Hyderabad (adapted from Srinivasa Rao et al. 2018)

Table 3.1 Worldwide network of established active FACE system from the countries other than India

Country	Locations	Name	Targeted crops/ ecosystem	Year of establishment
Australia	Horsum	AGFACE (Australian Grains Free Air CO_2 Enrichment) Facility	Wheat–pulse	2007
	Pontville, Tasmania	TasFACE	Native grassland	2002
	Walpeup	AGFACE	Wheat	2008
	Yabulu, Queensland	OzFACE	Natural tropical Savanna	2001
China	Wuxi Experiment Station, Jiangsu Province	China FACE	Rice–wheat	2001
Germany	Braunschweig, DEU	FAL FACE	Winter barley, winter wheat, corn, sugar beet, grass, cover crops	1999
	Giessen, DEU	GiFACE	Grassland	1998
India	Hyderabad	FACE		
Italy	Genomics Research Centre	Durum FACE	Durum wheat	2011
Japan	Tsukuba, Ibaraki	Tsukuba FACE	Rice	2010
New Zealand	Bulls	New Zealand FACE	Grazed pasture grassland	2008
Switzerland	Hofstetten, CHE	Web FACE	Mature temperate forest	2000
UK	Bangor	Bangor FACE	Deciduous forest	2004
USA	Minnesota	BioCON	Grassland	1998
	Cheyenne, WY	PHACE	Native rangeland (Prairie)	2006
	North Carolina	Duke FACE	Pine forest	1996
	California	Jasper Ridge Global Change Experiment	Grassland	1997
	Tennessee	ORNL FACE	Deciduous forest	1998
	Wisconsin	ASPEN FACE	Aspen forest	1997
	Illinois	SoyFACE	Soybean	2001
	Mojave Desert	Nevada Desert FACE	Desert scrub and C_3/ C_4 grasses	1997

For crop growth simulation studies, meta-analytic studies on elevated CO_2 fertilization effect between OTCs and FACE studies is useful to make comparable estimates of crop response. De Graaff et al. (2006) observed more aboveground biomass accumulation for similar level of atmospheric CO2 enrichment in OTC studies than FACE studies. However, the direct comparison between FACE and OTC is difficult (Bishop et al. 2014). However, under similar level of atmospheric CO_2 elevation, the yield increase from CO_2 fertilization effect was lower in wheat and rice grown under FACE system than OTCs (Wang and Feng 2013; Wang et al. 2015). However, increase in circle diameter for pipe installation reduces the reach of CO_2 exposure to the plant growing at center. Therefore, small diameter size is always preferable with more number of individual FACE system for conducting elevated CO_2 research. Even evidences showed that 24 h per day elevated CO_2 exposure has more yield stimulation in both FACE (Bunce 2014a) and OTC systems (Bunce 2014b).

6. *Temperature Gradient tunnels (TGTs) and Free Air Temperature Enrichment Technology (FATE)*: Temperature enrichment in our future climate is almost obvious with subsequent rise in atmospheric CO_2 enrichment. Temperature gradient tunnels (TGTs) are constructed for conducting variable temperature regime exposure experiments (Fig. 3.7). The facility is mainly constructed using UV transparent PVC film made of plastic green house with an air inlet at one end and an exhaust fan at other end. The inside air is generally heated with natural sunlight (greenhouse effect) and exhaust fans are used to create the temperature gradient within the crop canopy height. Thermocouples measures the temperature at frequent interval but the net temperature difference between air inlet and outlet varies between 4 and 5 °C. The system may be controlled by automated electronic systems in the recently developed field models. TGT should be installed in north-south orientation to obtain fairly uniform year around solar radiation in summer and avoid the effect of mutual shading from nearby chambers during winter months. However, imbalance in temperature and increased inside humidity often humidity in the modified the natural crop growth environment to a greater extent and gave biased over estimated results from TGT experiments.

Fig. 3.7 Temperature Gradient Tunnel at ICAR-IARI, New Delhi

a b

Fig. 3.8 Free air temperature enrichment facilities in India (**a**) ICAR-IARI, New Delhi (adapted from User Guide and Instruction Manual of Free Air Temperature Enrichment and Temperature Gradient tunnel, ICAR-IARI, New Delhi) (**b**) ICAR-CRIDA, Hyderabad (adapted from Srinivasa Rao et al. 2018)

Therefore, FATE has been designed that is comprised of additional ceramic IR heaters (with tungsten filament) to irradiate heat with 40° field of view; installed at 1.0–1.5 m height (Fig. 3.8a, b). The spatial and temporal coverage of IR heating in FATE is almost uniform over the crop growth stages. These temperature facilities are often integrated with previously mentioned elevated atmospheric CO_2 exposure facilities like FACE (Kimball et al. 2008; Godfree et al. 2011).

3.2.2 Climate Change Monitoring Techniques

3.2.2.1 Capturing GHG Using Eddy Covariance Techniques

United Nations General Assembly committed itself to the cause by drawing upon the resolution of Sustainable Development Goals (SDG) in the year 2015. One of the 17 goals mentions "to protect, restore and promote sustainable use of terrestrial ecosystems, sustainably manage forests, combat desertification, and halt and reverse land degradation and halt biodiversity loss" (goal 15, Anonymous 2018). Climate induced setback and calamities are a stern concern worldwide and considered as a threat against the above SDG. To achieve a healthy future and inclusive growth, agriculture must develop resilience towards the dilemmas of nature. The Intergovernmental Panel on Climate Change (IPCC) reported linear trend in global surface temperature during the period of 1906 and 2005 with a warming rate of 0.74 °C/year (range 0.56–0.92 °C/year) (Chatterjee and Saha 2018). However, a faster warming trend was observed over the last 50 years. The future projections reported a substantial warming of atmosphere in the range of 2 °C to 4 °C during the twenty-first century (IPCC 2007). Carbon dioxide (CO_2), methane (CH_4), nitrous oxide (N_2O), ozone (O_3), water vapor (H_2O), Chlorofluorocarbons (CFCs), and hydrochlorofluorocarbons (HCFCs) are the apprehended gases causing the greenhouse effect thereafter setting out global warming. Rapid urbanization and industrialization in the post-industrial revolution era caused an increase in emission of

Fig. 3.9 Eddy covariance system in ICAR-NRRI, Cuttack, India

greenhouse gases (GHGs) at an increasing rate in atmosphere (Chatterjee and Saha 2015). Greenhouse effect is caused due to the trapping of radiation in earth's atmosphere due to the presence of these gases. It is evident in today's scenario that the human interference has created a ruckus in the concentration of those particular gases which has intensified the global warming phenomenon.

Agriculture contributes about 10–12% of total GHG emissions worldwide in the form of CH_4, N_2O, and CO_2 (IPCC 2007). Rice field produces lots of water vapor whose greenhouse effect has not been estimated yet. To overcome the problem associated with climate change, the stakeholders are advised to follow climate smart agricultural practices. One of the prime focus of the climate resilient agriculture deals with the reduced net gas concentration from the field. Such objective requires the measuring of fluxes on a spatial as well as on real time basis. Greenhouse gas concentrations can be estimated by manual closed chamber method (collected at a particular point of time) and eddy covariance (EC) (collected on a real time basis) (Fig. 3.9).

EC has surfaced up as a viable alternative to tedious and limiting gas chamber method. The approval of EC within the past two decades has increased as it measures the direct gas fluxes along with net radiation exchange and temperature with respect to year wise, seasonal, and diurnal variation. It gives sensor based real time measurement of GHGs (Chatterjee et al. 2017). The atmosphere has trace amounts of gases which are in state of continuous turbulence due to the rotation of air eddies. Eddies move the air parcels in a haphazard manner thereby carrying the gases of interest. The EC monitors the gas and heat balances between the land surface and the canopy atmosphere. Flux is measured as the covariance between vertical wind velocity and the concentration of point of interest. EC estimates heat, H_2O, CH_4, and CO_2 and methane fluxes based on high frequency (10–20 Hz) real time data.

3.2.2.2 Capturing CO_2 Flux

EC monitors CO_2 exchange between vegetation and atmosphere continuously. Net ecosystem carbon dioxide (NEE) exchange is the resultant flux associated with CO_2 assimilation in photosynthesis and respiration (Swain et al. 2018b). The mean vertical flux density of CO_2 is obtained as the covariance between the Uz (vertical wind velocity) fluctuations with respect to the mixing ratio of CO_2 (Baldocchi 2003). In flooded rice, higher carbon sequestration capacity was observed due to higher gross primary productivity and slowing down of organic matter decomposition (Bhattacharyya et al. 2013). The NEE increases with the development in phenology and reaches its peak at flowering. The NEE value during the daytime is at large negative with lesser values during nighttime which corresponds to the resultant effect of daytime photosynthesis and nighttime respiration. Due to higher temperature and more incident solar radiation during daytime, there is considerable accumulation of C by plants, thus counterpoising the respiration process. Nighttime efflux of C to the atmosphere is due to the canopy, soil, and microbial respiration as an effect of nil photosynthesis.

The carbon dioxide budget is affected by the air temperature. Photosynthesis retard during the afternoon hours due to the increase in relative temperature (Jones 1992) which ultimately reduces the decrease in CO_2 uptake. Vapor pressure deficit (VPD) influences the photosynthetic activity of plants. Increasing VPD decreases the GPP as it results in partial stomatal closure thereby retarding photosynthesis (Pakoktom et al. 2009). Conversely, the air temperature is increased by VPD and thus influences the heterotrophic and autotrophic respiration (Alberto et al. 2009). The carbon fixation by plants in warmer temperatures diminishes as the heat stress inactivates the RuBisCO enzyme which initiates the C3 cycle or Calvin cycle. It lessens the CO_2 uptake which may be due to high evaporation or by invigorating photorespiration and usual respiration process. The warm season grasses have long growing season as well as high photosynthetic efficiency. Hence they have a profound impact on carbon behavior, thereby assessing the managed grasslands as source or sink for carbon sequestration. Carbon sequestered by grasses was quantified as 1.34 Mg/ha/year (Hiller et al. 2011).

3.2.2.3 Capturing Methane

About 25% of global warming is due to the CH_4 emission which is responsible for 18% of radiative forcing (Schubert et al. 2012). The soil microbes synthesize the methane under anaerobic conditions and release it to the atmosphere. The major sources are flooded paddies, natural wetlands, and digestion in ruminants. Methane exchange and production of CH_4 are dictated by the oxidation and redox potential of soil. Anaerobic condition accelerates methanogenesis (Neue 1993). CH_4 exchange between soil and atmosphere takes place by three processes, viz. ebullition, plant mediated transport, and diffusion (Cicerone and Shetter 1981).

Positive correlation between soil temperature and CH_4 flux has been established (Satpathy et al. 1997). With soil temperature, the growth rate of microbes and the change in microbial reactions are affected. Soil moisture content decreases the solubility of methane and thereby increases the diffusivity of CH_4 to the atmosphere. Ebullition creates CH_4 pockets which are directly affected by the wind velocity and pressure. High frictional velocity of air along with less ambient pressure causes a better mixing ratio of CH_4 (Xu et al. 2014).

Flooded condition in rice paddies is an important source of CH_4 emission due to the creation of anaerobic state. In flooded rice ecosystem, the net ecosystem methane exchange measured by EC is found to be controlled by several environmental variables, viz. soil temperature, air temperature, net radiation, and photosynthetically active radiation (PAR) (Swain et al. 2018c). Flood water slows down CH_4 transport when rice crop is under flooded condition and the surge in CH_4 escape is reported during the drainage (Miyata et al. 2000). High ebullition exchange is also altered due to the availability of high organic matter content in soil (Wassmann et al. 1996). Gross ecosystem productivity may be the driving force for slow production and transport of CH_4 and induce the diurnal variation of CH_4 (Ge et al. 2018). With development in phonological stages of rice, plant mediated transport by the aerenchyma becomes the major pathway for CH_4 transport.

3.2.2.4 Capturing Water Vapor Flux

Water vapor is the necessary evil for the climate change process happening throughout. Of the air volume, 1–4% is represented by the water vapor. It behaves as a sink of infrared radiation from the earth's surface, henceforth increasing the temperature of the planet. Warmer temperature induces the content elevation of water vapor which forms a positive loop. The EC has a technique to analyze the water vapor flux. It is estimated with covariance of Uz (vertical wind speed) and water vapor concentration. The main importance of measuring water vapor flux is to study the hydrological cycle between the terrestrial and atmospheric ecosystem. Cloud cover gives a distinct diurnal variation for water vapor content. Water vapor flux is highest in summer and lowest in winter (Li et al. 2006). Due to low soil evaporation and leaf transpiration during nighttime, the water vapor flux is nearly zero. Daily net radiation and temperature are positively correlated with the water vapor flux (Chatterjee et al. 2019b). Relative humidity and soil water content affect the water vapor content on days with cloud cover. The role of relative humidity and vapor pressure deficit turns out to be of immense importance on days of clear weather. Relative humidity plays a functional part for the water vapor transport. With increase in RH, evaporation and transpiration pathways show visibly decreasing trend. In forest ecology, no correlation is found in between the wind velocity and water vapor flux due to the weakening effect of vegetation on wind. Due to the differential temperature between the hydrological and terrestrial ecosystem, the wind speed shows diurnal variation which is an important component to measure the flux of the GHG.

3.3 Capturing and Monitoring Land Surface Energy Fluxes

Energy and mass transfer are two major biophysical processes that affect the energy balance of an ecosystem. In flooded rice ecology, the transfer of CO_2, CH_4, water vapor, and energy shows a close interrelation between the carbon cycle, the hydrological cycle, and the energy balance (Goosse et al. 2010).

3.3.1 Monitoring the Heat Fluxes

Heat flux monitoring Earth's surface energy is characterized primarily by four types of energy fluxes, i.e., net radiation flux (Rn), sensible heat flux (H), latent heat flux (LE), and soil heat flux (G) coming in or out of the soil or water (Chatterjee et al. 2019a). Throughout the day, the H is directed away from the surface, while at night it is in the opposite direction. The LE is the result of surface evaporation and evapotranspiration. The Rn is the result of the surface radiation balance, a product of upwell and downwell radiation. During the daytime Rn is directed towards the soil surface, while at night it is directed away from the soil surface. After some depth, the G at soil surface is dissimilar to the soil below (Masseroni et al. 2012, 2015). Rn is the key contributor to the surface during the day, while LE is the key receiver from the surface, whereas G and S are the key contributors of energy during the night, and then Rn and LE are the main recipients (Swain et al. 2018a, b).

3.3.2 Monitoring the Energy Balance

Determining a correct energy balance (EB) is a crucial prerequisite for understanding an agro-ecosystem that is heavily reliant on climatological factors such as rainfall, air temperature, radiation, moisture, and photosynthetically active radiation (PAR). Theoretically, heat fluxes control rainfall, which is one of the major inputs for agriculture to sustain its productivity and decrease farming costs (Gautam et al. 2019). Measurements of the energy balance over a lake in southern Finland using EC showed that the lake was a heat sink until July–August and the monthly closure of the energy balance ranged from 57% to 112% with an average of 72–82% (Nordbo et al. 2011). Unlike other crops, because of standing water in the field, the rice field has nearly the same situation micrometeorologically like the lake surface. Among all components of energy, latent heat (LE) is the dominant part of energy budget in flooded rice (Chatterjee et al. 2019a). Such differential nature in growing of rice may alter the area's surface runoff, groundwater storage, water cycle, surface energy budget, and cumulatively the region's microclimate (Simmonds et al. 1999). It is possible to estimate the energy balance closure (EBC) in three ways. The ordinary least square (OLS) relationship is established between turbulence heat flux (LE + H) and available heat flux (V), which is Rn–G, and linear regression coefficients (slope and intercept) (Wohlfahrt and Widmoser 2013). This method is perceived to be effective if the independent variable does not contain random errors. It is also possible to use the energy balance ratio (EBR) to assess its closure (Wilson et al. 2002). This is the ratio of aggregated LE + H and available heat flux over a period of time. In an experiment in Cuttack, it was observed that around 72% of the energy is balanced during the dry fallow after growing flooded rice (Tripathi et al. 2018). During the 1–3 days following the rainfall events, energy imbalance occurred. The energy is probably stored and advected in the freshwater. With the exception of a few cloudy days, the dry season and dry fallow were almost free of rainfall. Consequently, during the rain-free days, the energy components are almost balanced

(Chatterjee et al. 2019a). The third method, i.e., residual heat flux (R), quantifies the inconsistency between the available heat flux (V) and turbulence heat flux (LE + H) and provides information on the overestimation or underestimation of the LE + H measured by the EC system. When the EB is perfectly closed, the R should be zero. If $R > 0$, then the energy supply is higher than the energy loss; otherwise, the result is the other way around.

3.3.3 Monitoring Albedo and Bowen Ratio

Albedo is a crucial land characteristic that decides for the energy budget and influences the allocation of radiation energy in the earth-atmosphere system and thereby controls patterns of atmospheric circulation and hydrological cycle (Dickinson 1983; Grant et al. 2000). It depends largely on soil temperature and humidity (Breuer et al. 2003; Gascoin et al. 2009; Zheng et al. 2014). It is the downwell solar radiation flux density that is reflected by the surface and is controlled by the reflective properties of the surface and the spectral and angular distribution of downwell radiation (Grant et al. 2000). The close relationship of the albedo on surface characteristics implies it can represent as a variable by which alterations in land cover can be monitored, for example, in response to human activity on climate change. The Bowen ratio is an indirect method that has been extensively used to characterize the land in different environments (Hatton and Vertessy 1990). This ratio shows the relationship between sensible (H) and latent heat (LE) fluxes that can be used as a measure of evapotranspiration (Fuchs and Tanner 1970; Sinclair 1975).

3.3.4 Monitoring Air and Soil Temperature

Soil temperature can influence energy and it plays an important role in the energy balance. Change in soil temperature in the soil profile controls the microclimate of crop-soil-water continuum (Hillel 1998; Ghuman and Lal 1985). Soil thermal characteristics vary with soil water content, air temperature, porosity, saturated vapor pressure, and water vapor flux (Abu-Hamdeh 2003; Evett et al. 2012). During the day, the earth's surface is warmed more than the soil below, resulting in temperature gradient between topsoil–subsoil and earth surface–air layers. This triggers heat flow downward within the soil as a thermal wave whose intensity alters with depth. Soil heat flux (G) estimation from the soil temperature may provide a comprehension of the soil's gain or loss of heat (Chacko and Renuka 2002).

3.4 Conclusion

Monitoring and assessing the impact of climate change is necessary to maintain the sustainability of the agricultural production system. Climate change may adversely affect agriculture through decrease in crop yield, increase in the pest and disease incidence, reduction of the quality of the produce, increase in the number and degree of abiotic stress events to exert a negative impact on food security and nutritional security. To take a quick action, rapid response is required in the next few decades. Taking a quick response needed more precise measurements using sophisticated instruments. Moreover, the data generated by the impact assessment and monitoring tools may be used as an input for many of the models for future climate prediction.

References

Abu-Hamdeh NH (2003) Thermal properties of soils as affected by density and water content. Biosyst Eng 86:97–102

Alberto RCM, Wassmann R, Hirano T, Miyata A, Kumar A, Padre A, Amante M (2009) CO_2/heat fluxes in rice fields: comparative assessment of flooded and non-flooded fields in the Philippines. Agric For Meteorol 149:1737–1750

Allen LH, Pan D, Boote KJ, Pickering NB, Jones JW (2003) Carbon dioxide and temperature effects on evapotranspiration and water use efficiency of soybean. Agron J 95:1071–1081

Anonymous (2018). https://unstats.un.org/sdgs/report/2016/goal-15/. Accessed 04 Apr 2019

Baldocchi DD (2003) Assessing the eddy covariance technique for evaluating carbon dioxide exchange rates of ecosystem: past, present and future. Glob Chang Biol 9:479–492

Bhattacharyya P, Neogi S, Roy KS, Dash PK, Tripathi R, Rao KS (2013) Net ecosystem CO_2 exchange and carbon cycling in tropical lowland flooded rice ecosystem. Nutr Cycl Agroecosys 95:133–144

Bishop KA, Betzelberger AM, Long SP, Ainsworth EA (2014) Is there potential to adapt soybean (*Glycine max* Merr.) to future [CO_2]? An analysis of the yield response of 18 genotypes in free-air CO_2 enrichment. Plant Cell Environ 38:1683–1691

Breuer L, Eckhardt K, Frede HG (2003) Plant parameter values for models in temperate climates. Ecol Model 169:237–293

Bunce JA (2014a) Limitations to soybean photosynthesis at elevated carbon dioxide in free-air enrichment and open top chamber systems. Plant Sci 226:131–135

Bunce JA (2014b) CO_2 enrichment at night affects the growth and yield of common beans. Crop Sci 54:1744–1747

Bunce JA (2016) Response of soybeans and wheat to elevated CO_2 in free-air and open-top chamber systems. Field Crops Res 186:78–85

Chacko PT, Renuka G (2002) Temperature mapping, thermal diffusivity and subsoil heat flux at Kariavattom of Kerala. J Earth Syst Sci 111:79–85

Chatterjee D, Nayak AK, Chatterjee S, Bhattacharyya P, Kumar A, Lal B, Swain CK, Tripathi R (2017) Eddy covariance technique for micrometeorological measurements. In: Nayak AK, Sarkar RK, Chattopadhyay K, Reddy JN, Lal B, Chatterjee D (eds) Enhancing climate resilience in rice: abiotic stress tolerance and greenhouse gas mitigation. ICAR-National Rice Research Institute, Cuttack, pp 91–103

Chatterjee D, Saha S (2015) Climate change and agriculture: a multidimensional perspective. In: Chatterjee D, Sangma CBK, Kikon ZJ, Ray SK, Chowdhury P, Bordoloi LJ, Deka BC (eds) Resource conservation technologies in the context of climate change. ICAR Research Complex for NEH Region, Medziphema, pp 1–8

Chatterjee D, Saha S (2018) Response of soil properties and soil microbial communities to the projected climate change. In: Advances in crop environment interaction. Springer, Singapore, pp 87–136

Chatterjee D, Tripathi R, Chatterjee S, Debnath M, Shahid M, Bhattacharyya P, Swain CK, Tripathy R, Bhattacharya BK, Nayak AK (2019a) Characterization of land surface energy fluxes in a tropical lowland rice paddy. Theor Appl Climatol 136(1–2):157–168. https://doi.org/10.1007/s00704-018-2472-y

Chatterjee D, Nayak AK, Vijayakumar S, Debnath M, Chatterjee S, Swain CK, Bihari P, Mohanty S, Tripathi R, Shahid M, Kumar A (2019b) Water vapor flux in tropical lowland rice. Environ Monit Assess 191(9):550

Cicerone RJ, Shetter JD (1981) Sources of atmospheric methane: measurements in rice paddies and a discussion. J Geophys Res Oceans 86(C8):7203–7209

De Graaff MA, van Groenigen KJ, Six J, Hungate B, van Kessel C (2006) Interactions between plant growth and soil nutrient cycling under elevated CO_2: a meta-analysis. Glob Chang Biol 12:2077–2091. https://doi.org/10.1111/j.1365-2486.2006.01240.x

Dickinson RE (1983) Land surface processes and climate—surface albedos and energy balance. Adv Geophys 25:305–353

Evett SR, Schwartz RC, Casanova JJ, Heng LK (2012) Soil water sensing for water balance, ET and WUE. Agric Water Manag 104:1–9

Fuchs M, Tanner CB (1970) Error analysis of Bowen ratios measured by differential psychrometry. Agric Meteorol 7:329–334

Gascoin S, Ducharne A, Ribstein P, Perroy E, Wagnon P (2009) Sensitivity of bare soil albedo to surface soil moisture on the moraine of the Zongo glacier (Bolivia). Geophys Res Lett 36: L02405

Gautam P, Lal B, Nayak AK, Raja R, Panda BB, Tripathi R, Shahid M, Kumar U, Baig MJ, Chatterjee D, Swain CK (2019) Inter-relationship between intercepted radiation and rice yield influenced by transplanting time, method, and variety. Int J Biometeorol 63(3):337–349. https://doi.org/10.1007/s00484-018-01667-w

Ge HX, Zhang HS, Zhang H, Cai XH, Song Y, Kang L (2018) The characteristics of methane flux from an irrigated rice farm in East China measured using the eddy covariance method. Agric For Meteorol 249:228–238

Ghuman BS, Lal R (1985) Thermal conductivity, thermal diffusivity, and thermal capacity of some Nigerian soils. Soil Sci 139:74–80

Godfree R, Robertson B, Bolger T, Carnegie M, Young A (2011) An improved hexagon open-top chamber system for stable diurnal and nocturnal warming and atmospheric carbon dioxide enrichment. Glob Chang Biol 17:439–451

Goosse H, Barriat PY, Lefebvre W Loutre MF, Zunz V (2010) Introduction to climate dynamics and climate modeling. Online textbook. http://www.climate.be/textbook

Grant IF, Prata AJ, Cechet RP (2000) The impact of the diurnal variation of albedo on the remote sensing of the daily mean albedo of grassland. J Appl Meteorol 39(2):231–244

Hatton TJ, Vertessy RA (1990) Transpiration of plantation *Pinus radiata* estimated by the heat pulse method and the Bowen ratio. Hydrol Process 4:289–298

Hendrey GR, Kimball B (1990) Free-air carbon dioxide enrichment. Application to field-grown cotton. DOE FACE project brochure, national technical information service. U.S. Department of Commerce, Springfield

Hillel D (1998) Environmental soil physics: fundamentals, applications, and environmental considerations. Academic, Cambridge

Hiller RV, McFadden JP, Kljun N (2011) Interpreting CO_2 fluxes over a suburban lawn: the influence of traffic emissions. Bound-Layer Meteorol 138:215–230

IPCC (2007) Climate change 2007: the physical science basis, contribution of working group I to the fourth assessment report of the intergovernmental panel on climate change. Cambridge University Press, Cambridge

IPCC (2013) Summary for policy makers. In: Stocker TF, Qin D, Plattner G-K, MMB T, Allen SK, Boschung J, Nauels A, Xia Y, Bex V, Midgley PM (eds) Climate change 2013: the physical science basis. Contribution of working group I to the fifth assessment report of the intergovernmental panel on climate change. Cambridge University Press, Cambridge, pp 1–27

Jones AG, Scullion J, Ostle N, Levy PE, Gwynn-Jones D (2014) Completing the FACE of elevated CO_2 research. Environ Int 73:252–258

Jones HG (1992) Plants and microclimate: a quantitative approach to environmental plant physiology, 2nd edn. Cambridge University Press, Cambridge

Kimball BA, Conley MM, Wang S, Lin X, Luo C, Morgan J, Smith D (2008) Infrared heater arrays for warming ecosystem field plots. Glob Chang Biol 14:309–320

Li J, Liu Y, Yang X, Li J (2006) Studies on water-vapor flux characteristic and the relationship with environmental factors over a planted coniferous forest in Qianyanzhou Station. Acta Ecol Sin 26 (8):2449–2456

Masseroni D, Facchi A, Romani M, Chiaradia EA, Gharsallah O, Gandolfi C (2015) Surface energy flux measurements in a flooded and an aerobic rice field using a single eddy-covariance system. Paddy Water Environ 13:405–424. https://doi.org/10.1007/s10333-014-0460-0

Masseroni D, Ravazzani G, Corbari C, Mancini M (2012) Turbulence integral length and footprint dimension with reference to experimental data measured over maize cultivation in Po Valley, Italy. Atmósfera 25:183–198

Miyata A, Leuning R, Denmead OT et al (2000) Carbon dioxide and methane fluxes from an intermittently flooded paddy field. Agric For Meteorol 102(4):287–303

Müller J, Eschenröder A, Muller WD (2009) Through-flow chamber CO_2/H_2O canopy gas exchange system—construction, microclimate, errors, and measurements in a barley (*Hordeum vulgare* L.) field. Agric For Meteorol 149(2):214–229

Neue HU (1993) Methane emission from rice fields. Bioscience 43:466–475

Nordbo A, Launiainen S, Mammarella I, Leppäranta M, Huotari J, Ojala A, Vesala T (2011) Long term energy flux measurements and energy balance over a small boreal lake using eddy covariance technique. J Geophys Res 116:D02119. https://doi.org/10.1029/2010JD014542

Pakoktom T, Aoki M, Kasemsap P, Boonyawat S, Attarod P (2009) CO_2 and H_2O fluxes ratio in paddy fields of Thailand and Japan. Hydrol Res Lett 3:10–13

Pal M, Karthikeyapandian V, Jain V, Srivastava AC, Raj A, Sengupta UK (2004) Biomass production and nutritional levels of berseem (*Trifolium alexandrium*) grown under elevated CO_2. Agric Ecosyst Environ 101:31–38

Pal M, Rao LS, Srivastava AC, Jain V, Sengupta UK (2003) Impact of CO_2 enrichment and variable nitrogen supplies on composition and partitioning of essential nutrients of wheat. Biol Plant 47:227–231

Pérez-Priego O, Testi L, Orgaz F, Villalobos FJ (2010) A large closed canopy chamber for measuring CO_2 and water vapour exchange of whole trees. Environ Exp Bot 68:131–138

Phene CJ, Baker DN, Lambert JR, Parsons JE, McKinion JN (1978) SPAR – a soil-plant-atmosphere research system. Trans ASAE 21(5):924–938

Prasad KVV, Boote KJ, Allen LH Jr, Tomas JMG (2003) Supra-optimal temperatures are detrimental to peanut (*Arachis hypogaea* L.) reproductive processes and yield at both ambient and elevated carbon dioxide. Glob Chang Biol 9:1775–1787

Rodeghiero M, Niinemets U, Cescatti A (2007) Major diffusion leaks of clamp-on leaf cuvettes still unaccounted: how erroneous are the estimates of Farquhar et al. model parameters? Plant Cell Environ 30(8):1006–1022

Rogers HH, Heck WW, Heagle AS (1983) A field technique for the study of plant responses to elevated carbon dioxide concentration. J Air Pollut Control Assoc 33:42–44

Saha S, Chakraborty D, Lata PM, Nagarajan S (2011) Impact of elevated CO_2 on utilization of soil moisture and associated soil biophysical parameters in pigeon pea (*Cajanus cajan* L.). Agric Ecosyst Environ 142(3–4):213–221

Saha S, Chakraborty D, Sehgal VK, Nain L, Pal M (2016) Long-term atmospheric CO_2 enrichment impact on soil biophysical properties and root nodule biophysics in chickpea (*Cicer arietinum* L.). Eur J Agron 75:1–11. https://doi.org/10.1016/j.eja.2015.12.005

Saha S, Chakraborty D, Sehgal VK, Pal M (2015a) Potential impact of rising atmospheric CO_2 on quality of grains in chickpea (*Cicer arietinum* L.). Food Chem 187:431–436. https://doi.org/10.1016/j.foodchem.2015.04.116

Saha S, Chakraborty D, Sehgal VK, Pal M (2015b) Rising atmospheric CO_2: potential impacts on chickpea seed quality. Agric Ecosyst Environ 203:140–146

Saha S, Sehgal VK, Chakraborty D, Pal M (2015c) Atmospheric carbon dioxide enrichment induced modifications in canopy radiation utilization, growth and yield of chickpea (*Cicer arietinum* L.). Agric For Meteorol 202:102–111

Saha S, Sehgal VK, Nagarajan S, Pal M (2012) Impact of elevated atmospheric CO_2 on radiation utilization and related plant biophysical properties in pigeon pea (*Cajanus cajan* L.). Agric For Meteorol 158:63–70

Satpathy SN, Rath AK, Ramakrishnan B et al (1997) Diurnal variation in methane efflux at different growth stages of tropical rice. Plant Soil 195(2):267–271

Schubert CJ, Diem T, Eugster W (2012) Methane emissions from a small wind shielded lake determined by eddy covariance, flux chambers, anchored funnels, and boundary model calculations: a comparison. Environ Sci Technol 46(8):4515–4522

Simmonds I, Bi D, Hope P (1999) Atmospheric water vapor flux and its association with rainfall over China in summer. J Clim 12:1353–1367. https://doi.org/10.1175/1520-0442(1999)012<1353:AWVFAI>2.0.CO;2

Sinclair TC (1975) Photosynthate and nitrogen requirements for seed production by various crops. Science 189:565–567

Srinivasa Rao M, Vanaja M, Srinivas I, Nageshwara Rao CVK, Srinivas K, Maheswari M, Prabhakar M, Sreelakshmi P, Bhaskar S, Sammi Reddy K (2018) CTGC: a facility to study the interactive effects of CO_2 and temperature. Bulletin no.01/2018. ICAR-Central Research Institute for Dryland Agriculture, Hyderabad, 44 p

Swain CK, Bhattacharyya P, Nayak AK, Singh NR, Chatterjee D, Dash PK, Neogi S, Pathak H (2018a) Temporal variation of energy fluxes during dry season in tropical lowland rice. Mapan 33(3):241–251. https://doi.org/10.1007/s12647-018-0260-x

Swain CK, Nayak AK, Bhattacharyya P, Chatterjee D, Chatterjee S, Tripathi R, Singh NR, Dhal B (2018b) Greenhouse gas emissions and energy exchange in wet and dry season rice: eddy covariance-based approach. Environ Monit Assess 190:423. https://doi.org/10.1007/s10661-018-6805-1

Swain CK, Bhattacharyya P, Nayak AK, Singh NR, Neogi S, Chatterjee D, Pathak H (2018c) Dynamics of net ecosystem methane exchanges on temporal scale in tropical lowland rice. Atmos Environ 191:291–301

Taub DR, Miller B, Allen H (2008) Effects of elevated CO_2 on the protein concentration of food crops: a meta-analysis. Glob Change Biol 14:565–575

Tripathi R, Debnath M, Chatterjee S, Chatterjee D, Kumar A, Bhaduri D, Poonam A, Nayak PK, Shahid M, Satapathy BS, Panda BB, Nayak AK (2018) Assessing energy and water footprints for increasing water productivity in rice-based systems assessing energy and water footprints for increasing water productivity in rice based systems. In: Pathak H, Nayak AK, Jena M, Singh ON, Samal P, Sharma SG (eds) Rice research for enhancing productivity, profitability and climate resilience. ICAR-National Rice Research Institute, Cuttack, pp 242–257

Uprety DC, Garg SC, Bisht BS, Maini HK, Dwivedi N, Paswan G, Raj A, Saxena DC (2006) Carbon dioxide enrichment technologies for crop response studies. J Sci Ind Res 65 (11):859–866

Uprety DC, Mahalaxmi V (2000) Effect of elevated CO_2 and nitrogen nutrition on photosynthesis, growth and carbon–nitrogen balance in *Brassica juncea*. J Agron Crop Sci 184:271–276

Vanaja M, Jyothi M, Ratnakumar P, Vaghera P, Raghuram Reddy P, Jyothi N, Yadav SKL, Maheswari M, Venkateswarlu B (2008) Growth and yield responses of castor bean (*Ricinus communis* L.) to two enhanced CO_2 levels. Plant Soil Environ 54:38–46

Wang J, Wang C, Chen N, Xiong Z, Wolfe D (2015) Response of rice production to elevated [CO_2] and its interaction with rising temperature or nitrogen supply: a meta-analysis. Clim Change 130 (4):529–543

Wang L, Feng Z (2013) Effects of elevated atmospheric CO_2 on physiology and yield of wheat (*Triticum aestivum* L.): a meta-analytic test of current hypotheses. Agric Ecosyst Environ 178:57–63

Wang X, Nakatsubo T, Nakane K (2012) Impacts of elevated CO_2 and temperature on soil respiration in warm temperate evergreen *Quercus glauca* stands: an open-top chamber experiment. Ecol Res 27:595–602

Wassmann R, Neue HU, Alberto MCR et al (1996) Fluxes and pools of methane in wetland rice soils with varying organic inputs. Environ Monit Assess 42(1):163–173

Wilson K, Goldstein A, Falge E, Aubinet M, Baldocchi D, Berbigier P, Bernhofer C, Ceulemans R, Dolman H, Field C, Grelle A (2002) Energy balance closure at FLUXNET sites. Agric For Meteorol 113:223–243. https://doi.org/10.1016/S0168-1923(02)00109-0

WMO (2014). http://www.wmo.int/pages/mediacentre/press_releases/pr_991_en.html

Wohlfahrt G, Widmoser P (2013) Can an energy balance model provide additional constraints on how to close the energy imbalance? Agric For Meteorol 169:85–91

Xu L, Lin X, Amen J et al (2014) Impact of changes in barometric pressure on landfill methane emission. Glob Biogeochem Cycles 28(7):679–695

Zheng ZY, Wei ZG, Li ZC, Wei H, Liu H (2014) Study of parameterization of surface albedo of bare soil over the Gobi Desert in the Dunhuang region [in Chinese with English abstract]. Chin J Atmos Sci 38(2):297–308. https://doi.org/10.3878/j.issn.1006-9895.2013.13147

Advancement in Soil Testing with New Age Sensors: Indian Perspective

4

R. K. Swetha, Swagata Mukhopadhyay, and Somsubhra Chakraborty

Abstract

Soil is one of the crucial resources for maintaining a sustainable future of Indian agriculture for food security. It is a heterogeneous dynamic body influenced by natural and anthropogenic agents; hence, its spatial and temporal variability is inevitable. The soil health card scheme introduced by the Govt. of India aims to issue soil card to farmers which will carry crop-wise recommendations of nutrients and fertilizers for the individual farms for improved productivity through judicious use of inputs based on the soil health card for the area. All soil samples are to be tested under this scheme which includes about 121 million agricultural fields across India. The capacity of the soil testing laboratories far lags behind the requirement that is coming under the soil health card scheme. The Government aims to use this card to 14 crore farmers across India. In general, precise mapping of soil using conventional analysis is laborious and time consuming. Combining this challenge with the ambitious soil health card scheme makes it quite a challenging task to accomplish. Advanced sensing techniques such as portable X-ray fluorescence spectrometry (PXRF) and diffuse reflectance spectroscopy (DRS) can be used to help this challenge by developing soil spectral libraries. Soil spectral library contains spectral signatures of specific soil types that can be linked to a range of spectral properties. With the development of empirical models for different nutrients of benchmark soils, spectral libraries can help rapid analysis of thousands of samples in a short time. This chapter will give a basic overview of PXRF, DRS, and other commercially available soil sensors with their advantages and disadvantages.

Keywords

Soil Health Card · PXRF · VisNIR DRS · NixPro · Soil Sensor

R. K. Swetha · S. Mukhopadhyay · S. Chakraborty (✉)
Agricultural and Food Engineering Department, IIT Kharagpur, Kharagpur, India
e-mail: somsubhra@agfe.iitkgp.ac.in

A. Rakshit et al. (eds.), *Soil Analysis: Recent Trends and Applications*,
https://doi.org/10.1007/978-981-15-2039-6_4

4.1 Soil Testing in India

Soil testing is an important step for increasing agricultural production and raising farm income. Traditional soil testing methods are based on chemical methods carried out under laboratory conditions. These methods are generally time consuming, tedious, and involve elaborate sample preparation steps. On the other hand, the number of soil samples needed to be analyzed is large because of the small size of the landholdings in many parts of India. Consequently, even if soil samples are collected from different agricultural fields, timely testing of these samples is generally not possible and the test results often fail to reach farmers in a timely manner. This initiates a negative feedback loop creating a strong aversion to soil testing among our agricultural community.

India currently has about 137 million individual landholdings. Currently, there are about 1049 soil testing labs operating in the country with an annual analyzing capacity of only 10.7 million samples. In West Bengal, there are about 14 soil testing laboratories. As a part of the soil health card (SHC) mission, the national mandate for the year 2015–2016 was to generate 13 lakh SHCs. As of July 2016, 5.28 lakh samples have been collected out of which only 1.3 lakh could be tested. Thus, the analyzing capacity of soil testing labs simply lags far behind the requirement. Moreover, almost no efforts are made to monitor soil physical properties or soil water holding attributes in standard soil testing campaigns and, hence, water resource management in India is based on the distribution and supply of water instead of actual crop requirement. Furthermore, there is a need to repeat soil testing from time to time depending on soil types and crops. Hence, new technology has to be introduced to make soil testing-based nutrient management a reality.

4.2 Success Stories in India

Recently, the Government of Karnataka (GoK) has completed the "Bhoochetana" programme in collaboration with ICRISAT, Hyderabad, in which soil testing was done in 92,000 soil samples collected covering 3.6 million farmers over a period of 5 years (2009–2010 to 2013–2014). In this programme, soil testing-based fertilizer application was implemented with two simple recommendations: (a) full dose of recommended fertilizer was applied in those fields where soil test showed >50% nutrient deficiency and (b) the recommended dose was reduced to half where soil test showed <50% nutrient deficiency. These two simple recommendations resulted in 20% increase in yield with a net income of about INR 1267.6 crore and there was a huge amount of fertilizer saving. This programme has clearly demonstrated that technology intervention and a participatory approach have the potential to increase productivity at large scales.

4.3 Soil Health and Its Sensing: International Efforts

Agriculture is challenged with feeding an increasing population with limited land and water resources. Long-term declines in soil health due to unsustainable agricultural practices and environmental management currently threaten continued delivery of these critical ecosystem services, which has prompted researchers to place greater focus on properties related to soil health. Private and international initiatives have also emphasized the need to build healthy soils to provide the basis for sustainable and healthy food and fiber production. However, it has been difficult to understand the true impacts of healthy soils on agricultural productivity and environmental preservation due to the lack of common indicators of soil health.

The concept of soil health addresses the continued capacity of soil to perform ecosystem functions that support plant growth, habitat for organisms, and regulates environmental quality (Doran and Parkin 1996; Karlen et al. 1997). Several soil physical, chemical, and biological characteristics have been proposed that may be used as sensitive and consistent indicators of changes in soil health (Cano et al. 2018). For example, changes in physical properties such as soil macroaggregates, whose formation and stability are influenced by biological activity (Blankinship et al. 2016), can indicate potential changes in soil aeration, moisture holding capacity and availability, and water infiltration (Beare et al. 1994; Shaver et al. 2002). Soil carbon dynamics may be monitored over time by measuring both total soil organic C (SOC) stocks and soil organic matter (SOM), which indicate C storage potential for long-term sustainability as well as organic matter involved in improving soil water holding characteristics and nutrient cycling dynamics (Magdoff and Van Es 2000; Rawls et al. 2003). Finally, the size, composition, and function of soil microbial communities are important biological soil health indicators that reflect microbially mediated labile C pools and SOM formation, indicate shifts in relative abundances of bacteria and fungi that perform different ecosystem roles, and regulate transformations of C, N, P, and S involved in decomposition and nutrient availability to plants. Researchers rely on these measurements to assess changes in soil health due to agricultural management strategies and environmental change (e.g., Acosta-Martínez and Cotton 2017).

Researchers have long struggled with an effective method for quantifying soil health, especially considering the large number of chemical, physical, and biological indicator measurements needed to accurately assess soil health, for which specific needs and methods may vary by region and soil type, and the time and labor costs associated with this approach. Sensor-based approaches may provide a cost-effective, site-specific solution for soil health monitoring and management. Soil sensors with wireless connections in the fields can continuously monitor soil moisture, temperature, pH, electrical conductivity, and salinity. Emerging proximal sensor technologies such as diffuse reflectance spectroscopy (DRS) and portable X-ray fluorescence spectrometry (PXRF) can efficiently quantify soil salinity, total C/total N, and other soil properties. The PXRF spectrometer uses low power X-rays (10–40 keV) produced from an Rh X-ray tube to forcibly eject inner shell electrons of matter (Fig. 4.1).

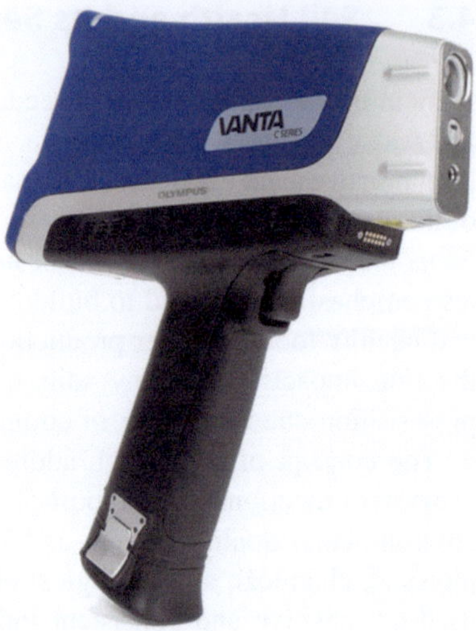

Fig. 4.1 An Olympus® VANTA portable X-ray fluorescence spectrometer

The specific energy identifies the element and the strength of emission enables quantification via silicon drift detector. This analysis can be done in the field, on-site, in 60 s with little to no sample preparation needed. PXRF analysis has successfully been applied for elemental quantification of solids (soil). Coupled with georeferencing, the combined use of DRS + PXRF enables us to predict multiple soil properties in a single day on-site with non-destructive scans (Aldabaa et al. 2015). Visible and near-infrared diffuse reflectance spectroscopy (VisNIR DRS) is a promising hyperspectral scanning technology that has become popular for rapidly quantifying and identifying multiple soil parameters simultaneously (Rossel et al. 2006) (Fig. 4.2). This hyperspectral technique has achieved wider acceptance in soil science, owing to its cost-effectiveness and advantages over other analytical spectroscopic and wet chemistry methods.

Conversely, beyond direct reporting of total elemental concentration of plant essential nutrients and heavy metals (e.g., Pb, Cd, Cr, As, etc.) (Weindorf et al. 2013a), PXRF-based elemental data combined with various regression techniques have been used to determine soil pH (Sharma et al. 2014), salinity (Swanhart et al. 2014), cation exchange capacity (Sharma et al. 2015), soil texture (Zhu et al. 2011), gypsum content (Weindorf et al. 2013b), calcium carbonate development (Chakraborty et al. 2017a), lithologic discontinuities (Weindorf et al. 2015), and base saturation percentage (Rawal et al. 2019). Studies have been conducted on natural soils, mine tailings (Koch et al. 2017), and areas rife with heavy metal pollution (Chakraborty et al. 2017a). Newer approaches have extended the application of PXRF to land use/land management characteristics (Chakraborty et al. 2019b), compost (Li et al. 2018; Weindorf et al. 2018), vegetation (McGladdery et al. 2018), and water (Pearson et al. 2017, 2018) analysis. To date, at least three reference methods are given for the use of PXRF for soil/sediment analysis (Weindorf and Chakraborty 2016; Soil Survey Staff 2014; US-EPA 2007).

Fig. 4.2 Field use of VisNIR DRS (photo: D. C. Weindorf)

Table 4.1 shows some of the currently available soil sensors and their advantages and disadvantages.

Remote sensing is a valuable tool to monitor and evaluate relationships between land management, water, and crops at various spatial and temporal scales. Remote sensing through satellites or aircraft has been increasingly used in agriculture because it is noninvasive and provides a detailed measure of soil or plant growth conditions at relatively low cost (Varvel et al. 1999). The spectral properties of soil are influenced by soil constituents, soil water content, organic matter content, soil texture, and aggregate size. For example, soils typically appear darker when wet than when dry, due to a decreased reflectance in the visible region of the spectrum (Escadafal et al. 1989). Soil organic matter content has a strong influence on soil reflectance, especially when the organic matter content exceeds 2.0% (Mulders 1987). Generally, fine textures show a high reflectance than coarse textures. Soil reflectance also decreases with increasing diameter of aggregates, yet for large diameters (>2.5 mm) there may be no decrease in reflectance (Mulders 1987). Therefore, we can potentially assess soil characteristics within the ranges above using remote sensing imagery. With remote sensing imagery, we also can infer information about soil condition by observing vegetation growth. Vegetation indices such as the normalized difference vegetation index (NDVI) have been used as an indicator of green vegetation cover and plant biomass. Total NDVI accumulation within a year and ground cover (GC) can provide useful information about agricultural productivity and changes in soil health (Sheffield and Morse-McNabb 2015). Quantifying production and soil health indicators over the long term using satellite

Table 4.1 A comparison of currently available soil sensors

Sl no.	Device	Advantage	Disadvantage
1.	VisNIR DRS	1. Commercially available 2. Cost-effective 3. Can model OC, clay, sand, available N and P 4. Rapid	1. High cost of the instrument 2. Produce better results with dry samples
2.	PXRF	1. Commercially available 2. Cost-effective 3. Portable 4. Rapid 5. Can model available K, Ca, Fe, Cu, Zn, and Mn using total nutrient content 6. Showed potential for predicting available P 7. Can be used to measure plant nutrients 8. Can be used for dissolved elements in water	1. Cannot work in sub ppm level 2. Better accuracy for higher Z elements
3.	NixPro™ sensor	1. Cheap 2. Portable 3. Measure color indices 4. Can be used to predict OC and other soil properties	Requires calibration
4.	Penetrometer + EC sensor	Measure BD	Measurement is difficult, small-scale variability, influence of soil moisture and soil texture on sensor readings
5.	Electromagnetic induction (EMI) sensor and galvanic contact resistivity (GCR) sensor	Well established, fast, mechanically robust, no security issues, EMI is light-weighted, GCR is cheap, large sample support, detect soil layering by depth sounding, different frequencies might give additional info	Ambiguous relationships to soil properties of interest, some EMI instruments tend to drift, EMI instruments are very sensitive to metal, GCR are heavy, GCR do not work well on dry soils
6.	Gamma ray spectrometry	Acknowledged by scientists, fast, direct relationship to K content and geology, indirect relationship to clay and others	Requires careful calibration by reference sampling, high cost
7.	Ion-selective electrodes	Direct relationship to target parameters (pH, NO_3^-, K^+, etc.), well established, no security issues	Not very robust (besides metal electrodes), sensitive to interfering ions, slow measurement, delayed response, drift, expensive (other electrodes than pH),

(continued)

Table 4.1 (continued)

Sl no.	Device	Advantage	Disadvantage
			does not work well for other ions besides H^+ (e.g., no PO_4^- electrodes)
8.	Ion-selective field effect transistors (ISFET)	Can be made cheap (chip technology) several options for ion-selective membranes	Currently only a few ions can be detected (more R&D required), mechanical sensitivity of membranes, drift, flow injection of soil solution

imagery can help build a system of soil health and productivity across the landscape. Emerging unmanned aerial system (UAS) technology provides images with centimeter resolutions to measure plant growth conditions, such as plant height, leaf area index, plant vigor, and biomass. Information derived from UAS images can be used as an in-situ measurement that may be integrated into Landsat or Sentinel-2 images to monitor crop growth and soil health at a regional scale.

4.4 Advancement in Modeling Exercises

One of the challenges for sensor-based data modeling is that it involves high-dimensional low sample size (HDLSS) datasets. A major part of the data are spectra, therefore it is HDLSS. Most of the least square methods suffer from the ill-posed estimation problems on HDLSS data. In addition, overfitting, model instability, black-box algorithms that lack interpretation, computation, and data visualization are common problems for high-dimensional data analysis. Scientists have applied modern statistical modeling, programming, and machine learning techniques to address these problems. For example, in the functional estimation, they have used regularized regression methods to overcome the ill-posed problems and overfitting for HDLSS data. They also used ensemble methods and model averaging to improve the prediction accuracy. For interpretation, researchers have utilized modern interpretation tools such as relative variable importance and partial dependence plot to interpret the black-box algorithm and shed lights into the underlying data generation mechanism. To improve computation efficiency, scientists have taken advantages of recent developments in high-performance computing, parallel programming, and GPU programming. For data visualization, they have used PCA/biplot and multidimensional scaling (MDS). Recently, scientists are currently using some non-linear dimensional reduction tools, such as ISOMAP, to explore the intrinsic geometry of the data based on local patterns on the manifold.

Another challenge for sensor modeling is that the data come from multiple sources. Hence, it is necessary to integrate these data from different sources to produce a more accurate and consistent prediction of soil health indicators.

Researchers have applied additive modeling and ensemble learning to address this problem. They have also applied data fusion and the hierarchical mixture of experts to further improve the accuracy in this study.

Calibrations to relate the soil health and the other factors were developed using several multivariate techniques, such as regularized linear regression, partial least squares regression (PLSR), principal component regression, penalized spline regression, support vector machine, random forest regression, and boosted regression trees. Model accuracy assessment was executed using root mean squared error (RMSE), residual prediction deviation (RPD), the coefficient of determination (R^2), the ratio of performance to interquartile range (RPIQ), bias, and Lin's concordance correlation coefficient (CCC).

Selecting the most feasible variable selection method is a crucial step in DRS-based real-time spatial variability analysis. Essentially, the optimal selection of the spectral variables may improve the predictive model accuracy. Recently, Raj et al. (2018) evaluated the potential of merging variable indicator-based DRS spectral channels and geostatistical interpolations for rapid production of spatial maps of several soil properties collected from Romanian Transylvanian plain (Fig. 4.3). Also, Chakraborty et al. (2019a) have utilized novel external parameter orthogonalization (EPO) technique to remove the masking effect of moisture while using DRS spectra for soil property prediction.

Recently, a collaboration from researchers of Texas Tech University (USA) and India has developed a patented methodology for combining PXRF and DRS to improve the predictive accuracy relative to either technique in isolation on the analyte of interest. In fact, the sensors have complementary strengths: DRS is very sensitive to moisture and carbon-based compounds; PXRF is highly sensitive to elements heavier than Cr, with acceptable quantification from Mg to V. The results of their testing proved successful, so much so that an US patent (US10107770B2) was awarded titled "Portable Apparatus for Soil Chemical Characterization." In this approach, the DRS spectrum is used as the primary modeling dataset. However, PXRF elemental data is fed into the model as auxiliary input data, essentially giving the model more information to resolve and predict the parameter of interest. Relative to laboratory methods, implementation of the combined data approach generally results in a higher coefficient of determination (R^2), lower root mean square error (RMSE), and higher residual prediction deviation (RPD = RMSE/std. error). Examples of the approach have been documented for hydrocarbons (Chakraborty et al. 2015), salinity (Aldabaa et al. 2015), and total C total N (Wang et al. 2015). Beyond the research group of Weindorf and Chakraborty, independent research from the University of Sydney (Horta et al. 2015) has verified the significance of the approach.

4.5 Precision Agriculture via Soil Sensors: Research in India

Precision agriculture (PA) has emerged as a management option that maximizes agricultural productivity and minimizes the risk of environmental pollution. PA employs a site-specific management approach to crop production, hence the term

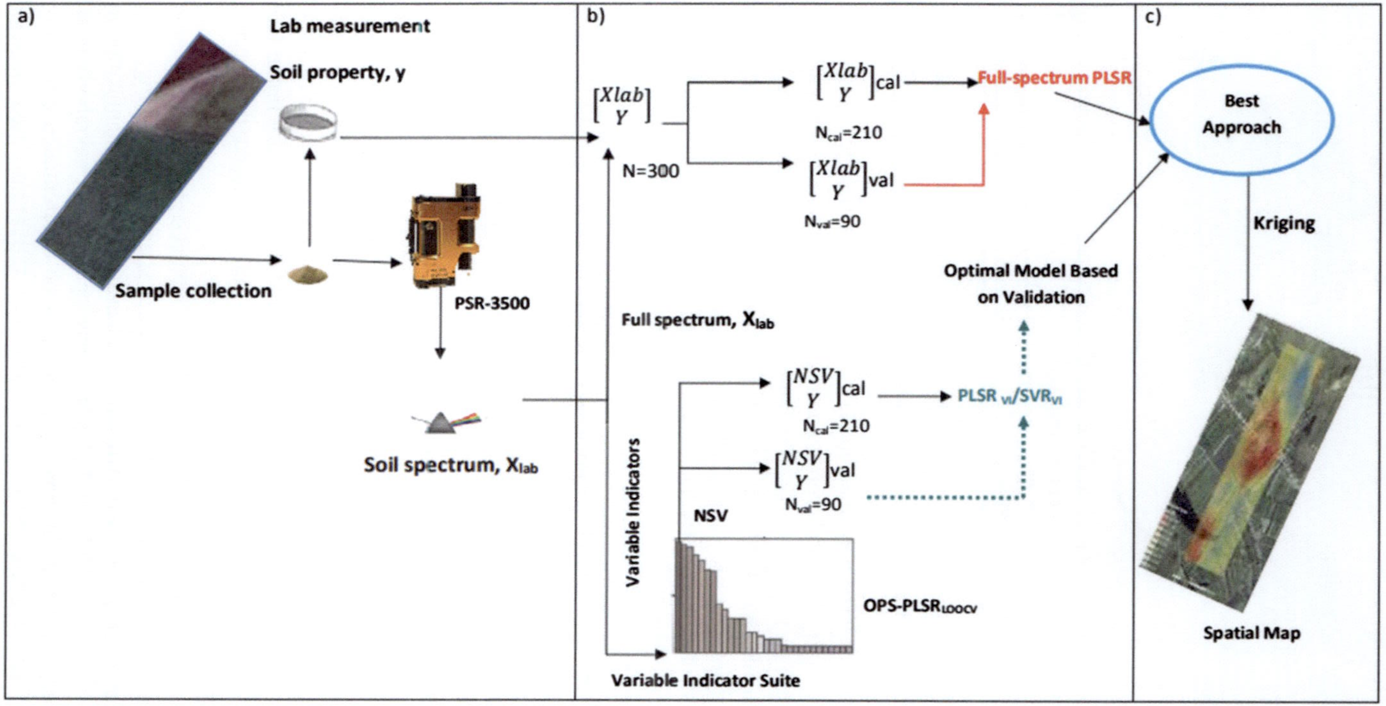

Fig. 4.3 Flow chart summarizing the variable indicator-based DRS modeling of soil properties and subsequent spatial variability mapping (Raj et al. 2018)

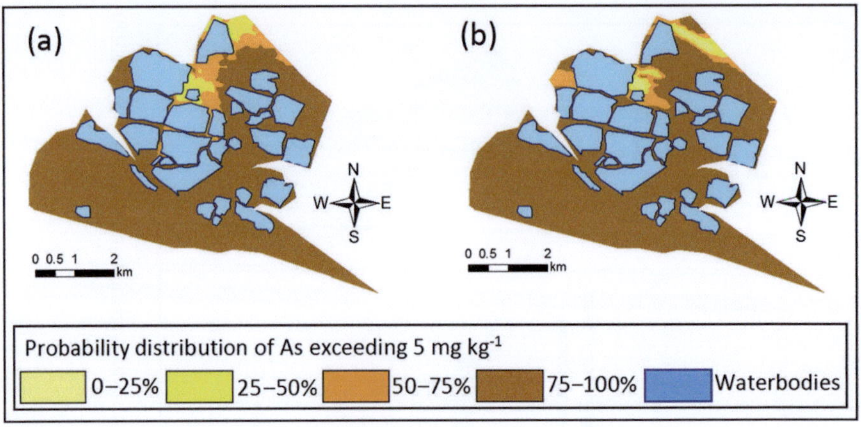

Fig. 4.4 Soil As contamination mapping via VisNIR DRS in Dhapa, Kolkata (Chakraborty et al. 2017c)

"site-specific farming." The implementation of PA requires knowledge of the spatial distribution of soil properties and crop conditions in order to properly tailor application of agrichemicals to crop needs. Traditionally, the spatial variability of soil properties is obtained via field grid sampling: farmers or agricultural specialists collect soil samples that are later analyzed for individual constituents such as nitrogen (N) and phosphorus (P). This conventional procedure is labor-intensive, time-consuming, and expensive.

Currently, most soil analyses in India are done through laboratory chemical analysis. There are about 1049 soil testing laboratories operating in the country with an annual analyzing capacity of 10.7 million samples. Distribution of soil health cards to individual farmers is the dream scheme of Indian Prime Minister and country has approximately 137 million agricultural fields and the capacity of soil testing labs simply lags far behind the requirement.

In India, very few studies have focused on developing advance sensor-based soil test expert system for reliable prediction of soil physicochemical attributes. Vibhute et al. (2018) used VisNIR DRS spectra of 74 soil specimens which were agglomerated by farming sectors of Phulambri Tehsil of the Aurangabad region of Maharashtra, India. Furthermore, the quantitative analysis of VisNIR spectrum was done. Srivastava et al. (2017) used DRS to rapidly predict soil salinity in the IGP with around 80% accuracy. They have also successfully validated the spectral models for other salinity parameters.

Chakraborty et al. (2017b, c) used DRS to rapidly predict soil As contamination and several pools of soil As in a landfill agricultural site of Kolkata (Fig. 4.4). Chakraborty et al. (2015) also used MIR DRS to measure soil Pb contamination in India soils. The first use of PXRF in Indian soil was reported by Chakraborty et al. (2019b) who concluded that different land uses can be differentiated effectively based on soil PXRF data (Fig. 4.5). Chakraborty et al. (2014) also used DRS for effectively monitoring compost enzymatic activity.

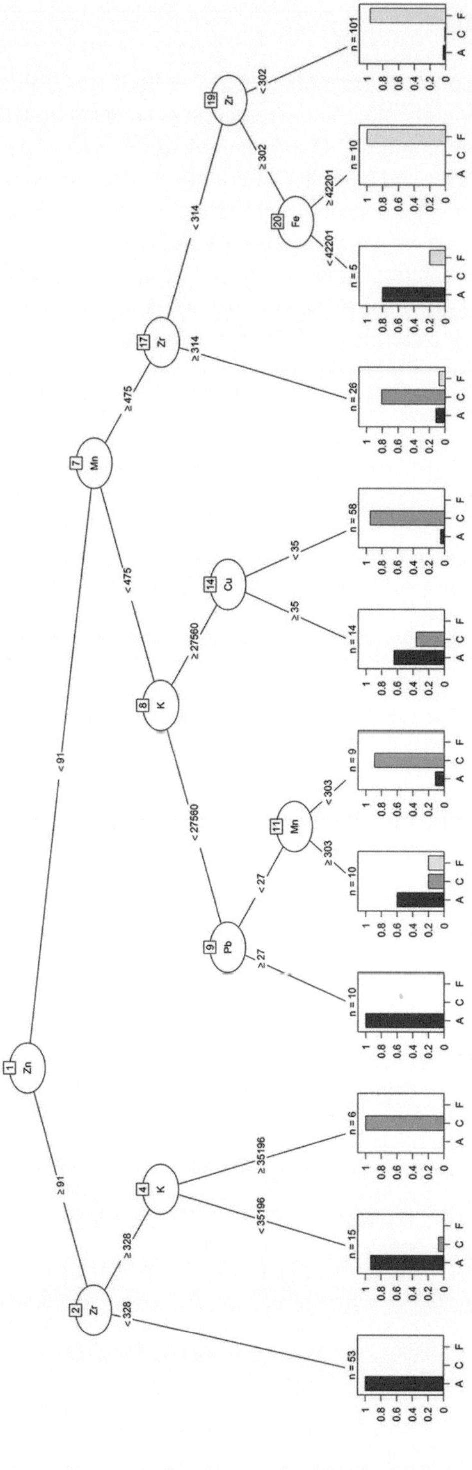

Fig. 4.5 A classification and regression tree (CART) model using PXRF elements to separate soil samples from three different land uses (Chakraborty et al. 2019b)

4.6 New Age Soil Color Sensors

Since soil OC content influences soil color (Baumgardner et al. 1969), it is possible to create spectral prediction models for soil color. Many researchers have calibrated DRS-based spectra for rapid prediction of soil OC (Morgan et al. 2009). Nevertheless, due to high cost of spectroradiometer and complex nature of spectral modeling researchers have proposed some cheap color sensors to classify soils (Rossel et al. 2006). NixPro is another cheap color sensor which has been utilized recently as a colorimeter for developing soil OC prediction models (Stiglitz et al. 2016). This rechargeable and portable sensor measures color indices like RGB, CYMK, and LAB and can replace the use of Munsell color chart. Moreover, research is going on for using this sensor for rapid estimation of plant nutrient status.

References

Acosta-Martínez V, Cotton J (2017) Lasting effects of soil health improvements with management changes in cotton-based cropping systems in a sandy soil. Biol Fertil Soils 53:533–546. https://doi.org/10.1007/s00374-017-1192-2

Aldabaa AAA, Weindorf DC, Chakraborty S, Sharma A, Li B (2015) Combination of proximal and remote sensing methods for rapid soil salinity quantification. Geoderma 239-240:34–46. https://doi.org/10.1016/j.geoderma.2014.09.011

Baumgardner MF, Kristof S, Johannsen CJ, Zachary A (1969) Effects of organic matter on the multispectral properties of soils. Purdue University Agr Exp Station J 3939:413–422

Beare MH, Hendrix PF, Coleman DC (1994) Water-stable aggregates and organic matter fractions in conventional- and no-tillage soils. Soil Sci Soc Am J 58:777–786

Blankinship JC, Fonte SJ, Six J, Schimel JP (2016) Plant versus microbial controls on soil aggregate stability in a seasonally dry ecosystem. Geoderma 272:39–50

Cano A, Nunez A, Acosta-Martinez V, Schipanski M, Ghimire R, Rice C, West C (2018) Current knowledge and future research directions to link soil health and water conservation in the Ogallala Aquifer region. Geoderma 328:109–118. https://doi.org/10.1016/j.geoderma.2018.04.027

Chakraborty S, Das BS, Ali N, Li B, Sarathjith MC, Majumder K, Ray DP (2014) Rapid estimation of compost enzymatic activity by spectral analysis method combined with machine learning. Waste Manag 34:623–631

Chakraborty S, Weindorf DC, Paul S, Ghosh B, Li B, Ali MN, Ghosh RK, Ray DP, Majumdar K (2015) Diffuse reflectance spectroscopy for monitoring lead in landfill agricultural soils of India. Geoderma Reg 5:77–85

Chakraborty S, Weindorf DC, Weindorf CA, Das BS, Li B, Duda B, Pennington S, Ortiz R (2017a) Semi-quantitative evaluation of secondary carbonates via portable X-ray fluorescence spectrometry. Soil Sci Soc Am J 81:844–852

Chakraborty S, Li B, Deb S, Paul S, Weindorf DC, Das BS (2017b) Predicting soil arsenic pools by visible near infrared diffuse reflectance spectroscopy. Geoderma 296:30–37

Chakraborty S, Weindorf DC, Deb S, Li B, Paul S, Choudhury A, Ray DP (2017c) Rapid assessment of regional soil arsenic pollution risk via diffuse reflectance spectroscopy. Geoderma 289:72–81

Chakraborty S, Li B, Weindorf DC, Morgan CLS (2019a) External parameter orthogonalisation of Eastern European VisNIR-DRS soil spectra. Geoderma 337:65–75

Chakraborty S, Li B, Weindorf DC, Deb S, Acree A, De P, Panda P (2019b) Use of portable X-ray fluorescence spectrometry for classifying soils from different land use land cover systems in India. Geoderma 338:5–13

Doran JW, Parkin TB (1996) Quantitative indicators of soil quality: a minimum data set. In: Doran JW, Jones AJ (eds) Methods for Assessing Soil Quality, vol 49. Soil Science Society of America, Special Publication, Madison, pp 25–37

Escadafal R, Girard M, Courault D (1989) Munsell soil color and soil reflectance in the visible spectral bands of Landsat MSS and TM data. Remote Sens Environ 27:37–46

Horta A, Malone B, Stockmann U, Minasny B, Bishop TFA, McBratney AB, Pallasser R, Pozza L (2015) Potential of integrated field spectroscopy and spatial analysis for enhanced assessment of soil contamination: a prospective review. Geoderma 241–242:180–209

Karlen DL, Mausbach MJ, Doran JW, Cline RG, Harris RF, Schuman GE (1997) Soil quality: a concept, definition, and framework for evaluation. Soil Sci Soc Am J 61:4–10

Koch J, Chakraborty S, Li B, Moore-Kucera J, van Deventer P, Daniell A, Faul C, Man T, Pearson D, Duda B, Weindorf CA, Weindorf DC (2017) Proximal sensor analysis of mine tailings in South Africa: an exploratory study. J Geochem Explor 181:45–57

Li B, Chakraborty S, Godoy M, Kusi NYO, Weindorf DC (2018) Compost cation exchange capacity via portable X-ray fluorescence (PXRF) spectrometry. Compost Sci Util 26:271. https://doi.org/10.1080/1065657X.2018.1522280

Magdoff F, Van Es H (2000) Building soils for better crops. Sustainable Agriculture Network, Beltsville, p 230

McGladdery C, Weindorf DC, Chakraborty S, Li B, Paulette L, Podar D, Pearson D, Kusi NYO, Duda B (2018) Elemental assessment of vegetation via portable X-ray fluorescence (PXRF) spectrometry. J Environ Manag 210:21–225

Morgan CLS, Waiser TH, Brown DJ, Hallmark CT (2009) Simulated in situ characterization of soil organic and inorganic carbon with visible near-infrared diffuse reflectance spectroscopy. Geoderma 151(3–4):249–256

Mulders M (1987) Remote sensing in soil science, Development in soil science. Elsevier, Amsterdam, p 379

Pearson D, Chakraborty S, Duda B, Li B, Weindorf DC, Deb S, Brevik E, Ray DP (2017) Water analysis via portable X-ray fluorescence spectrometry. J Hydrol 544:172–179

Pearson D, Weindorf DC, Chakraborty S, Li B, Koch J, Van Deventer P, de Wet J, Yaw Kusi N (2018) Analysis of metal-laden water via portable X-ray fluorescence spectrometry. J Hydrol 561:267–276

Raj A, Chakraborty S, Duda BM, Weindorf DC, Li B, Roy S, Sarathjith MC, Das BS (2018) Soil mapping via diffuse reflectance spectroscopy based variable indicators: an ordered predictor selection approach. Geoderma 314:146–159

Rawal A, Chakraborty S, Li B, Lewis K, Godoy M, Paulette L, Weindorf DC (2019) Determination of base saturation percentage in agricultural soils via portable X-ray fluorescence spectrometer. Geoderma 338:375–382

Rawls WJ, Pachepsky YA, Ritchie JC, Sobecki TM, Bloodworth H (2003) Effect of soil organic carbon on soil water retention. Geoderma 116:61–76

Rossel RAV, Walvoort DJJ, McBratney AB, Janik LJ, Skjemstad JO (2006) Visible, near infrared, mid infrared or combined diffuse reflectance spectroscopy for simultaneous assessment of various soil properties. Geoderma 131:59–75

Sharma A, Weindorf DC, Man T, Aldabaa A, Chakraborty S (2014) Characterizing soils via portable X-ray fluorescence spectrometer: 3. Soil reaction (pH). Geoderma 232-234:141–147

Sharma A, Weindorf DC, Wang D, Chakraborty S (2015) Characterizing soils via portable X-ray fluorescence spectrometry: 4. Cation exchange capacity (CEC). Geoderma 239-240:130–134

Shaver TM, Peterson GA, Ahuja LR, Westfall DG, Sherrod LA, Dunn G (2002) Surface soil physical properties after twelve years of dryland no till management. Soil Sci Soc Am J 66:1296–1303

Sheffield K, Morse-McNabb E (2015) Using satellite imagery to asses trends in soil and crop productivity across landscapes. IOP Conf Ser Earth Environ Sci 25:012013. https://doi.org/10.1088/1755-1315/25/1/012013

Soil Survey Staff (2014) Kellogg soil survey laboratory methods manual. Soil Survey Investigations Report No. 42. Version 5.0. USDA-NRCS, Lincoln, NE

Srivastava R, Sethi M, Yadav RK, Bundela DS, Singh M, Chattaraj S (2017) Visible-near infrared reflectance spectroscopy for rapid characterization of salt-affected soil in the Indo-Gangetic Plains of Haryana, India. J Indian Soc Remote Sensing 45(2):307–315

Stiglitz R, Mikhailova E, Post C, Schlautman M, Sharp J (2016) Evaluation of an in expensive sensor to measure soil color. Comput Electron Agric 121:141–148

Swanhart S, Weindorf DC, Chakraborty S, Bakr N, Zhu Y, Nelson C, Shook K, Acree A (2014) Soil salinity measurement via portable X-ray fluorescence spectrometry. Soil Sci 179:417–423

US-EPA (2007) Method 6200: field portable X-ray fluorescence spectrometry for the determination of elemental concentrations in soil and sediment. https://www.epa.gov/sites/production/files/2015-12/documents/6200.pdf (verified 12 Dec 2018)

Varvel GE, Schlemmer MR, Schepers JS (1999) Relationship between spectral data from an aerial image and soil organic matter and phosphorus levels. Precis Agric 1:291–300

Vibhute AD, Kale KV, Mehrotra SC, Dhumal RK, Nagne AD (2018) Determination of soil physicochemical attributes in farming sites through visible, near-infrared diffuse reflectance spectroscopy and PLSR modeling. Ecol Process 7:26

Wang D, Chakraborty S, Weindorf DC, Li B, Sharma A, Paul S, Ali N (2015) Synthesized use of VisNIR DRS and PXRF for soil characterization: total carbon and total nitrogen. Geoderma 243-244:157–167

Weindorf DC, Chakraborty S (2016) Portable X-ray fluorescence spectrometry analysis of soils. Methods Soil Anal. Available online at https://dl.sciencesocieties.org/publications/msa/abstracts/1/1/methods-soil.2015.0033 (verified 17 Aug 2018)

Weindorf DC, Herrero J, Bakr N, Swanhart S (2013a) Direct soil gypsum quantification via portable X-ray fluorescence spectrometry. Soil Sci Soc Am J 77:2071–2077

Weindorf DC, Paulette L, Man T (2013b) On-site assessment of metal contamination via portable X-ray fluorescence spectroscopy: Zlatna, Romania. Environ Pollut 182:92–100

Weindorf DC, Chakraborty S, Aldabaa A, Paulette L, Corti G, Cocco S, Michéli E, Wang D, Li B, Man T, Sharma A, Person T (2015) Lithologic discontinuity identification via proximal sensors. Soil Sci Soc Am J 79(6):1704–1716

Weindorf DC, Chakraborty S, Li B, Deb S, Singh A, Kusi NY (2018) Compost salinity assessment via portable X-ray fluorescence (PXRF) spectrometry. Waste Manag 78:158–163

Zhu Y, Weindorf DC, Zhang W (2011) Characterizing soils using a portable X-ray fluorescence spectrometer: 1. Soil texture. Geoderma 167-168:167–177

Isotopes and Tracer Techniques for Soil Analysis

5

Manoj Shrivastava, P. C. Srivastava, and S. F. D'Souza

Abstract

Natural and manmade radio and stable isotopes are very useful research tool for soil scientists. Since the nutrient requirements of crops vary considerably with soil type, climatic conditions and plant species, precise knowledge is required on the type, amount, method and time of application of fertilizer materials best suited for specific soil–crop combinations. The availability of isotopes for many plant nutrients such as ^{13}C, ^{2}H, ^{15}N, ^{32}P, ^{35}S, ^{59}Fe, ^{54}Mn, ^{65}Zn, etc. makes it possible for the researchers to explore investigations on soil fertility, nutrient and water use efficiency of plant and soil erosion and degradation and hence provides essential data to develop effective soil and water management strategies to protect soil health and water quality for sustainable crop production.

Keywords

A-value · E-value · Isotope exchange kinetics · L-value · Radio-isotopes · Stable isotopes · Soil · Tracers

5.1 Introduction

Soil is an important part of the ecological environment, and is a major resource contributing to human survival and development. Climate change, urbanization, industrialization, mineral resource development, intensive agriculture and land use

M. Shrivastava (✉)
Centre for Environment Science and Climate Resilient Agriculture, Indian Agricultural Research Institute, New Delhi, India

P. C. Srivastava
Department of Soil Science, G. B. Pant University of Agriculture and Technology, Pantnagar, India

S. F. D'Souza
Bhabha Atomic Research Centre, Mumbai, India

A. Rakshit et al. (eds.), *Soil Analysis: Recent Trends and Applications*,
https://doi.org/10.1007/978-981-15-2039-6_5

change causing soil degradation. Therefore, an effective measure of soil heath and degradation and the control and management of excessive emission of GHG and soil pollutant are both important means to improve the soil productivity. Plant mineral nutrition involves studies on nutrient uptake, translocation, cellular partitioning and utilization. The other important factors controlling plant mineral nutrition are (re)-translocation of the nutrients once they are taken by the plant that is an important adaptive mechanism under condition of nutrient deficiency. Isotopes and isotopic techniques play an important role in agriculture, in general, and soil science research, in particular (Ramachandran et al. 2007, 2009). Since the nutrient requirements of crops vary considerably with the soil type, climatic conditions and plant species, precise knowledge is required on the type, amount, method and time of application of fertilizer materials best suited for specific soil–crop combinations. Use of isotope labeled fertilizer materials permits such determinations and reveals directly within weeks, information which otherwise takes long period and that too with elaborate field studies.

The availability of stable and radio-isotopes for many of plant nutrients such as ^{15}N, ^{32}P, ^{35}S, ^{59}Fe, ^{54}Mn, ^{65}Zn, etc. makes it possible for the researchers to investigate on soil fertility, and availability of plant macro- and micro-nutrients. Agriculture currently accounts for about 70% of global freshwater use and the Food and Agriculture Organization (FAO) of the United Nations (UN) forecasts that, by 2050, global water requirements for agriculture will increase by 50% in order to meet the increased food demands of a growing population. With an increasing scarcity of freshwater, due to indiscriminate use and a changing climate with extreme weather events of droughts and flooding, there is an urgent need to improve the management of this resource. Isotopic and nuclear techniques are useful and effective tools to assess the soil water status, particularly in the immediate vicinity of crop roots, to trace soil water movement and to identify hot spots of land degradation that deliver sediments and affect downstream water quality.

5.2 Overview of Theory and Use of Isotopes for Soil Analysis

5.2.1 Isotopic Tracer Methodology

Use of isotopic tracer methodology in soil analysis is mainly based on isotope dilution principle. This principle can be explained by the following equation:

$$S = s\left[\left(a_i/a_f\right) - 1\right] \tag{5.1}$$

where $S = g$ or moles of the test substance, $s =$ amount of test substance associated with added isotopic tracer, $a_i =$ specific activity before equilibrium, $a_f =$ specific activity after equilibrium

Table 5.1 Important radio-isotopes used in soil study

Element	Radio-isotope	Emission	Half-life	Uses
Carbon	^{14}C	Beta	5720 years	Photosynthesis, organic matter
Phosphorus	^{32}P	Beta	14.3 days	Soil fertility, root distribution
Potassium	^{40}K	Beta	1.3×10^9 years	Soil fertility, K balance
Calcium	^{45}Ca	Beta	165 days	Ion uptake, soil exchangeable Ca
Magnesium	^{28}Mg	Beta, gamma	21.3 h	Movement in plants
Sulphur	^{35}S	Beta	87 days	Soil availability, plant uptake
Iron	^{55}Fe ^{59}Fe	Beta Gamma, beta	2.6 years 45.6 days	Soil erosion, foliar nutrition, soil availability
Manganese	^{54}Mn	Gamma, beta	314 days	Foliar nutrition, soil availability
Copper	^{64}Cu	Gamma, beta	12.8 h	Soil and plant movement
Zinc	^{65}Zn	Gamma, beta	245 days	Soil fertility, soil movement
Molybdenum	^{99}Mo	Gamma, beta	66.7 h	Plant movement
Sodium	^{22}Na	Gamma, beta	2.6 years	Salt tolerance, cell permeability
Chlorine	^{36}Cl	Beta	3.08×10^5 years	Salt tolerance, solute movement in soil

5.2.1.1 Radioactive Isotopes as Tracers

Radioactive isotopes can be used to follow a particular element through various pathways and quantitative measurements may be made. They have the advantage of behaving in the same way that their stable counterparts do, but they can be readily traced. Radioactive isotopes can be likened to a coloured dye. They have a wide range of uses and are particularly valuable in plant nutrition research. The physical properties of a radioactive nuclide determine its usefulness as a tracer. The three most important are half-life, mode of decay and decay energy. If the half-life of a nuclide is very short, any compound labeled with it will be difficult to prepare, use and measure within the time of decay. The mode and energy of decay determine how the nuclide will be measured. Some of the important radio-isotopes used in soil studies are given in Table 5.1.

Measurement of Radio-Isotopes

Radioactive decay is a spontaneous reaction occurring when there is nuclear instability. Nuclides vary considerably in their instability and unstable nuclei emit subatomic particles like alpha (α), beta (β) and electromagnetic radiation like gamma (γ) through a series of decay reactions. By detecting these particles and the radiation by various instruments radioactivity can be measured. In soil study mainly β and γ emitting radio-isotopes are being used.

Fig. 5.1 Energy spectra for ^{32}P and ^{35}S (source: IAEA-TCS 14 2001)

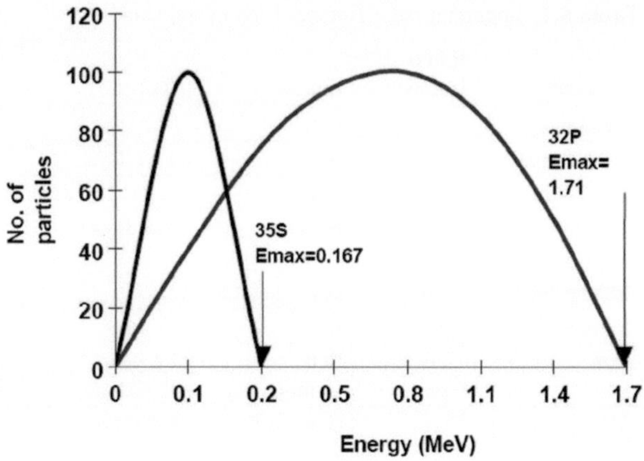

Beta (β) Counting

When β emitter radio-isotopes are used in soil plant system, they can be measured by various instruments. Following techniques are being used to measure the β emitting radio-isotopes:

Liquid Scintillation Counting (LSC)

Beta decay occurs when an unstable nucleus with an excess of neutrons returns to stability equilibrium through the conversion of a neutron to a proton with the emission of an electron and an antineutrino as follows:

$$\text{Neutron (n)} \rightarrow \text{Proton (p+)} + \text{Electron (e−)} + \text{Antineutrino } (^{\vee})$$

This process occurs within the nucleus of an isotope like phosphorus-32, which has 15 protons and 17 neutrons but as a result of beta decay, transforms into stable sulphur with a nucleus of 16 protons and 16 neutrons. The total energy of beta decay, Emax, consists of the combined energy of the emitted the beta and antineutrino particles. Very few of the emitted beta particles have maximum energy, as energy is shared between the beta particle and the antineutrino. Most of the emitted beta particles have an average energy of approximately one third of Emax (Fig. 5.1). The windows can be set on the liquid scintillation counter LSC to capture β particles of particular energies. These window settings can also be used to simultaneously count two or more isotopes that have different energy spectra such as the ^{32}P and ^{35}S isotopes in Fig. 5.1.

Cerenkov Counting

When β particles are emitted they leave the nucleus at speeds approaching that of light in a vacuum. However, in the surrounding medium the speed of light is lower and consequently the passage of particles through the medium causes shock waves from which light photons are emitted. This light can be counted to give a measure of the radioactivity present. The minimum energy required to produce light in an aqueous solution is 0.263 MeV. This means that only those radionuclides with high-energy emissions, such as ^{32}P, may be counted by this method.

Geiger–Müller (GM) Counting

In contrast to the scintillation counter, Geiger–Müller counter is a device for measuring ionization. Its efficiency is usually low and its use has declined markedly since the introduction of sophisticated scintillation counters. However, sample preparation is very simple; no solvent is used and the radiation from the sample interacts directly with the ionization chamber. As the source material does not interfere with the operation of the GM tube, this technique can be readily used on soil providing the samples are uniformly treated.

Gamma (γ) Counting

Gamma rays are electromagnetic radiation similar to light but with much shorter wavelength. Both have well defined energies. Gamma emissions are mono-energetic so they are not like beta emission where there is a broad spread of energies. Most gamma emitters produce gamma particles with a single energy but some like ^{137}Cs emit two or even more gamma rays of different energies, which arise from different decay processes. The energy of the gamma photon is three to six orders of magnitude higher than that of light photon. The energies measured in a gamma counter are in the range 20 keV–2 MeV. The gamma counter has two detector elements: a single crystal of thallium activated sodium iodide NaI (Tl) or more efficient high purity germanium HpGe crystal and the photomultiplier. In the energies below 2 MeV interaction of gamma rays with the crystal may take place by two principal mechanisms. In the photoelectric effect a gamma photon disappears and a (photo) electron ejected from one of the atomic electron shells with kinetic energy, which is the difference between the gamma photon and the orbital electron binding energy. The photoelectric absorption is used in the measurement of the energy of a gamma photon. Solid and liquid samples can be counted. The sample is introduced into the counter alongside the crystal and photomultiplier and the gamma rays emitted from the sample counted.

Radiotracer Techniques to Evaluate Phosphorus Dynamic in Soil Plant System

^{32}P and ^{33}P can be used in soil fertility and P cycling studies. ^{32}P emitting the β^- radiation of high energy (1.71 MeV) can be readily detected and counted by Cerenkov counting on a liquid scintillation counter (L'Annunziata 1987) as well as by GM counter. The main limitation with the use of ^{32}P is its short half-life (14.3 days), which limits the duration of crop growth experiments. In addition, the high-energy β^- radiation emitted by ^{32}P poses a considerable external radiation hazard. On the other hand ^{33}P has longer half-life (24.4 days) which allows longer plant growth experiments. The β^- radiation emitted by ^{33}P has a lower energy level (0.248 MeV) than that of ^{32}P, and is thus less hazardous for external exposure. However, ^{33}P is more expensive than ^{32}P and since the β^- particle energy from ^{33}P is lower than the threshold level (0.263 MeV for aqueous samples) required for Cerenkov counting, ^{33}P can only be counted by liquid scintillation, which requires the addition of fluorescent reagents (scintillation cocktails) (Simonnet 1990). Further the availability of both radioactive P isotopes (^{32}P and ^{33}P) makes it possible to use double-labeling techniques. Counting of ^{32}P and ^{33}P dual-labeled samples can be easily carried out on a liquid scintillation counter with dual channels (L'Annunziata 1987).

Determination of Exchangeable (or Labile) P in the Soil

Exchangeable P in the soil refers to P in soil solution and on soil surfaces that undergoes exchanges with ^{32}P phosphate introduced into the system. Its measurement using isotopes is based on the principle that if an aliquot containing labeled P is introduced into a soil system, the labeled P will exchange with and be diluted by the isotopically distinguishable, but otherwise identical form of P in the soil. The radioactivity per unit weight of phosphate, known as the specific activity (SA), in a sample taken from the equilibrated system will reflect the relative quantities of the two isotopes (ratio of $^{32}P/^{31}P$). If the initial specific activity of the labeled P added to the soil is SA_1, defined as:

$$SA_1 = q_o/Q_1 \tag{5.2}$$

where q_o is the radio-isotope activity introduced to the system and Q_1 is the quantity of the labeled P, after the labeled is diluted by the soil exchangeable P (Q_2), the final specific activity (SA_2) becomes

$$SA_2 = q_o/(Q_1 + Q_2) \tag{5.3}$$

The amount of isotopically exchangeable P in the soil (Q_2) is calculated by combining Eqs. (5.2) and (5.3) as:

$$Q_2 = Q_1\ (SA_1/SA_2 - 1) \tag{5.4}$$

If the isotope is used carrier-free, Q_2 is then calculated as:

$$Q_2 = q_o/SA_2 \tag{5.5}$$

The above equations represent the isotopic dilution principle.

Two isotopic methods have been developed using the isotopic dilution principle as to determine the amount of exchangeable P in the soil. The values from these methods are known as the *E*- and *L*-values. Though *E*- and *L*-values have similar conceptual basis, the *E*-value is measured after exchange times from a few hours to several days, in contrast to several months for *L*-values.However, the *E*-value is measured in a soil suspension, whereas the *L*-value is obtained under plant growing conditions; plant root-induced processes (e.g., acidification, mycorrhizal effects) may significantly increase the amount P available for plant uptake in the rhizosphere (Di et al. 1997).

E-value

The *E*-value is a measure of soil exchangeable P obtained by shaking soil with a solution containing ^{32}P and measuring the specific activity of the solution after a period of equilibration (McAuliffe et al. 1948; Russell et al. 1954). The solution may be distilled water, dilute carrier-P solution or dilute electrolyte containing ^{32}P. A solution to solid ratio of 10–20 is used for convenience. The period of shaking is kept in the range of 1–24 h. The *E*-value is measured in dilute soil suspension, a situation that is very different from that under which plants grow, the ability of *E* to reflect soil P supply to plants may be limited, particularly in soils with high P retention capacity (Amer et al. 1969).

Isotope Exchange Kinetics and E_{piet} Value

Fardeau et al. (1985) introduced the concept of isotope exchange kinetics and the value is known as $E_{piet.}$ When $^{32}PO_4$ ions are introduced carrier-free in a soil suspension, the change of radioactivity with time could be described by the following equation for exchange times between 30 s and 4 months

$$R_t/R_0 = \left\{R_1/R_0\left[t + (R_1/R_0)^{(1/n)}\right]^{-n}\right\} + (R_\infty/R_0) \tag{5.6}$$

where *t* is the isotopic exchange time (minutes), R_0 the total radioactivity (MBq) introduced, R_t the radioactivity remaining in the solution after *t* min, R_1 the radioactivity remaining in the solution after 1 min, R_∞ the quantity of radioactivity remaining in the solution after an infinite exchange period and *n* an experimental factor ranging between 0 and 0.5. The value R_∞/R_0 is the maximum possible dilution of the added $^{32}PO_4$ ions by $^{31}PO_4$ ions present in the soil/solution system, and can be approximated by the ratio of water-soluble P to the mineral P content of the soil (Di et al. 1997):

$$R_\infty/R_0 = (10)\ (\mathrm{C}/\min\ \mathrm{P}) \tag{5.7}$$

where R_∞ and R_0 have the same meaning as in Eq. (5.6), min P is the mineral P (mg P kg^{-1}) content of the soil and C is the concentration of P in the soil solution (mg L^{-1}). The factor 10 arises from the fact that 1 g soil is suspended in 10 mL water. The decrease of radioactivity in the solution is attributed to exchanges between the introduced $^{32}PO_4$ ions and the $^{31}PO_4$ ions located on the solid phase of the soil. The specific activity of phosphate in the solution is assumed to be identical to that of the exchanged system:

$$R_0/E_{piet} = R_t/(10\,\mathrm{C}) \tag{5.8}$$

$$E_{piet} = 10\,\mathrm{C}(R_0/R_t) \tag{5.9}$$

where E_{piet} is the quantity of isotopically exchangeable P.

In this approach, soil mineral P is not divided into two pools, one containing the available P and the other containing unavailable forms of P; rather the exchange reactions are regarded as continuous processes which ultimately could result in the exchange of all the mineral P in the soil. The amount of exchangeable P therefore depends on the exchange time. A useful index developed from this approach is the amount of soil P exchangeable after 1 min equilibration, E_{pie1} (Tran et al. 1988; Salcedo et al. 1991):

$$E_{\mathrm{pie1}} = 10\,\mathrm{C}\,(R_0/R_1) \tag{5.10}$$

The exchangeable P pool measured at 1 min is regarded as a homogeneous pool of ions including phosphate in soil solution and phosphate on soil surfaces with the same mobility.

Soil exchangeable P divided into four pools as measured after different exchange times, and interpreted these P pools in terms of plant uptake patterns (a) P exchangeable between 1 min and 24 h (time period for root uptake of P from soil), (b) between 24 h and 3 months (growth period of the root system of an annual crop), (c) between 3 months and 1 year and (d) periods longer than 1 year (Di et al. 1997)

Soil P Buffer Capacity (PBC)

Soil P buffer capacity (PBC) is another index which is associated with exchangeable P and soil P supply to plants. It describes the overall relationship between the concentration of P in soil solution (intensity, I) and the quantity of exchangeable P on soil surfaces (quantity, Q) (Ozanne 1980). The value of R_0/R_1 (as defined in Eq. (5.10)) is used to provide a measure of the soil PBC. The R_0/R_1 values are shown to vary widely (about 1–30) in a range of soils, and were primarily influenced by clay and aluminium oxide.

L-value Technique

The *L*-value is measured by labeling soil with carrier-free ^{32}P, and using plants to sample the specific activity of soil exchangeable P. Thus the *L*-value effectively represents the fraction of soil P that is exchangeable with added isotope P as measured by plant uptake. The *L*-value method is based on the assumption of equilibrium between the added ^{32}P with exchangeable P in the soil (Larsen 1952, 1967). To obtain specific activity to derive *L*-value, plants are harvest as early as practically possible, recover the plants completely from the soil (including both tops and roots) and determine the total ^{32}P and ^{31}P content. The amount of P contained in the seeds is subtracted from the readily exchangeable P calculated from Eq. (5.5) (where Q_2 is L).

A-value

The *A*-value is a measure of plant available P in soil compared with an isotopically labeled fertilizer standard. The *A*-value does not refer to the amount of exchangeable P in the soil (as is the case with E and L) but the amount of plant available P compared with a fertilizer standard. The actual values of A for the same soil may vary, depending on the type of fertilizer standard, application rate, method of application, crop species and the stage of plant growth (Fried and Dean 1952; Shrivastava et al. 2007). Its measurement is based on the assumption that if a plant is presented with two sources of a nutrient, it will take up the nutrient from the two sources in direct proportion to the respective quantities available. *A*-value techniques may be divided in two categories; (a) labeling the applied P source (Srivastava and D'Souza 2007; Shrivastava et al. 2007; Srivastava et al. 2009) or (b) labeling of soil P (Shrivastava et al. 2011; Shrivastava and D'Souza 2014).

Labeling the Applied P Source

An inorganic or organic P source is labeled with ^{32}P or ^{33}P phosphate and then applied to the soil; the release rate of the added P is determined by monitoring the specific activity of P in the plants growing in the treated soil. The temporal variation pattern in the amount of radioactive isotope (usually in proportion to total isotope applied) taken up by plants, or specific activity, may indicate the rate at which P is released from the added source. An alternate to labeling the fertilizer under study is to label a standard or reference fertilizer and measure the *A*-value of soils treated with different fertilizer sources (Fried and Dean 1952; Kucey and Bole 1984; Shrivastava et al. 2007).

A-value is calculated by the following equation:

$$A = B\,(1 - Y)/Y \tag{5.11}$$

where B is the amount of fertilizer nutrient standard applied and Y is the proportion of nutrient in the plant derived from the fertilizer which is calculated as follows:

$$Y = \mathrm{SA}_2/\mathrm{SA}_1 \tag{5.12}$$

where SA_1 and SA_2 are the specific activities of the fertilizer standard and the plant grown, respectively. Combining Eqs. (5.11) and (5.12) gives

$$A = B\,(\mathrm{SA}_1/\mathrm{SA}_2 - 1) \tag{5.13}$$

Labeling of Soil P

Naturally occurring sparingly soluble organic and inorganic P sources like phosphate rocks, organic manure, sewage sludge, crop residues and guano materials are difficult to label. Instead of labeling these P sources, the soil to which the P source is applied can be labeled with ^{32}P or ^{33}P. The amount of P released from the P source is determined by the extent to which the radio-isotope is diluted. If the soil is uniformly labeled with ^{32}P and mixed completely with a P source, then the amount of P released is assessed by the increase in the exchangeable P.

If the specific activity of plants grown in the soil which received the fertilizer is SA_{s+f}, and that of plants grown in a control (i.e., without fertilizer application) is SA_s, then the amount of exchangeable P in the fertilized soil (Q_{s+f}) and in the control (Q_s) is calculated as:

$$Q_{s+f} = q_0/\mathrm{SA}_{s+f} \tag{5.14}$$

$$Q_s = q_0/\mathrm{SA}_s \tag{5.15}$$

where q_0 is the activity of ^{32}P introduced into the system.

The difference between Q_{s+f} and Q_s indicates the amount of P available from the fertilizer. The proportion of P derived from the soil exchangeable P (rather than from fertilizer P) in the plants grown in the fertilized soils (Z) is

$$Z = \mathrm{SA}_{s+f}/\mathrm{SA}_s \tag{5.16}$$

and the proportion of P in the plants derived from the fertilizer (Y) is

$$Y = 1 - SA_{s+f}/SA_s \tag{5.17}$$

A-value technique is used extensively in soil fertility research. In this technique, the different isotopic parameters, namely percent phosphorus (P) derived from fertilizer (%Pdff), percent P derived from soil (%Pdfs), *A*-value (available P from soil) and percent fertilizer P utilization (FPU) are computed.

$$\%\text{Pdff} = \left[\text{Specific activity of plant}\left(\text{dpm mg P}^{-1}\right)\right]/\left[\text{Specific activity of fertilizer}\left(\text{dpm mg P}^{-1}\right)\right] \times 100 \tag{5.18}$$

$$A - \text{value}\left(\text{mg P pot}^{-1}\right) = (100 - \%\text{Pdff})/(\%\text{Pdff}) \times \text{Applied P rate}\left(\text{mg pot}^{-1}\right) \tag{5.19}$$

$$\%\text{FPU} = \left[\%\text{Pdff} \times \text{total P uptake}\left(\text{mg pot}^{-1}\right)\right]/\text{Fertilizer P added}\left(\text{mg pot}^{-1}\right) \tag{5.20}$$

The *A*-value expresses the availability of P in soil system relative to that of the ^{32}P-SSP in units of carrier (mg P pot^{-1}). Same way *A*-value for other nutrients like N, S, Zn, Fe, Mn and Cu can be evaluated by using respective tracer (isotopes).

5.2.1.2 Stable Isotopes as Tracers

Stable isotopes are used in the same way as radioactive isotopes in soil/plant studies. Whereas radioactive isotopes emit particles which are captured in photomultiplier tubes and counted stable isotopes are separated from each other by passing a gas containing them through a strong magnetic field, which deflects them differentially according to their mass.

Stable isotopes are helpful to understand the elemental cycles in soil-plant-atmospheric continuum. Stable isotope data can contribute both source-sink (tracer) and process information. The elements C, N, S, H and O all have more than one isotope, and isotopic compositions of natural materials can be measured with great precision with a mass spectrometer. The most common stable isotope used is ^{15}N but a large number of other stable isotopes are produced which are increasingly being used in soil studies. Some of the important stable their characteristics and uses are shown in Table 5.2. Many investigations have been carried out, namely plant uptake studies of macro- and micronutrients, fertilizer use efficiency of inorganic and organic fertilizers, uptake, mobility and adsorption studies of different micronutrients, heavy metal pollutants and fungicides in different soil types for various crops, utilizing the various stable isotopes.

With respect to natural abundance of C isotopes in the environment, 98.9% of C exists as ^{12}C, 1.1% as ^{13}C and 10^{-8}% as the cosmogenic radioactive ^{14}C. Basis for the work with stable isotopes (^{12}C and ^{13}C) at natural abundance is the fact that during kinetic and thermodynamic processes, such as biochemical reactions, phase changes or diffusion, heavier isotopes are normally discriminated against the lighter counterparts because of the higher kinetic energy of the latter. As a consequence, the

Table 5.2 Important stable isotopes used in soil fertility research

Sr. no.	Element	Isotope	Abundance (%)	Uses
1	Hydrogen	^{1}H ^{2}H (D)	99.985 0.015	Water movement, biochemical studies, water cycling
2	Carbon	^{12}C ^{13}C	98.89 1.11	C-12 enriched (C-13 depleted). Organic matter reaction mechanisms work Soil organic matter studies in ecosystems, photosynthesis, carbon translocation, carbon cycling, carbon sequestration
3	Nitrogen	^{14}N ^{15}N	99.63 0.37	Fertilizer N use efficiency, biological nitrogen fixation, N balance, N transformation in soils, N availability from organic-materials, animal nutrition studies
4	Oxygen	^{16}O ^{17}O ^{18}O	99.759 0.037 0.204	Photosynthesis, respiration, soil organic matter studies, ecological studies, hydrology
5	Sulphur	^{32}S ^{33}S ^{34}S ^{36}S	95.00 0.76 4.22 0.014	Potentially useful for fertilizer use efficiency, environmental pollution, ecological and medical research

lighter isotopes accumulate relative to the heavier ones in the reaction products. For easier handling, the small numbers achieved when calculating the ^{13}C:^{12}C ratio and comparing the ratios of different samples, the isotope composition is expressed as δ-value, which was introduced by Craig (1953) and which can be calculated as below according to Eq. (5.21),

$$\delta X(\{U\}\{wasy\}h) = \{\frac{R_{\text{Sample}} - R_{\text{Standard}}}{R_{\text{Standard}}}\} \times 1000(\text{per mil}) \tag{5.21}$$

where X is ^{13}C, ^{15}N or ^{34}S, and R is the corresponding ratio $^{13}C/^{12}C$, $^{15}N/^{14}N$ or $^{34}S/^{32}S$.

For example, Eq. (5.22)

$$\delta^{13}C(‰) = \left\{\frac{R_{\text{Sample}} - R_{\text{Standard}}}{R_{\text{Standard}}}\right\} \times 1000 \text{ (per mil)} \tag{5.22}$$

where R is the molar ratio of the heavy to light isotope, i.e., $^{13}C/^{12}C$. The international reference standard for carbon was a limestone, Pee Dee Belemnite (PDB), which had a $\delta^{13}C$ being equal to 0.0112372.

The δ-values are measures of the amounts of heavy and light isotopes in a sample. Increases in these values denote increases in the amount of the heavy isotope components. Standard reference materials are carbon in the PeeDee limestone, nitrogen gas in the atmosphere and sulphur from the Canyon Diablo meteorite. The precision of the measurements is typically ±0.2‰ or better (Peterson and Fry 1987). A mass spectrometer is required for accurate detection of small differences in

δ-values and gaseous samples are required for the isotopic determinations. Sample preparation for isotopic analysis using isotope ratio mass spectrometer (IRMS) requires a complete conversion of sample to gas. High temperature sealed tube combustion to convert carbon and nitrogen compounds to CO_2 and N_2 can be used. N_2 can also be prepared from Kjeldahl digestions or ammonia (Minagawa et al. 1984). Sulphur-containing materials are converted to sulphates and sulphides, which are in turn converted to SO_2 (Fritz et al. 1974; Halas et al. 1982; Yanagisawa and Sakai 1983). Pure CO_2, N_2 and SO_2 are separated from one another and from water, using various cold traps that allow differential volatilization and trapping under high vacuum conditions. A pure gas is then introduced into a dual or triple collector isotope ratio mass spectrometer, and its isotopic composition measured relative to a known standard.

Measurement of Stable Isotopes

Isotopes have identical chemical properties but some slightly different physical properties. Detection methods use one of these properties such as mass, emission spectrum, IR absorption. The most common and most precise method to measure stable isotopes is mass spectrometry. For the determination of ^{15}N emission spectrometry can also be used, but with much less precision.

Mass Spectrometry

Mass spectrometry (MS) is an analytical technique in which atoms or molecules from a sample are ionized, separated according to their mass-to-charge ratio (m/z), and then recorded. There is a wide range of mass spectrometers for different types of samples with different ionization and separation methods. For determining the isotope ratios of light element stable isotopes (H, C, N, O and S) isotope ratio mass spectrometers (IRMS) is used. The sample has to be converted to a gas (N_2, CO_2, H_2 and SO_2) by means of a suitable preparation system. This gas is fed into the mass spectrometer where the ratios of the isotopes of interest are determined.

Emission Spectrometer

Emission spectrometers are much simpler than mass spectrometer. They can be maintained much easier but they can be used only to determine $^{15}N/^{14}N$.

5.3 Application of Isotopes in Soil and Water Studies

5.3.1 Fertilizer Use Efficiency

Fertilizer use efficiency is one of the key parameters of long-term soil fertility management. Isotope-aided research has established that ammonium polyphosphate (APP), a fertilizer source of phosphorus and a carrier of micronutrients is equal or superior to the orthophosphate fertilizer (DAP) in diverse soil-crop regimes (Yadav and Mistry 1985; D'Souza and Yadav 1993). Further, fertilizer use efficiency of

^{65}Zn-APP was significantly higher than that of ^{65}Zn-DAP for major crops in vertisol and was equally effective in ultisol.

Radioisotopic tracer studies on cotton crop (Srivastava and D'Souza 2007) indicated that the agronomic efficiency of single super phosphate (SSP) and di ammonium phosphate (DAP) is comparable and is superior to nitro phosphate (NP) in vertisol under greenhouse conditions. It was found from radiotracer studies (Shrivastava et al. 2007) that the fertilizer use efficiency of Purulia rock phosphate (PRP) was better than that of Mussoorie rock phosphate (MRP) in two acid soils.

Studies on ^{32}P and ^{65}Zn-labeled P- and Zn-enriched biosludges, from molasses based distillery (Indian Patent Nos. 238485 and 239929), applied to a mollisol, separately (Srivastava et al. 2008, 2009) indicated that application of P-enriched post-methanation biosludge to a rice crop grown under greenhouse conditions significantly enhanced the dry matter yield as compared to SSP application; greenhouse experiments using Zn-enriched post-methanation biosludge revealed that the dry matter yield and total uptake of Zn by rice were statistically similar to those rice crop fertilized with zinc sulphate heptahydrate. Both P- and Zn-enriched biosludges applied to a rice crop had a significantly higher residual effect on the dry matter yield of subsequently grown wheat crop. Application of P- and Zn-enriched biosludges resulted in significant increase in wheat yield under field conditions as well.

Studies on the fertilizer use efficiency of the crop combinations (Kotur et al. 2010), namely capsicum, onion—watermelon, radish—okra, French bean using ^{32}P-labeled fertilizer indicated that P use efficiency of all the crop combinations was intermediate between that of the respective sole crops.

Studies on the mobility of labeled micronutrients, ^{65}Zn, ^{54}Mn and ^{59}Fe in black clay loam and laterite soils revealed that synthetic chelating agent like EDTA significantly induced mobility of these micronutrients by the formation of micronutrient-EDTA complexes; further, uptake of these micronutrients was more in laterite than in black soil and accumulation of Zn and Mn in aerial tissues was more than that of Fe (D'Souza and Mistry 1979).

Zinc-65-aided research on mycorrhizal systems has shown that the Zn use efficiency in maize plants significantly increased in mycorrhizal (M^+) as compared to non-mycorrhizal (M^-) treatments. Further, the water-soluble plus exchangeable Zn (WSEX-Zn) and organically bound Zn (OC-Zn) in M^+ soil were increased by 52 and 82%, respectively, over uninoculated conditions, thus increasing the pool of available Zn in soil. The Zn in crystalline oxide and residual fractions of M^+ soil were reduced by 20–30% indicating thereby the efficiency of mycorrhizal fungus to exploit highly fixed fractions of Zn (Subramanian et al. 2007).

5.3.2 Soil Organic Matter

Soil organic matter is considered a key factor in maintaining soil quality and it is very essential for long-term soil fertility (Wallace 1994). It provides a reservoir of plant nutrients and improves soil structure. The fate of carbon and nutrients released during organic matter decomposition is an important determinant of the short- and long-term soil fertility. Hence, maintenance of soil organic matter must be considered as one of the main objectives of sustainable land-management of productive agriculture. Stable isotopic tracers like ^{15}N, ^{13}C, ^{11}B, ^{34}S, ^{18}O and ^{25}Mg and radioactive tracers such as ^{3}H, ^{14}C, ^{32}P, ^{35}S and ^{86}Rb are very useful in nutrient dynamics in soil. The difference in ^{13}C between C3 and C4 plants has been used to study carbon turnover rates. The different carbon assimilation pathways in C3 and C4 plants result in different ratios of ^{12}C and ^{13}C between these two groups of plants (Cadisch and Giller 1996). Using ^{25}Mg, a stable isotope in their study, Jentschke et al. (2000) found that ectomycorrhiza was capable of enhancing the Mg supply to Norway spruce seedlings.

5.3.3 Root Distribution in Soil

Soil fertility research is also related to root distribution of nutrients and assessment of soil erosion and sedimentation. Intercropping involves simultaneous cultivation of more than one plant species in the same field, which can improve the use of plant growth resources in space and time. Isotopes are very useful to elucidate root distribution and competition of nutrients in intercropped plant species (Jensen 1996).

5.3.4 Soil Erosion Studies

Soil erosion and associated land degradation are serious problems, which affects the soil fertility and productivity. There is a need to quantify the soil erosion for the development of effective soil conservation. Radio-isotopes such as ^{137}Cs, ^{210}Pb and ^{7}Be are useful in assessing soil erosion losses and sedimentation rates (Blake 1999; Mabit et al. 2008).

Cesium-137 is an anthropogenic radio-isotope released from atmospheric atomic bomb tests primarily during 1950s and 1960s. Lead-210 is a naturally occurring geogenic radionuclide derived from the decay of gaseous ^{222}Rn, a daughter in the ^{238}U decay series. Be-7 is a naturally occurring cosmogenic radio-isotope produced by spallation of O and N atoms in the upper atmosphere. The three radionuclides are deposited to earth surfaces mainly in the form of precipitation. To date the three tracers have been accepted and used by the erosion research community to estimate point soil redistribution rates (Matisoff and Whiting 2011). The retrospective estimation of long-term mean soil redistribution for individual points on a landscape is deemed a core advantage of the tracing technique (Zapata 2010).

5.3.5 Compound-Specific Isotope Analyses Techniques

Compound-specific stable isotope (CSSI) signatures of inherent soil organic biomarkers allow discriminating and apportioning the source of soil contribution from different land uses. Plant communities label the soil where they grow by exuding organic biomarkers. Although all plants produce the same biomarkers, the stable isotopic signature of those biomarkers is different for each plant species. For agri-environmental investigations, the CSSI technique is based on the measurement of carbon-13 (^{13}C) natural abundance signatures of specific organic compounds such as natural fatty acids (FAs) in the soil. By linking fingerprints of land use to the sediment in deposition zones, this approach has been shown to be a useful technique for determining the source of eroded soil and thereby identifying areas prone to soil degradation.

5.3.6 Isotopic Techniques for Soil Moisture Study

Nuclear and isotopic techniques play an important and sometimes unique role in providing information essential to developing strategies aimed at improving agricultural water use efficiency, and hence in providing solutions to mitigate the increasing water scarcity. The soil moisture neutron probe (SMNP) is ideal for the measurement of soil water in the immediate vicinity of the crop roots, and providing accurate data on the accessibility to the crop of available water to establish optimal irrigation schedules. The SMNP is currently the most suitable instrument to accurately measure soil moisture under saline conditions. It is also widely used to calibrate conventional moisture sensors for direct use in farmers' fields. Both oxygen and hydrogen are components of water. The use of the isotopic signatures of oxygen (^{18}O) and hydrogen (^{2}H) in water vapour taken from field crops can facilitate the quantification of crop water uptake, i.e., plant transpiration, and water lost through soil evaporation. It therefore provides information on factors affecting transpiration and evaporation, essential for improving the water use efficiency of crops. Carbon (C) is an important building component of plants. Green plants assimilate carbon from atmospheric carbon dioxide through the process of photosynthesis. Carbon dioxide is composed of two stable isotopes, the less abundant ^{13}C and the lighter ^{12}C. During photosynthesis the plant discriminates against the heavier isotope in favour of the lighter one. The extent of this discrimination depends on environmental factors, such as water availability and salts in the soil. The variation in the relative abundance of the carbon isotopes can therefore be used as a surrogate marker of water stress, water use efficiency and crop tolerance to drought and salinity.

5.4 Conclusion

The isotopes, both stable and radioactive have proved to be very useful in soil analysis. They play a major role in the advancement of soil fertility and degradation studies and water management. Their availability and advantages of using them in agricultural research as a tool would definitely result in increased food production and security.

References

Amer F, Mahdi S, Alradi A (1969) Limitations in isotopic measurements of labile phosphate in soils. J Soil Sci 20:91–100

Blake WH (1999) Fallout beryllium-7 as a tracer in soil erosion investigations. Appl Radiat Isot 51:599–605

Cadisch G, Giller KE (1996) Estimating the contribution of legumes to soil organic matter build up in mixed communities of C3/C4 plants. Soil Biol Biochem 28:823–825

Craig H (1953) The geochemistry of the stable carbon isotopes. Geochim Cosmochim Acta 3(2–3):53–92

D'Souza TJ, Mistry KB (1979) Uptake, distribution and metabolic fate of ^{59}Fe, ^{58}Co, ^{54}Mn and ^{65}Zn in plants and their mobility and availability to crops in typical black and laterite soils. In: Proceedings of the Symposium on Isotopes and Radiations in Research on Soil Plant Relationships, IAEA, Vienna, 1979, pp 407–425

D'Souza TJ, Yadav VB (1993) Evaluation of ammonium polyphosphate fertilizer as a carrier of zinc. In: Proceedings of the FAO/IAEA Seminar for Asia and Pacific on Nuclear Related Methods in Soil-Plant Aspects of sustainable Agriculture, Colombo, Sri Lanka, April 1993, pp 209–217

Di HJ, Condron LM, Frossard E (1997) Isotope techniques to study phosphorus cycling in agricultural and forest soils: a review. Biol Fertil Soils 24:1–12

Fardeau JC, Morel C, Jappé J (1985) Cinétique d'échange des ions phosphate dans les systémes sol-solution. Vérificatiou expérimentale de l'équation théorique. C R Seances Acad Sci III 8:371–376

Fried M, Dean LA (1952) A concept concerning the measurement of available soil nutrients. Soil Sci 73:263–271

Fritz P, Drimmie RJ, Nowicki VK (1974) Preparation of sulfur dioxide fbr mass spectrometer analyses by combustion of sulfides with copper oxide. Anal Chem 46:164–166

Halas S, Shakur A, Krouse HR (1982) A modified method of SO_2 extraction from sulphates for isotopic analysis using $NaPO_3$. Isotopenpraxis 18:433–435

IAEA-TCS-14 (2001) Use of isotope and radiation methods in soil and water management and crop nutrition. IAEA, Vienna, p 247

Jensen ES (1996) Grain yield, symbiotic N_2 fixation and interspecific competition for inorganic N in pea-barley intercrops. Plant Soil 182:25–38

Jentschke G, Brandes B, Khun AJ, Shroder WH, Becker JS, Godbold DL (2000) The mycorrhizal fungus *Paxillus involutus* transports magnesium to Norway spruce seedlings. Evidence from stable isotope labeling. Plant Soil 220:243–246

Kotur SC, Anjaneyulu K, Ramesh PR, Ramachandran V (2010) Assessment of fertilizer use efficiency under enhanced crop intensity of vegetables due to inter-cropping using ^{15}N and ^{32}P labeled fertilizers. Indian J Agri Sci 80:325–326

Kucey RMN, Bole JB (1984) Availability of phosphorus from 17 rock phosphates in moderately and weakly acidic soils as determined by ^{32}P dilution, A value and P uptake methods. Soil Sci 138:180–188

L'Annunziata MF (1987) Radionuclide tracers, their detection and measurement. Academic, London

Larsen S (1952) The use of ^{32}P in studies on the uptake of phosphorus by plants. Plant Soil 4:1–10

Larsen S (1967) Soil phosphorus. Adv Agron 19:151–210

Mabit L, Benmansour M, Walling DE (2008) Comparative advantages and limitations of the fallout radionuclides ^{137}Cs, $^{210}Pbex$ and ^{7}Be for assessing soil erosion and sedimentation. J Environ Radioact 99:1799–1807

Matisoff G, Whiting PJ (2011) Measuring soil erosion rates using natural (7Be, 210Pb) and anthropogenic (137Cs, 239Pu, 240Pu). In: Baskaran M (ed) Handbook of environmental isotope geochemistry, Advances in Isotope Geochemistry, vol 1. Springer, Berlin, pp 487–519

McAuliffe CD, Hall NS, Dean LA, Hendricks SB (1948) Exchange reactions between phosphates and soil: hydroxylic surfaces of soil minerals. Soil Sci Soc Am Proc 12:119–123

Minagawa M, Winter DA, Kaplan IR (1984) Comparison of Kjeldahl and combustion methods for measurement of nitrogen isotope ratios in organic matter. Anal Chem 56:1859–1861
Ozanne PG (1980) Phosphate nutrition of plants—a general treatise. In: Khasawneh FE, Sample EC, Kamprath EJ (eds) The role of phosphorus in agriculture. ASA-CSSA-SSA, Madison, pp 559–589
Peterson BJ, Fry B (1987) Stable isotopes in ecosystem studies. Annu Rev Ecol Syst 18(1):293–320
Ramachandran V, Shrivastava M, D'Souza SF (2007) Role of isotopes in soil-plant, water and fertilizer research. IANCAS Bull 6:341–353
Ramachandran V, Shrivastava M, Vasu K, Kotur SC, Srivastava PC, Subramanian KS, D'Souza SF (2009) Radioisotopes in soil fertility and soil pollution research. In: International conference on peaceful use of atomic energy, New Delhi, 29 September–1 October 2009, pp 86–91
Russell RS, Rickson JB, Adams SN (1954) Isotope equilibria between phosphates in soil and their significance in the assessment of fertility by tracer methods. J Soil Sci 5:85–105
Salcedo IH, Bertino F, Sampaio EVSB (1991) Reactivity of phosphorus in northeastern Brazilian soils assessed by isotopic dilution. Soil Sci Soc Am J 55:140–145
Shrivastava M, D'Souza SF (2014) Bio-solubilization of rock phosphate and plant growth promotion by Aspergillus niger TMPS1 in Ultisol and Vertisol. In: Proceeding-international symposium on managing soils for food security and climate change adaptation and mitigation. Food and Agriculture Organization of the United Nations, Rome, pp 73–77
Shrivastava M, Bhujbal BM, D'Souza SF (2007) Agronomic efficiency of Indian rock phosphates in acidic soils employing radiotracer A-value technique. Commun Soil Sci Plant Anal 38:461–471
Shrivastava M, Kale SP, D'Souza SF (2011) Rock phosphate enriched post-methanation bio-sludge from kitchen waste based biogas plant as P source for mungbean and its effect on rhizosphere phosphatase activity. Eur J Soil Biol 47:205–212
Simonnet G (1990) The scintillation counter. In: Slater RJ (ed) Radioisotopes in biology, a practical approach. Oxford University Press, Oxford, pp 31–85
Srivastava M, D'Souza SF (2007) Assessment of agronomic efficacy of phosphatic fertilizers for cotton crop in vertisol using ^{32}P radiotracer technique. J Radioanal Nucl Chem 274:397–401
Srivastava PC, Singh AP, Kumar S, Ramachandran V, Shrivastava M, D'Souza SF (2008) Desorption and transformation of zinc in a mollisol and its uptake by plants in a rice-wheat rotation fertilized with either zinc enriched bio-sludge from molasses or with inorganic zinc. Biol Fertil Soils 44:1035–1041
Srivastava PC, Singh AP, Kumar S, Ramachandran V, Shrivastava M, D'Souza SF (2009) Efficacy of phosphorus enriched post-methanation bio-sludge from molasses based distillery as P source to rice and wheat crops grown in a Mollisol: I. Laboratory and greenhouse evaluation with ^{32}P labeled sources. Geoderma 149:312–317
Subramanian KS, Virgine Tenshia JS, Ramachandran V, Jayalakshmi K, Rajarajan A (2007) Influence of mycorrhizal inoculation on utilization of fertilizer zinc by maize. J Nucl Agric Biol 36:133
Tran TS, Fardeau JC, Giroux M (1988) Effects of soil properties on plant-available phosphorus determined by the isotopic dilution phosphorus-32 method. Soil Sci Soc Am J 252:1383–1390
Wallace A (1994) Soil organic matter must be restored to near original levels. Commun Soil Sci Plant Anal 25:29–35
Yadav VB, Mistry KB (1985) Evaluation of ammonium ortho- and polyphosphates as carriers of zinc for maize-bean sequence. J Nucl Agric Biol 14:102–106
Yanagisawa F, Sakai H (1983) Thermal decomposition of barium sulfate-vanadium pentoxide-silica glass mixtures for preparation of sulfur dioxide in sulfur isotope ratio measurements. Anal Chem 55:985–987
Zapata F (2010) Handbook for the assessment of soil erosion and sedimentation using environmental radionuclides. Kluwer Academic Publisher, Dordrecht, p 219

Protocols for Determination and Evaluation of Organic Carbon Pools in Soils Developed Under Contrasting Pedogenic Processes and Subjected to Varying Management Situations

6

Rakesh S, Deepranjan Sarkar, Shikha, Ardith Sankar, Abhas Kumar Sinha, Prabir Mukhopadhyay, and Amitava Rakshit

Abstract

The fundamental role of soil organic carbon (SOC) in maintaining soil quality and regulating Earth's carbon cycle generates renewed interests of scientists to track its status in the ecosystem. Changes in SOC dynamics is a resultant of soil's complex interactions with vegetation, climate, and land-use practices. Even a small variation in SOC content could bring a significant impact on the atmospheric concentration of carbon dioxide. The fluxes of carbon (C) in soil are mainly dominated by its different pools rather than the total SOC. Therefore, profiling of SOC pools is very important for developing sound management practices that can sustain crop productivity and soil fertility and also reduce C emissions or mitigate climate change. However, accuracy, time requirement, and cost-effectiveness of the present analytical methods direct the need for advancing towards the standard protocols. A systematic appraisal and critical investigation of the different SOC determination methods applied by diverse research groups in contrasting agroecosystems and management conditions will improve our understanding and bridge the gaps in selecting the right protocols.

Rakesh S · A. K. Sinha · P. Mukhopadhyay
Department of Soil Science and Agricultural Chemistry, Uttar Banga Krishi Viswavidyalaya, Cooch Behar, West Bengal, India

D. Sarkar (✉) · A. Rakshit
Department of Soil Science and Agricultural Chemistry, Institute of Agricultural Sciences, Banaras Hindu University, Varanasi, Uttar Pradesh, India

Shikha
Krishi Vigyan Kendra, Ranichauri, Veer Chandra Singh Garhwali Uttarakhand University of Horticulture and Forestry, Tehri Garhwal, Uttarakhand, India

A. Sankar
Department of Agronomy, Institute of Agricultural Sciences, Banaras Hindu University, Varanasi, Uttar Pradesh, India

A. Rakshit et al. (eds.), *Soil Analysis: Recent Trends and Applications*,
https://doi.org/10.1007/978-981-15-2039-6_6

Keywords
Soil organic carbon pools · Carbon stocks · Assessment methods · Land-use systems · Management practices

6.1 Introduction

Soil acting as a source and sink of carbon (C) requires special attention from being a grievously misused natural resource. The important source of C in soil is soil organic matter (SOM), which is also known as the basis of soil fertility. Soil organic carbon (SOC) and SOM terms are frequently used interchangeably to describe soil functioning. Soil organic matter represents a key index of soil quality because it has a profound influence on the physical, chemical, and biological properties of soil. Increasing the levels of SOM in the soil will contribute significantly in achieving sustained production systems. The distribution of SOC is influenced by factors like climate, soil (texture, mineralogy), topography, land-use systems (native forest vegetation, agroforestry, cropping systems, grasslands), and agricultural management practices (mulching, tillage, application of organic amendments and mineral fertilizers) (Parton et al. 1987; Giri et al. 2007; Bandyopadhyay and Lal 2015). However, declining trends of SOM are reported with the adoption of destructive land-use and management practices, e.g., deforestation, intensive use of tillage and chemicals, biomass burning, or reduction in the use of organic inputs. All such activities not only lead to rapid mineralization and soil disturbance but also increase the concentration of atmospheric carbon dioxide (CO_2) and global warming. Therefore, in recent times, good knowledge is required for quantifying SOC with a standard protocol to maintain the quality and productivity of soils besides offsetting the greenhouse gas emissions.

Short- and medium-term changes in quantity and quality of SOC is not reflected in total organic C (TOC) content of soil as these variations are slow, spatial, and temporal; and soil may have high background C concentrations varying very little in its mineral associations (Bosatta and Ågren 1994; Russell et al. 2004; Lal 2006). Interestingly, different pools of SOC are found to act as sensitive or quick indicators of the changes in soil quality (Guo and Gifford 2002; Tan et al. 2007; Chen et al. 2009). These pools are developed on the basis of their composition and susceptibility towards oxidative forces (Baldock and Nelson 2000; Leggett and Kelting 2006). Further fractionation of SOC has been sketched by physical, chemical, and biological means (Fig. 6.1). Understanding the relative proportion of each fraction is very imperative to control the flux dynamics. Moreover, the SOC pools are also involved in several ecosystem functions such as maintaining agronomic productivity, nutrient cycling, water quality, biodiversity, and mitigating climate change (Lal 2015).

Global carbon budgeting under the influence of the Kyoto Protocol has gained considerable interest in scientists and policy makers in reducing the amount of greenhouse gases in the atmosphere. Developing countries residing in tropics and subtropics (e.g., India) can generate prospective revenue from C trading as soils are

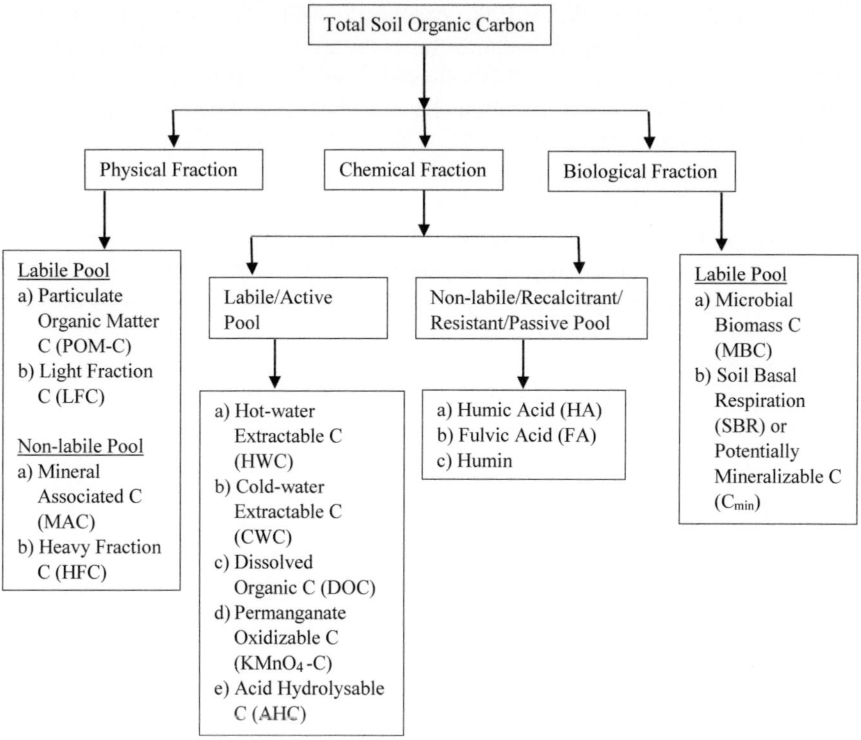

Fig. 6.1 Scheme for estimation of different soil organic carbon pools

starving for C sequestration and ample sunlight is available for photosynthesis (Mandal 2011). However, the potential of an ecosystem to serve as a source or sink of CO_2 is dependent upon the balance between C inputs (gains) and outputs (losses). There is an urgent need to adopt the best interventions that can sequester more C and promote the build-up of SOC.

The chapter was attempted to outline the methods used for determining the various pools of SOC, comparing and highlighting their suitability in different land-use systems. Conceptual background about the SOC pools will help in apprais-ing the best land-use systems, characterizing C sequestration and/or C credits, and providing the technical know-how for developing sound methodologies.

6.2 Soil Organic Carbon Stocks

Carbon in soil is existing in organic and inorganic forms; combining these two components forms the total soil C (Nieder and Benbi 2008). The organic component is derived from the decayed plant and animal remains, while the inorganic form is lithogenic and pedogenic in origin, present as carbonates [e.g., calcium carbonate ($CaCO_3$)]. The soil is the largest reservoir of terrestrial C, stocking ~1500 Pg (1 Pg = 1015 g = 1 billion tons) of organic C in the upper layer (up to 1 m depth) of soil at the global level (Batjes 1996; Amundson 2001). This is roughly twice the amount of C in the atmosphere and three times the amount of C in the above-ground vegetation (Scharlemann et al. 2014), comprising approximately two-thirds of the C storage of terrestrial ecosystems (Schimel et al. 1994). The global soil C reservoir is equivalent of ~300 times as much C as the annual emissions from fossil fuel combustion (Schulze and Freibauer 2005).

Soils found in warm, humid, and per humid climates are of major concern because they are inherently low in SOC. India gifted with diverse group of soils and ecoregions have a wide range of organic C stocks. The first estimation of Indian SOC stock using the databank of 48 soil series by Gupta and Rao (1994) revealed it to be around 24.3 Pg in surface and subsurface depth ranging from 44 to 186 cm. Inventorying the SOC stocks in five main physiographic regions of India, Bhattacharyya et al. (2000) found the total SOC stock to be 21 Pg and 63.2 Pg for 0.3 m and 1.5 m depth, respectively. Srinivasarao et al. (2009) have studied organic C stocks in different soil orders and diverse agro-ecological regions across the country up to 1.5 m depth. The SOC stocks varied from 20.1 to 95.9 Mg ha^{-1}, and trailed the order: Vertisols > Inceptisols > Alfisols > Aridisols; soils receiving high rainfall had greater stocks of organic C than soils of low rainfall regions. Agricultural soils of the Indo-Gangetic Plains have very low SOC concentration of 8.5–15.2 Mg ha^{-1} in 40 cm depth and 12.4–22.6 Mg ha^{-1} in 1 m depth (Singh et al. 2011). It can be generalized that Indian soils are low in organic C content having stock in the range of 20–25 Pg to top 1 m depth. This current stock can be increased to 35 Pg by sequestering more C in soils with appropriate management strategies (Singh et al. 2014). The size of the SOC pool (stock) is computed by multiplying SOC concentration with bulk density and depth of soil.

6.3 Soil Organic Carbon Pools

The total SOC is composed of several fractions of varying lability, i.e., active and passive pools (Fig. 6.1). To understand the C dynamics, we must study different compartments (physical, chemical, and biological) of its complex structure. Identification of each component (pool) and their specific roles will bridge our knowledge gaps regarding the stability of SOM and various transformation processes. Soil organic C is generally determined by wet oxidation (digestion) method of Walkley and Black (1934) using an acid dichromate solution. The SOM is oxidized with standard potassium dichromate ($K_2Cr_2O_7$) solution (1.0 N) and concentrated sulphuric acid (H_2SO_4). The unused dichromate is back titrated with 0.5 N ferrous ammonium sulfate (FAS). In case of TOC, the soil is digested in digestion tubes at 150 °C for 45 min using 0.4 N $K_2Cr_2O_7$ and 18.0 N H_2SO_4, and thereafter titrated against 0.2 N FAS (Tiessen and Moir 1993). Total SOC is also calculated indirectly by subtracting total soil inorganic C (estimated from $CaCO_3$ content) from total soil C (analyzed by CHN Elemental Analyzer). Each method has its own advantages and disadvantages. Walkley–Black method is time-consuming and generates environmental toxic dichromate waste. The correction factor (for estimating actual C) used in this method varies with location of soil samples as the recovery percentages (assuming incomplete oxidation) changes with texture, land uses, and analytical methods (Chatterjee et al. 2009). Accuracy of estimation through elemental analyzer is high, but involves costly investments and separate analysis of inorganic C. Recently, in situ methods (e.g., spectral reflectance technology) are gaining popularity in C estimation due to their speed, economics, reliability, and non-destructive nature. In this chapter, fractionation approaches have been mainly focused.

6.3.1 Labile Pools

The labile pools of SOC have a fast turnover rate or short residence time (<10 years) as they serve as the primary source of energy for the soil microbial assemblages, responding very fast to ecological changes (Cheng and Kimble 2001; McLauchlan and Hobbie 2004; Benbi et al. 2012). Enlisting the fractions as mineralizable C should be done with proper care as they are often used as a measure of sustainability. Besides nourishing the microbiota, they also play a great role in the nutrient availability in soil. Some examples of the bio-reactive pools are microbial biomass C (MBC), potential mineralizable C (C_{min}), hot-water extractable C (HWC), cold-water extractable C (CWC), permanganate oxidizable C ($KMnO_4$-C), dissolved organic C (DOC), and particulate organic matter C (POM-C). Such groups of C are made of carbohydrates (hemicellulose, starch, soluble sugars), amino acids, proteins, and other organic compounds (Rovira and Vallejo 2002). A variety of methods are applied to determine the labile C pools; however, acid hydrolysis and water extraction are very commonly used approach in chemical fractionation (Ahn et al. 2009). For separating the SOM chemically, the extractants are chosen according to solubility and affinity of the organic compounds (Cheng and Kimble 2001; Nieder and Benbi 2008). Physical fractions are mainly categorized based on their density or distribution of aggregate size (Cambardella and Elliott 1993; Tan et al. 2007). In biological separation, these pools are indirectly estimated by measuring the CO_2 evolved during mineralization of SOM by microbes and directly by estimating their pool size (microbial biomass) (McLauchlan and Hobbie 2004).

Particulate organic matter C estimation involves disaggregation of soil (<2 mm) using 0.5% sodium hexametaphosphate solution (1:3 ratio) and shaking in a reciprocal shaker for a period of 18 h (Sollins et al. 1999). The soil suspension is passed through a 53 μm sieve. The material retained on the sieve is oven dried at 60 °C for 48 h, weighed, and subjected to combustion (muffle furnace) at 550 °C for 4 h to determine the mass of sand free POM-C. This pool of SOC contains about 18–39% of TOC (Cambardella and Elliott 1992). Analyzing POM-C we get information regarding organic matter developed from recent incorporation of residues (plant origin) rather than the organic material of microbial origin linked with the mineral fractions (silt and clay) of soil (Cheshire and Mundie 1981; Puget et al. 1999). Density fractionation is another physical method for separating SOM into light and heavy fractions. Generally, liquids having specific gravity (SG) of 1.6–2.0 g cm^{-3} are used in separation schemes (Crow et al. 2007; Benbi et al. 2012). Light fraction (LF) representing recent and partially decomposed organic matter (less stable) or plant-like particulate organic matter can be easily separated from the more stable fraction (heavy fraction) by using solutions of low density (SG: 1.6–1.8 g cm^{-3}). Isolation of LF from the soil is carried using sodium iodide solution (SG = 1.72 g cm^{-3}) (Janzen et al. 1992). After gentle shaking (30 min) and equilibration (48 h), the suspended matter (LF) is shifted to a suction filtration unit. The LF under suction is repeatedly washed with three aliquots of 0.01 M calcium chloride ($CaCl_2$) and three aliquots of distilled water. The material is weighed after an overnight drying (50 °C), and the entire process is repeated to determine the residual LF in the settled material of equilibration period. As the process of isolation is completed, the material is analyzed for total C using elemental analyzer.

Chemical fractionation involves application of different extracting solutions made of water, oxidizing agents, salts, acids, and bases. Hot-water extractable C is estimated by shaking soil (3 g) with distilled water (30 mL) for 30 min at 20 °C (Ghani et al. 2003). The soil is centrifuged (3000 rpm) for 20 min. Then, again 30 mL of distilled water is added to the residue and a 16 h extraction is followed at 80 °C (hot-water bath). After centrifugation, the supernatant is filtered (0.45 μm cellulose nitrate membrane) and analyzed for total C. The targeted HWC is determined by subtracting the inorganic component from total HWC. For estimating CWC, the heating treatment (extraction in hot-water bath) in the aforementioned procedure of HWC is skipped. Water-soluble organic fractions are composed of organic acids, carbohydrates (monosaccharides, polysaccharides), and nitrogenous compounds (Pansu and Gautheyrou 2006). Among carbohydrates, simple sugars (glucose, fructose) and polysaccharides (starch) are present in CWC and HWC, respectively. Dissolved organic C is extracted from field-moist soil using 2 M potassium chloride (KCl), and centrifuged or filtered (Whatman Grade 42) for quantification (Jones and Willett 2006). Finely ground soil is oxidized with 0.333 M potassium permanganate ($KMnO_4$), shaken (1 h), and centrifuged for determining the readily oxidizable organic C (Blair et al. 1995). The diluted supernatant is measured in a spectrophotometer at 565 nm. Carbon content in the extracts of different pools can be quantified either through wet oxidation or by colorimetric methods. Acid hydrolysis studies mostly use hydrochloric acid (HCl) and sulfuric acid (H_2SO_4) for isolation of hydrolysable C. However, H_2SO_4 is popular as a better predictor of SOM quality (Rovira and Vallejo 2002; Belay-Tedla et al. 2009). Acid hydrolysable C (AHC) is analyzed by treating soil sample with 12 M H_2SO_4 for 16 h at room temperature (Puget et al. 1999). The hydrolysate is diluted to 1 M H_2SO_4 with distilled water and oven dried (100 °C) for 5 h.

The living component (microbial biomass C) of SOM is estimated in field-moist samples by chloroform fumigation–extraction method (Vance et al. 1987). One set of soil is fumigated (24 h) with ethanol-free chloroform, and thereafter, both fumigated and non-fumigated samples are extracted with 0.5 M potassium sulfate (1:4 soil:solution ratio) in a shaker (30 min). Organic C is determined in the filtered extract by chromic acid digestion method. The proportion of MBC is only 1–3% of TOC (Devi and Yadava 2006). Soil basal respiration (SBR) or C_{min} involves measurement of CO_2 released by soil microorganisms under incubated conditions using acid-base titration (Anderson 1982). The evolved CO_2 trapped in sodium hydroxide is titrated with dilute HCl. In general, C_{min} is an indicator of biological activity in soil, reflecting the metabolism of active microbes involved in decomposition of organic matter, and the total amount of CO_2 production during depletion of SOM represents the readily mineralizable pool of SOC (Hopkins 2007).

6.3.2 Recalcitrant Pools

Organic matter located within mineral aggregates is shielded against any enzyme attack, and thus form the biologically stable (nonavailable) organic C (Jastrow and Miller 1998). Thoroughly degraded organic matter or microbially processed material associated with mineral surfaces form the heavy fraction (HF). The mineral and organo-mineral material or HF is determined from the material that settled after the removal of LF and residual LF by passing it through a 53 μm sieve, washing with distilled water, and weighing the retained material of sieve after drying at 50 °C (Benbi et al. 2012). The passed slurry consisting of silt and clay particles and the organic matter associated with these separates is further used for determination of mineral associated C (MAC). This is achieved by centrifuging, drying, and weighing. Carbon content in the above-mentioned fractions is analyzed with elemental analyzer.

Humic substances (HS), one of the largest fraction of SOM are studied to know the stability of SOC rather than deciphering its transformation dynamics (Cheng and Kimble 2001). These are heterogeneous, colloidal, amorphous, high-molecular-weight, dark colored (yellow to black), and polymeric material formed by secondary synthesis reactions (humification) (Stevenson 1994). According to their solubility characteristics, they are further categorized into three factions, viz., humic acid (HA), fulvic acid (FA), and humin. The HA is alkali soluble fraction having dark brown to black color. Fulvic acid (yellow to brownish-yellow) is soluble in water as well as in acid and alkaline solutions. Humin (black) is insoluble fraction of SOM. These fractions are separated by alkaline hydrolysis with 0.5 M sodium hydroxide (NaOH) or sodium carbonate (Na_2CO_3). The soluble part is further treated with acid (HCl) to extract the precipitated HA (acid insoluble). After extraction and purification, HS are subjected to characterization (functional groups, total acidity, molecular weight, elemental analysis, morphology, and crystallinity) for their qualitative estimation. The mean residence time of these substances varies from 100 to 2000 years (Wang and Chang 2001). They provide very little evidence about the changes in their chemical composition due to the continuation of management practices.

6.4 Soil Carbon Dynamics

6.4.1 Land-Use Systems

A wide variation is noted in SOC pools between natural ecosystems and agroecosystems in the existing literature, indicating the requirement of systematic documentation of the changes occurring in C pool size using the standard methods (Table 6.1). Generally, forests and grasslands are characterized by a higher amount of C content than croplands. In a north-eastern state (Meghalaya) of India, although the carbon stock (1 m depth) was found in the order: mixed forests (13.8 kg C m^{-2}) > grasslands (12.7 kg C m^{-2}) > croplands (11.0 kg C m^{-2}), but organic C content in surface soils (0–0.3 m) followed the order: grasslands (8.5 kg C m^{-2}) > croplands (5.2 kg C m^{-2}) > mixed forests (4.0 kg C m^{-2}) (Dutta et al. 2013). The increased surface accumulation in grasslands is explained by enhanced root biomass, and in croplands factors like deeper root systems and fertilization might be the reasons for such a trend. Gami et al. (2009) reported mean depletion in SOC stocks of 0–15 (55%), 15–30 (32%), 30–45 (17%), and 45–60 (10%) cm depths in the eastern Indo-Gangetic Plains due to land cleaning (removal of forests) for agricultural practices (rice-wheat system). Forest soils are characterized by high FA content, but peat and grassland soils are known for high HA content (Stevenson 1994; Nieder and Benbi 2008).

Bandyopadhyay and Lal (2015) noted significantly higher POM-C and MAC in the forest soil (12.7 and 23.6 g kg^{-1}) than the cultivated soil (5.8 and 14.3 g kg^{-1}) collected (0–10 cm depth) from a long-term field experiment (16 years) in Ohio (USA). The effect of horticultural land uses (plantation of fruit trees) on TOC stock (20 cm) was found to be higher in guava (28.8 Mg C ha^{-1}) and jamun (27.3 Mg C ha^{-1}) plantation followed by litchi (25.7 Mg C ha^{-1}) and mango trees (19.2 Mg C ha^{-1}) in the sodic soils of Karnal (Haryana, north-west India) (Datta et al. 2015). *Eucalyptus*-based agroforestry system (managed for 6 years) showed greater TOC stock and MBC than monocropped sugarcane-sugarcane system (managed for 2 years) in semi-arid region of Haryana (Kumar et al. 2018). This improvement is attributed to increased C inputs (litterfall, root exudates, dead roots, bark decomposition) and higher biological activities of soil. The effect of rice ecosystem type, viz., young terrace land (YTLRF), mature terrace land (MTLRF), slope land rice ecosystem (SLRF), and lowland rice ecosystem (LLRF) on C pools (DOC and MBC) in hill agriculture was studied by Sangma et al. (2016). Variable response was observed among the rice ecosystems: LLRF (1485 and 1234 μg g^{-1}) > MTLRF (1176 and 822 μg g^{-1}) > YTLRF (1073 and 709 μg g^{-1}) > SLRF (669 and 557 μg g^{-1}). An investigation of 24 study sites in Europe showed increase in mean SOC stocks (0–30 cm depth) with land-use changes from cropland to grassland (18 ± 11 Mg ha^{-1}) and cropland to forest (21 ± 13 Mg ha^{-1}), while changes from grassland to cropland (−19 ± 7 Mg ha^{-1}) and grassland to forest (−10 ± 7 Mg ha^{-1}) resulted a negative trend (Poeplau and Don 2013). Fractionation of SOC revealed that POM-C was more sensitive to these changes than other pools like MAC and DOC.

Table 6.1 Distribution of global SOC pools as influenced by land-use and/or management practices

Location	Soil type	Climate	Land-use/management practices	Depth (cm)	Method	Response of SOC pools	Reference
Physical fractions							
Melfort, Saskatchewan, Canada	Silty clay loam	Humid continental	Crop rotation [fallow + spring wheat + spring wheat (FWW) and sweet clover + spring wheat + spring wheat (GWW)]	0–7.5	LF	292 g kg^{-1} (FWW); 300 g kg^{-1} (GWW)	Janzen et al. (1992)
Boigneville, Essonne, France	Loess	Oceanic	NT and CT	2–3 (NT) and 0–30 (CT)	POM-C	73.8 g kg^{-1} (NT); 13.1 g kg^{-1} (CT)	Puget et al. (1999)
Coshocton, Ohio, USA	Silt loam	Oceanic	Maize (NT and CT) and forest (white oak, red oak, and yellow poplar)	0–5	LF and HF	88.0, 41.0, and 147 g kg^{-1} (LF); 19.4, 7.5; 28.6 (HF)	Tan et al. (2007)
Chenghuang, Linfen, China	Sandy loam	Semi-arid, semi-humid, continental	Winter wheat [conventional tillage with no residue (CT), shallow tillage with residue cover (STR), and no-tillage with residue cover (NTR)]	0–15 and 15–30	POM-C	4.2, 7.1, and 5.9 g kg^{-1} (0–15); 2.6, 2.9, and 3.1 g kg^{-1} (15–30)	Chen et al. (2009)
Ludhiana, Punjab, India	Sandy loam	Semi-arid, subtropical	Rice-wheat cropping system [no organics (NO), farmyard manure (FYM), rice straw (RS), FYM + RS (FYMRS)]	0–7.5	Coarse POM-C (250–2000 μm), fine POM-C (53–250 μm), LF, HF, and MAC	145, 771, 371, and 785 mg kg^{-1} (coarse POM-C); 1006, 1540, 1120, and 1589 mg kg^{-1} (fine POM-C); 256, 929, 425, and 1238 mg kg^{-1} (LF); 1127, 1830,	Benbi et al. (2012)

(continued)

Table 6.1 (continued)

Location	Soil type	Climate	Land-use/management practices	Depth (cm)	Method	Response of SOC pools	Reference
						1299, and 2563 mg kg^{-1} (HF); 3475, 3557, 4392, and 4461 (MAC)	
Fargo, North Dakota, USA	Silty clay	Humid continental	Corn-soybean-sugarbeet rotation (ST, NT, and CT)	0–15	Coarse POM-C (53–2000 μm)	7.3 g kg^{-1} (ST); 7.7 g kg^{-1} (NT); 5.5 g kg^{-1} (CT)	Awale et al. (2013)
Ohio, USA	Silt loam	Humid continental	Hardwood forest and agroecosystem (maize)	0–10	POM-C and MAC	12.7 and 23.6 g kg^{-1} (forest); 5.8 and 14.3 g kg^{-1} (cropland)	Bandyopadhyay and Lal (2015)
New Delhi, India	Sandy clay loam	Hot semi-arid	Rice-wheat cropping system (mungbean residue + direct-seeded rice–zero-till wheat + rice residue–zero-till summer mungbean)	0–15	POM-C	3.6 g kg^{-1}	Dey et al. (2016)
Four study sites in the Argentine Pampas (Bengolea, Monte Buey, Pergamino, and Viale), Argentina	Sandy loam (Bengolea), silty loam (Monte Buey and Pergamino), and silty clay loam (Viale)	Temperate semi-humid (Bengolea and Monte Buey) and temperate humid (Pergamino and Viale)	Natural systems and agroecosystems	0–20	Coarse POM-C (105–2000 μm) and fine POM-C (53–105 μm)	1.8, 2.3, 1.8, and 2.6 g kg^{-1} (coarse POM-C); 3.1, 2.8, 2.3, and 1.8 g kg^{-1} (fine POM-C)	Duval et al. (2018)

Chemical fractions							
Boigneville, Essonne, France	Loess	Oceanic	NT and CT	2–3 (NT) and 0–30 (CT)	AHC and HWC	4.7 and 0.6 mg g^{-1} (NT); 1.6 and 0.2 mg g^{-1} (CT)	Puget et al. (1999)
North Island, Waikato, New Zealand	Sandy loam	Temperate	Dairy pastures, sheep and cattle grazed pastures, cropping sites (maize and sweetcorn), market gardening sites, and native bush sites	0–7.5	HWC	3000 μg g^{-1} (dairy pastures); 3400 μg g^{-1} (sheep/beef pastures); 1000 μg g^{-1} (cropping); 850 μg g^{-1} (market gardening); 4000 μg g^{-1} (native)	Ghani et al. (2003)
Wales, UK	Dystric cambisol (unimproved grassland) and Leptic podzol (coniferous forest)	Temperate, oceanic	Unfertilized grassland (sheep's fescue + common bent) and coniferous forest (Sitka spruce + European larch)	2–15 (grassland) and 0–5 (forest)	DOC	47 ± 2 mg l^{-1} (grassland); 2925 ± 67 mg l^{-1} (forest)	Jones and Willett (2006)
Chenghuang, Linfen, China	Sandy loam	Semi-arid, semi-humid, continental	Winter wheat [conventional tillage with no residue (CT), shallow tillage with residue cover (STR), and no-tillage with residue cover (NTR)]	0–15	DOC, HWC, and $KMnO_4$-C	90.5 g kg^{-1}, 375.2 mg kg^{-1}, and 1.4 g kg^{-1} (CT); 118.3 g kg^{-1}, 554.2 mg kg^{-1}, and 2.7 g kg^{-1} (STR);	Chen et al. (2009)

(continued)

Table 6.1 (continued)

Location	Soil type	Climate	Land-use/management practices	Depth (cm)	Method	Response of SOC pools	Reference
						102.3 g kg^{-1}, 514.2 mg kg^{-1}, and 3.7 g kg^{-1} (NTR)	
Santa Fe River Watershed, Florida, USA	Ultisols	Humid subtropical	Croplands, rangelands, and urban areas	0–30	AHC and HWC	3.1 and 0.5 g kg^{-1} (croplands); 4.6 and 0.7 g kg^{-1} (rangelands); 3.5 and 0.7 g kg^{-1} (urban areas)	Ahn et al. (2009)
New Delhi, India	Inceptisol	Semi-arid, subtropical	Maize-wheat cropping system (50% of NPK fertilizers + vermicompost @ 5 Mg ha^{-1})	0–15	$KMnO_4$-C	628 mg kg^{-1} (after maize); 931 mg kg^{-1} (after wheat)	Basak et al. (2012)
Fargo, North Dakota, USA	Silty clay	Humid continental	Corn-soybean-sugarbeet rotation (ST, NT, and CT)	0–15	$KMnO_4$-C	560 mg kg^{-1} (ST); 555 mg kg^{-1} (NT); 542 mg kg^{-1} (CT)	Awale et al. (2013)
New Delhi, India	Sandy clay loam	Hot semi-arid	Rice-wheat cropping system (mungbean residue + direct-seeded rice–zero-till wheat + rice residue–zero-till summer mungbean)	0–15	$KMnO_4$-C	0.4 g kg^{-1}	Dey et al. (2016)
Saiden, Ri-Bhoi, Meghalaya, India	Clay loam	Warm to hot moist humid	Lowland rice ecosystems	0–15	DOC	1485 μg g^{-1}	Sangma et al. (2016)
Four study sites in the Argentine	Sandy loam (Bengolea), silty loam	Temperate semi-humid (Bengolea	Natural systems and agroecosystems	0–20	AHC, HWC, and $KMnO_4$-C	2.0, 0.3, and 0.7 g kg^{-1} (Bengolea); 3.0,	Duval et al. (2018)

Pampas (Bengolea, Monte Buey, Pergamino, and Viale), Argentina	(Monte Buey and Pergamino), and silty clay loam (Viale)	and Monte Buey) and temperate humid (Pergamino and Viale)				0.7, and 0.9 g kg^{-1} (Monte Buey); 2.6, 0.4, and 0.6 g kg^{-1} (Pergamino); 5.5, 0.5, and 0.8 g kg^{-1} (Viale)	
Biological fractions							
Chenghuang, Linfen, China	Sandy loam	Semi-arid, semi-humid, continental	Winter wheat [conventional tillage with no residue (CT), shallow tillage with residue cover (STR), and no-tillage with residue cover (NTR)]	0–15 and 15–30	MBC	347.4, 446.0, and 422.8 mg kg^{-1} (0–15); 209.7, 228.8, and 231.7 mg kg^{-1} (15–30)	Chen et al. (2009)
Santa Fe River Watershed, Florida, USA	Ultisols	Humid subtropical	Croplands, rangelands, and urban areas	0–30	C_{min}	4.4, 6.0, and 5.8 mg kg^{-1}	Ahn et al. (2009)
New Delhi, India	Inceptisol	Semi-arid, subtropical	Maize-wheat cropping system (50% of NPK fertilizers + vermicompost @ 5 Mg ha^{-1})	0–15	MBC and C_{min}	281 mg kg^{-1} and 12 μg CO_2 g^{-1} h^{-1} (after maize); 259 mg kg^{-1} and 12.4 μg CO_2 g^{-1} h^{-1} (after wheat)	Basak et al. (2012)
Ludhiana, Punjab, India	Sandy loam	Semi-arid, subtropical	Rice-wheat cropping system [no organics (NO), farmyard manure (FYM), rice straw (RS), FYM + RS (FYMRS)]	0–7.5 and 7.5–15	Cumulative C_{min} (36 days)	185, 261, 246, and 355 mg kg^{-1} (0–7.5); 259, 311, 301, and 358 mg kg^{-1} (7.5–15)	Benbi et al. (2012)
Fargo, North Dakota, USA	Silty clay	Humid continental	Corn-soybean-sugarbeet rotation (ST, NT, and CT)	0–15	C_{min} (30 days) and	90 and 450 mg kg^{-1} (ST); 85 and	Awale et al. (2013)

(continued)

Table 6.1 (continued)

Location	Soil type	Climate	Land-use/management practices	Depth (cm)	Method	Response of SOC pools	Reference
					cumulative C_{min} (30 days)	433 mg kg^{-1} (NT); 70 and 370 mg kg^{-1} (CT)	
Saiden, Ri-Bhoi, Meghalaya, India	Clay loam	Warm to hot moist humid	Lowland rice ecosystems	0–15	MBC	1234 μg g^{-1}	Sangma et al. (2016)
Umiam, Meghalaya, India	Sandy loam	Mixed subtropical	Maize (50% of NPK fertilizers + FYM @ 5 t ha^{-1} + lime @ 0.5 t ha^{-1})	0–15	MBC	576 mg kg^{-1}	Verma et al. (2017)
Kurukshetra, Haryana, India	Sandy loam	Subtropical continental monsoon	*Eucalyptus*-based agroforestry system	Averaged across 0–105	MBC	15–23% higher MBC	Kumar et al. (2018)

6.4.2 Management Practices

Soil and crop management practices influence the C budget in the ecosystem. Adoption of appropriate techniques that can sequester C in soils rather than causing its depletion and enhancing the net emissions are very crucial in forming the positive and flexible budget. Strategic interventions affecting the composition of SOM are presented in Table 6.1. Basak et al. (2012) found that conjoint application of 50% of the recommended nitrogen (N), phosphorus (P), and potassium (K) fertilizers and value-added manures (e.g., vermicompost, compost, or farmyard manure) is effective in maintaining higher SOC pools (e.g., $KMnO_4$-C, MBC, and C_{min}). Significant improvement in SOC, MBC, and TOC was observed in an acid soil of Meghalaya (hilly regions) by applying integrated soil management involving good combination or choice of soil amendments (Verma et al. 2017). Conservation agricultural practices, e.g., zero tillage, brown manuring, green manuring, and residue incorporation are effective in increasing labile SOC pools than the intensive cropping practices of north-western Indo-Gangetic Plains (Dey et al. 2016). In a corn-sugarbeet-soybean cropping sequence, conservation tillage practices like strip-till (ST) and no-till (NT) resulted considerably higher SOC content (3.9 and 6.6%), SOC stock (11.9 and 8.7%), coarse POM-C (33 and 45%), and C_{min} at 30 days (34 and 28%) than conventional till (CT) (Awale et al. 2013). Among the three tillage practices, the maximum amount of $KMnO_4$-C was found under ST which resulted 3.3% higher $KMnO_4$-C than CT, but no significant difference was found between NT and CT. The effect of tillage practices on SOC fractions was observed in the order: coarse POM-C (physical) > cumulative C_{min} (biological) > $KMnO_4$-C (chemical). Ghani et al. (2003) concluded measurement of HWC as a sensitive indicator of soil quality because it can even reflect variations within an ecosystem, e.g., impact of grazing intensities or fertilization on pastures. While assessing the soil quality through the labile SOC fractions, Duval et al. (2018) reported SOC as well as $KMnO_4$-C were mainly affected by climate and soil conditions (types or depths), but POM-C, AHC, and HWC were more sensitive to management practices or land-use types. Eleven years of conservation tillage practices, viz., shallow tillage with residue cover and NT with residue cover significantly affected the quantity and quality of SOM (Chen et al. 2009). Both of these practices resulted significantly higher SOC stock, labile SOC pools (POM-C, HWC, DOC, $KMnO_4$-C, and MBC), and amount of macroaggrates as compared to CT. Long-term application (11 years) of organic amendments (farmyard manure, rice straw, and farmyard manure + rice straw) in a rice-wheat cropping sequence had a profound influence on SOC stocks and the relative proportion of different SOC pools (Benbi et al. 2012). Sensitiveness to the management practices was found in the order: LF > HF > MAC.

6.5 Carbon Sequestration

Removal of atmospheric C and capturing it in soil is dependent on addition of C inputs (crop-mediated and external) and environmental factors (climate, soil, and plant). Nutrient addition (fertilization) in soil besides improving the productivity increase C content in soil by producing the residual biomass. The measurement of soil C sequestration is usually performed by conducting long-term experiments. Such studies include direct approaches of measuring C dynamics over a time period or indirectly measured by calculating the net balance between the gains and losses of C. Singh and Sharma (2012) observed higher potential of *shisham* in accumulating organic C in soil than other seventeen years old leguminous tree species (*khair*, *subabul*, and *kikar*). Analyzing one long-term fertilizer experiment, Singh (2016) concluded that soybean-wheat cropping system was better in net C sequestration potential and sorghum-wheat system was superior in atmospheric C assimilation potential. Comparing the C emission and sequestration potential of different crop residues in soil, Sarma et al. (2013) noted that CO_2 evolution was higher in rice, wheat, and sesamum residues, but C sequestration was greater in horse gram and buckwheat residues. Conversion of conventional tillage to no-till in dominant cropping system (rice-wheat) of India is estimated to sequester (net C) 244–359 kg C ha^{-1} $annum^{-1}$, while such adaptation in maize-wheat and cotton-wheat systems can show sequestration rates of 219–231 kg C ha^{-1} $annum^{-1}$ (Grace et al. 2012). The C sequestration potential is highly affected by soil texture as rice-wheat soils were found to store more C with increase in silt+clay fraction of soil (Gami et al. 2009). Benbi et al. (2012) predicted continuing rice-wheat cropping system without maintaining an annual C input of 11.8 Mg ha^{-1} may severely cause reduction in SOC stocks of the Indo-Gangetic Plains.

6.6 Conclusions

Development of robust, faster, accurate, cheaper, transparent, non-destructive, and user-friendly methods for measuring the SOC pool is of high priority to maneuver the C cycle and legitimate the C trading at a broad scale. Selection of quantification methods is yet dependent on laboratory facility, economics, analysts (conception and accuracy in determination), safety issues, and standardization of protocols. Simple testing kits with field-based protocols will be helpful for the farmers to manage SOM. As SOC has wide variations across soil types, climatic zones, and landscapes, more data should be extrapolated at regional and national levels to draw some valid conclusions regarding its behavior (quantitative and qualitative changes). It is very difficult to interpret the SOC changes with the help of any single fraction of SOM; however, their integration (physical, chemical, and biological) has been found more informational to predict the land-use induced changes (natural systems as well as agroecosystems). Enriching C in the soil will not only curb global warming but also ensure food security, biodiversity, and environmental quality.

References

Ahn MY, Zimmerman AR, Comerford NB, Sickman JO, Grunwald S (2009) Carbon mineralization and labile organic carbon pools in the sandy soils of a North Florida watershed. Ecosystems 12:672–685

Amundson R (2001) The carbon budget in soils. Annu Rev Earth Planet Sci 29:535–562

Anderson JPE (1982) Soil respiration. In: Page AL, Miller RH, Keeney DR (eds) Methods of soil analysis, Part 2. Chemical and microbiological properties, 2nd edn. American Society of Agronomy and Soil Science Society of America, Madison, pp 831–871

Awale R, Chatterjee A, Franzen D (2013) Tillage and N-fertilizer influences on selected organic carbon fractions in a North Dakota silty clay soil. Soil Tillage Res 134:213–222

Baldock JA, Nelson PN (2000) Soil organic matter. In: Sumner ME (ed) Handbook of soil science. CRC Press, Boca Raton, pp 25–84

Bandyopadhyay KK, Lal R (2015) Effect of long-term land use management practices on distribution of C and N pools in water stable aggregates in Alfisols. J Indian Soc Soil Sci 63:53–63

Basak BB, Biswas DR, Rattan RK (2012) Comparative effectiveness of value-added manures on crop productivity, soil mineral nitrogen and soil carbon pools under maize-wheat cropping system in an inceptisol. J Indian Soc Soil Sci 60:288–298

Batjes NH (1996) Total carbon and nitrogen in the soils of the world. Eur J Soil Sci 47:151–163

Belay-Tedla A, Zhou X, Su B, Wan S, Luo Y (2009) Labile, recalcitrant, and microbial carbon and nitrogen pools of a tallgrass prairie soil in the US Great Plains subjected to experimental warming and clipping. Soil Biol Biochem 41:110–116

Benbi DK, Toor AS, Kumar S (2012) Management of organic amendments in rice-wheat cropping system determines the pool where carbon is sequestered. Plant Soil 360:145–162

Bhattacharyya T, Pal DK, Mandal C, Velayutham M (2000) Organic carbon stock in Indian soils and their geographical distribution. Curr Sci 79:655–660

Blair GJ, Lefroy RDB, Lisle L (1995) Soil carbon fractions based on their degree of oxidation, and the development of a carbon management index for agricultural systems. Aust J Agric Res 46:1459–1466

Bosatta E, Ågren GI (1994) Theoretical analysis of microbial biomass dynamics in soils. Soil Biol Biochem 26:143–148

Cambardella CA, Elliott ET (1992) Particulate soil organic-matter changes across a grassland cultivation sequence. Soil Sci Soc Am J 56:777–783

Cambardella CA, Elliott ET (1993) Methods for physical separation and characterization of soil organic matter fractions. Geoderma 56:449–457

Chatterjee A, Lal R, Wielopolski L, Martin MZ, Ebinger MH (2009) Evaluation of different soil carbon determination methods. Crit Rev Plant Sci 28:164–178

Chen H, Hou R, Gong Y, Li H, Fan M, Kuzyakov Y (2009) Effects of 11 years of conservation tillage on soil organic matter fractions in wheat monoculture in Loess Plateau of China. Soil Tillage Res 106:85–94

Cheng HH, Kimble JM (2001) Characterization of soil organic carbon pools. In: Lal R, Kimble JM, Follett RF, Stewart BA (eds) Assessment methods for soil carbon. Lewis, Boca Raton, pp 117–129

Cheshire MV, Mundie CM (1981) The distribution of labelled sugars in soil particle size fractions as a means of distinguishing plant and microbial carbohydrate residues. J Soil Sci 32:605–618

Crow SE, Swanston CW, Lajtha K, Brooks JR, Keirstead H (2007) Density fractionation of forest soils: methodological questions and interpretation of incubation results and turnover time in an ecosystem context. Biogeochemistry 85:69–90

Datta A, Basak N, Chaudhari SK, Sharma DK (2015) Effect of horticultural land uses on soil properties and organic carbon distribution in a reclaimed sodic soil. J Indian Soc Soil Sci 63:294–303

Devi NB, Yadava PS (2006) Seasonal dynamics in soil microbial biomass C, N and P in a mixed-oak forest ecosystem of Manipur, North-east India. Appl Soil Ecol 31:220–227

Dey A, Dwivedi BS, Bhattacharyya R, Datta SP, Meena MC, Das TK, Singh VK (2016) Conservation agriculture in a rice-wheat cropping system on an alluvial soil of north-western Indo-Gangetic plains: effect on soil carbon and nitrogen pools. J Indian Soc Soil Sci 64:246–254

Dutta D, Bandyopadhyay S, Baruah U, Sarkar D (2013) Use of taxonomic approach to estimate total soil organic carbon (SOC) pool sizes in Meghalaya state, India. J Indian Soc Soil Sci 61:237–243

Duval ME, Galantini JA, Martínez JM, Limbozzi F (2018) Labile soil organic carbon for assessing soil quality: influence of management practices and edaphic conditions. Catena 171:316–326

Gami SK, Lauren JG, Duxbury JM (2009) Influence of soil texture and cultivation on carbon and nitrogen levels in soils of the eastern Indo-Gangetic Plains. Geoderma 153:304–311

Ghani A, Dexter M, Perrott KW (2003) Hot-water extractable carbon in soils: a sensitive measurement for determining impacts of fertilisation, grazing and cultivation. Soil Biol Biochem 35:1231–1243

Giri JD, Singh SK, Singh RS, Shyampura RL (2007) Carbon stock and its distribution in soils of Ajmer district and management strategies for carbon sequestration. Agropedol 18:21–32

Grace PR, Antle J, Aggarwal PK, Ogle S, Paustian K, Basso B (2012) Soil carbon sequestration and associated economic costs for farming systems of the Indo-Gangetic Plain: a meta-analysis. Agric Ecosyst Environ 146:137–146

Guo LB, Gifford RM (2002) Soil carbon stocks and land use change: a meta analysis. Glob Change Biol 8:345–360

Gupta RK, Rao DLN (1994) Potential of wastelands for sequestering carbon by reforestation. Curr Sci 66:378–380

Hopkins DW (2007) Carbon mineralization. In: Carter MR, Gregorich EG (eds) Soil sampling and methods of analysis, 2nd edn. CRC Press, Boca Raton, pp 589–606

Janzen HH, Campbell CA, Brandt SA, Lafond GP, Townley-Smith L (1992) Light-fraction organic matter in soils from long-term crop rotations. Soil Sci Soc Am J 56:1799–1806

Jastrow JD, Miller RM (1998) Soil aggregate stabilization and carbon sequestration: feedbacks through organo-mineral associations. In: Lal R, Kimble JM, Follett RF, Stewart BA (eds) Soil processes and the carbon cycle. CRC Press, Boca Raton, pp 207–224

Jones DL, Willett VB (2006) Experimental evaluation of methods to quantify dissolved organic nitrogen (DON) and dissolved organic carbon (DOC) in soil. Soil Biol Biochem 38:991–999

Kumar P, Mishra AK, Chaudhari SK, Basak N, Rai P, Singh K, Singh R, Pandey CB, Sharma DK (2018) Carbon pools and nutrient dynamics under *Eucalyptus*-based agroforestry system in semi-arid region of north-west India. J Indian Soc Soil Sci 66:188–199

Lal R (2006) Soil carbon sequestration in Latin America. In: Lal R, Cerri CC, Bernoux M, Etcheves J, Cerri E (eds) Carbon sequestration in soils of Latin America. Food Products Press, New York, pp 49–64

Lal R (2015) Soil carbon sequestration in agroecosystems of India. J Indian Soc Soil Sci 63:125–143

Leggett ZH, Kelting DL (2006) Fertilization effects on carbon pools in loblolly pine plantations on two upland sites. Soil Sci Soc Am J 70:279–286

Mandal B (2011) The 29th Professor JN Mukherjee-ISSS Foundation lecture–soil organic carbon research in India—a way forward. J Indian Soc Soil Sci 59:S9–S22

McLauchlan KK, Hobbie SE (2004) Comparison of labile soil organic matter fractionation techniques. Soil Sci Soc Am J 68:1616–1625

Nieder R, Benbi DK (2008) Soil organic matter characterization. In: Carbon and nitrogen in the terrestrial environment. Springer, Dordrecht, pp 81–111

Pansu M, Gautheyrou J (2006) Handbook of soil analysis: mineralogical, organic and inorganic methods. Springer, Berlin

Parton WJ, Schimel DS, Cole CV, Ojima DS (1987) Analysis of factors controlling soil organic matter levels in Great Plains Grasslands. Soil Sci Soc Am J 51:1173–1179

Poeplau C, Don A (2013) Sensitivity of soil organic carbon stocks and fractions to different land-use changes across Europe. Geoderma 192:189–201

Puget P, Angers DA, Chenu C (1999) Nature of carbohydrates associated with water-stable aggregates of two cultivated soils. Soil Biol Biochem 31:55–63

Rovira P, Vallejo VR (2002) Labile and recalcitrant pools of carbon and nitrogen in organic matter decomposing at different depths in soil: an acid hydrolysis approach. Geoderma 107:109–141

Russell AE, Camberdella CA, Ewel JJ, Parkin TB (2004) Species, rotation, and life-form diversity effects on soil carbon in experimental tropical ecosystems. Ecol Appl 14:47–60

Sangma CB, Thakuria D, Biam DED (2016) Rice ecosystems in hill agriculture: effect on soil biological pools of carbon, nitrogen and phosphorus. J Indian Soc Soil Sci 64:391–401

Sarma UJ, Chakravarty M, Bhattacharyya HC (2013) Emission and sequestration of carbon in soil with crop residue incorporation. J Indian Soc Soil Sci 61:117–121

Scharlemann JP, Tanner EV, Hiederer R, Kapos V (2014) Global soil carbon: understanding and managing the largest terrestrial carbon pool. Carbon Manag 5:81–91

Schimel DS, Braswell BH, Holland EA, McKeown R, Ojima DS, Painter TH, Parton WJ, Townsend AR (1994) Climatic, edaphic, and biotic controls over storage and turnover of carbon in soils. Global Biogeochem Cycles 8:279–293

Schulze ED, Freibauer A (2005) Environmental science: carbon unlocked from soils. Nature 437:205–206

Singh M (2016) 20th Dr BV Mehta Memorial Lecture–Impact of nutrient management and changing climate on productivity and carbon sequestration: a case study from long-term fertilizer experiment. J Indian Soc Soil Sci 64:S67–S76

Singh B, Sharma KN (2012) Depthwise distribution of soil organic carbon and nutrients under some tree species after seventeen years of plantation. J Indian Soc Soil Sci 60:198–203

Singh H, Pathak P, Kumar M, Raghubanshi AS (2011) Carbon sequestration potential of Indo-Gangetic agroecosystem soils. Trop Ecol 52:223–228

Singh A, Kumar S, Singh YV, Bhatia A (2014) Organic carbon dynamics in soils amended with different organic manures and tillage practices in rice-wheat system. J Indian Soc Soil Sci 62:344–350

Sollins P, Glassman C, Paul EA, Swanston C, Lajtha K, Heil JW, Elliott ET (1999) Soil carbon and nitrogen: pools and fractions. In: Robertson GP, Coleman DC, Sollins P, Bledsoe CS (eds) Standard soil methods for long-term ecological research. Oxford University Press, New York, pp 89–105

Srinivasarao C, Vittal KPR, Venkateswarlu B, Wani SP, Sahrawat KL, Marimuthu S, Kundu S (2009) Carbon stocks in different soil types under diverse rainfed production systems in tropical India. Commun Soil Sci Plant Anal 40:2338–2356

Stevenson FJ (1994) Humus chemistry: genesis, composition, reactions, 3rd edn. Wiley, New York

Tan Z, Lal R, Owens L, Izaurralde RC (2007) Distribution of light and heavy fractions of soil organic carbon as related to land use and tillage practice. Soil Tillage Res 92:53–59

Tiessen H, Moir JO (1993) Total and organic carbon. In: Carter ME (ed) Soil sampling and methods of analysis. Lewis, Boca Raton, pp 187–199

Vance ED, Brookes PC, Jenkinson DS (1987) An extraction method for measuring soil microbial biomass C. Soil Biol Biochem 19:703–707

Verma BC, Choudhury BU, Kumar M, Hazarika S, Ramesh T, Bordoloi LJ, Moirangthem P, Bhuyan D (2017) Soil organic carbon fractions and enzymes activities as affected by organic and inorganic amendments in an acid soil of Meghalaya. J Indian Soc Soil Sci 65:54–61

Walkley A, Black IA (1934) An examination of the Degtjareff method for determining soil organic matter, and a proposed modification of the chromic acid titration method. Soil Sci 37:29–38

Wang MC, Chang SH (2001) Mean residence times and characteristics of humic substances extracted from a Taiwan soil. Can J Soil Sci 81:299–307

Analytical Strategies for Arsenic Estimation 7

Arghya Chattopadhyay, Amitava Rakshit, A. P. Singh, Deepranjan Sarkar, Nitesh Kumar Singh, and Aditi Chourasia

Abstract

Arsenic became a serious problem in various countries, affecting millions of people as it has major exposure root through drinking water. Arsenic is a non-threshold carcinogen. Many a time, its threat is overlooked only because of little knowledge about its estimation techniques. Various estimation techniques and instruments are available for risk quantification of arsenic. In many undeveloped countries where instrumental facilities are not available, a rapid arsenic testing technique through color development principle is beneficial there. Modern instruments mainly vary in their level of sensitivity. Arsenic estimation of groundwater samples required pre acid treatment before analysis to prevent oxidation of arsenic. Olsen reagent is most widely used for extracting plant available arsenic from soil. Digestion is preferred when we want to know the total amount of arsenic in samples. Among the various instruments available, hydride generation-atomic absorption spectroscopy (HG-AAS) is most widely used for estimating total arsenic and inductively couple plasma-mass spectroscopy (ICP-MS) hyphenated high pressure liquid chromatography (HPLC) is used for arsenic speciation analysis.

Keywords

Arsenic · Suitable extractant · Speciation · Detection limit · Sensitivity of instrument

A. Chattopadhyay (✉) · A. Rakshit · A. P. Singh · D. Sarkar · N. K. Singh · A. Chourasia
Department of Soil Science and Agricultural Chemistry, Institute of Agricultural Sciences, Banaras Hindu University, Varanasi, India

A. Rakshit et al. (eds.), *Soil Analysis: Recent Trends and Applications*,
https://doi.org/10.1007/978-981-15-2039-6_7

7.1 Introduction

Arsenic is widely dispersed and ubiquitous in environment. Average concentration of As in earth's crust is approximately 5 mg kg^{-1}. Under oxidizing conditions such as those prevailing in surface waters, the principal species is pentavalent arsenic and under reduce condition trivalent arsenic is dominant. Arsenic, the king of poison, is a non-threshold carcinogen and 20th most abundant element in the earth's crust. In the field of arsenic research, there are also significant advances in analytical chemistry to open new areas. Around 200 years back, Dr. Marsh in 1830s firstly started the arsenic analysis process. Last few decades, arsenic research grew continuously in accord with the discovery of new arsenic species and the importance of their environmental and biological activities. Recently, the growing interest of arsenic is increased because of its carcinogenic and leukemic effects on human body due to the drinking of groundwater (Mahfuzar 2007). A wide variety of methods to determine arsenic have been used: ultraviolet spectrometry, atomic absorption spectroscopy methods (AAS) coupled with hydride generation (HG-AAS), electrothermal AAS in graphite furnace (ETAAS), atomic fluorescence spectrometry (AFS), atomic emission spectrometry (AES), usually coupled with inductively coupled plasma (ICP-MS), X-ray spectrometry, neutron activation analysis (NAA), capillary electrophoresis, collision induced dissociation (CID), gas chromatography (GC), size exclusion chromatography (SEC), high-performance liquid chromatography (HPLC), Fourier transform ion cyclotron resonance-mass spectroscopy (FTICR-MS), stripping potentiometry, electroanalytical detection on gold plate, and gold film electrode preparation for anodic stripping voltammetric determination of arsenic. Methods involved in these techniques require expensive instrumentation, complicated procedures, and special sample pretreatment. Overall, all these methods are essentially sensitive to total arsenic. Growing interests in the determination of different arsenic species in groundwaters are caused by the fact that toxic effects of arsenic are solely connected with its chemical forms and oxidation states. The toxicity and bioavailability of arsenic can only be determined if all its forms can be identified and quantified. Several techniques including high-performance liquid chromatography separation joined with inductively coupled mass spectrometry, hydride generation atomic spectrometry and electrospray mass spectrometry detection have shown most powerful methods for arsenic speciation in environmental and biological matrices. These methods provide strong reliability on understanding of arsenic metabolism and biological cycling. In this review, we are trying to include recent developments and applications of analytical methods for the detection and speciation of groundwater arsenic.

7.2 Techniques for Arsenic Extraction

Estimation of arsenic level in soil sample comprises two major steps, one is extraction followed by determination. Extraction methodologies varies according to the fraction of arsenic we concern; like plant available, total, oxide bound or organically bound etc., the extraction methodologies varies accordingly.

Table 7.1 List of some commonly used acid mixture for wet digestion technique

Reference	Acid used
United States Environmental Protection Agency (US-EPA)206–5, 1974	H_2SO_4–HNO_3
US-EPA 7060A, 1994	H_2O_2–HNO_3
US-EPA 3050B, 1996	HNO_3–HCl
United States Geological Survey (USGS), 1999	HNO_3– H_2O_2, H_2SO_4–HF–HCl
United States Department of Agriculture, 2001 (CLG-ARS.03)	HNO_3–HCl

Hudson-Edwards et al. (2004)

7.2.1 Extraction of Total Soil Arsenic

For total arsenic estimation from soil or sediment sample, wet digestion technique is adopted. In this technique, various acid mixtures are used (Table 7.1) to achieve complete destruction of all As-bearing phases. The digestion methods can be carried out using a hotplate or microwave-digestion ovens to eliminate loss of volatile arsenic during the extraction.

7.2.2 Extracting Plant Available Arsenic from Soils and Sediments

Extraction of available arsenic is based on solubility product principle. Choice of method will depend on the types of soils and sediments being analyzed. Reduction of such As-bearing amorphous Fe oxides releases As to water systems. Acid ammonium oxalate is used for this purpose (Hudson-Edwards et al. 2004). Hydroxylamine hydrochloride is also used to extract Fe oxide-associated As (Montperrus et al. 2002). Extraction of plant available As from soil is done on a routine basis by Olsen extractant (0.5M NaHCO3, pH 8.5) (Table 7.2).

7.3 Techniques for Arsenic Determination

A variety of analytical techniques have been already applied for arsenic species determination. In the following paragraphs, we will summarize the major methods for the determination of arsenic.

7.3.1 Rapid Arsenic Test

First step reduction: Reducing agent is "nascent" hydrogen which is generated through the reaction of zinc metal and hydrochloric acid. The reduction may be accelerated by adding a small amount of potassium iodide and stannous chloride.

Table 7.2 List of some reagents used for soil extractable available arsenic

Soil extractants	References
Water	Reed and Sturgis (1936), Vandecaveye et al. (1936), Rosenfels and Crafts (1939), Deuel and Swobods (1972)
0.1, 0.5, and 1N NH_4OAc	Vandecaveye et al. (1936), Jacobs et al. (1970), Johnston and Barnard (1979)
0.1 and 0.5M NH_4NO_3	Vandecaveye et al. (1936)
0.1N KNO_3	Johnston and Barnard (1979)
0.5M $NaHCO_3$, 0.5M Na_2CO_3, and 0.5M $(NH_4)_2CO_3$	Woolson et al. (1971), Johnston and Barnard (1979)
0.5M NaHCO3 at pH $\geq$ 6.5 0.05N HCl + 0.025N H2SO4 at pH < 6.5	Woolson et al. (1973)
Bray P-1 (0.03N NH_4F + 0.025N HCl)	Jacobs et al. (1970)
Mixed acid (0.05N HCl + 0.025 N H_2SO_4),0.5N HCl, 0.05 and 0.5M KH_2PO_4 and 0.5M $(NH4)_2SO_4$	Woolson et al. (1971), Woolson (1973), Johnston and Barnard (1979)
0.1M solutions of Na2HPO4 (pH 9.1), Na2HPO4/NaH2PO4 (3:2, pH 7), Na_2HPO_4 (pH 4.5) and H_3PO_4 (pH 1.6),	Yamamoto (1975)
Modified Chang Jackson procedure	Johnson and Hitbold (1969), Jacobs et al. (1970), Woolson et al. (1971), Johnston and Barnard (1979)

$$Zn + 2\ HCl = ZnCl_2 + H_2$$

Second step checking interference: Sulfur is most ubiquitous in nature and causes serious interference in arsine gas generation by forming hydrogen sulfide. One way of checking this interference is by passing the gas stream through a filter impregnated with lead acetate which will form insoluble lead sulfide. An alternative way is to use cupric chloride in combination with ferric chloride. Ferric salts (FeC13) enhance the arsenic evolution and compensate the suppression effect of CuC12 (Cherukuri and Anjaneyulu 2005).

Third step volatilization: Both the tri and penta valent species of arsenic [As (V) and As (III)] generated arsine gas by the reduction with reducing agent (zinc dust) under acidic conditions (hydrochloric acid). The reactions that could occur are as follows:

$$As_2O_3 + 6\ Zn + 12\ HCl \rightarrow 2\ AsH_3 + 6\ ZnCl_2 + 3\ H_2O$$

$$H_3AsO_4 + 4\ Zn + 8\ HCl \rightarrow AsH_3 + 4\ ZnCl_2 + 4H_2O$$

Fourth step color development: Color stripe may be of two types, one is mercuric bromide ($HgBr_2$) and another is silver nitrate ($AgNO_3$) (Das et al. 2014). When the arsine (AsH_3) gas reacted with mercuric bromide ($HgBr_2$) a yellow-to-brown colored compound is formed (depending upon arsenic concentration) (Fig. 7.1). Reaction on the paper strip is as follows:

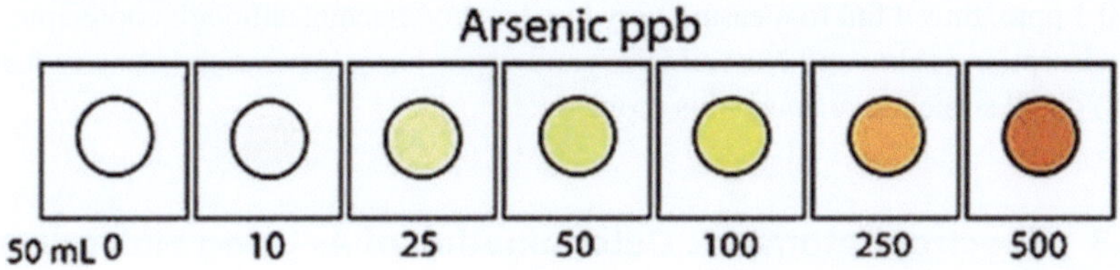

Fig. 7.1 Color chart for arsenic in mercuric bromide method. Color chart Kearns and Tyson (2012)

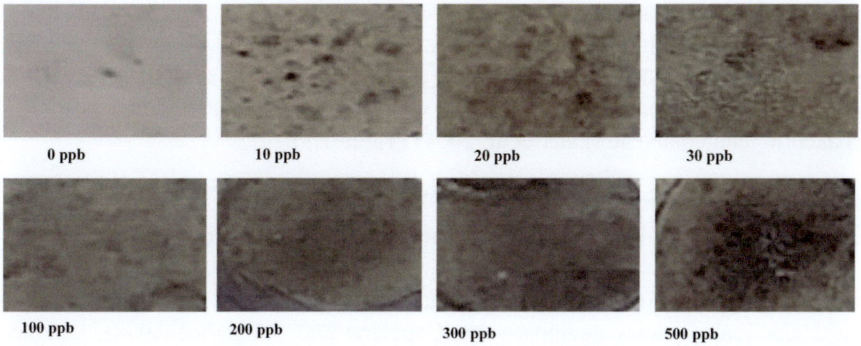

Fig. 7.2 Color chart for arsenic in silver nitrate method. Authors tested color chart

$$AsH_3 + HgBr_2 \rightarrow AsH(HgBr)_2 \text{ (yellow) } As(HgBr)_3 \text{ (brown)}$$

In case of silver nitrate method, the arsine (AsH3) when reacted with silver nitrate (AgNO3) formed a grey-to-black to a silver-black colored complex (depending upon arsenic concentration) (Fig. 7.2). Reaction on the paper strip is as follows:

$$AsH_3 + 3\ Ag+ \rightarrow Ag_3As \text{ (silver} - \text{black colored)} + 3H+$$

7.3.2 Spectrophotometric Determination of Arsenic by Silver Diethyldithiocarbamate

In this method, arsenic present in solution is reduced to arsine (AsH_3) by reducing agent. Then, the arsine gas bubble absorbed through 0.5% silver diethyldithiocarbamate which produces red color, and the intensity of the red color is measured at 540 nm. If the sample inherently contains sulfur, then it will produce sulfide gas before the formation of arsine and this interference can be removed by passing AsH_3 through lead acetate saturated glass wool. Contamination is the chief source of error of this process and this can be checked by rinsing the sample with 4% HNO_3. The sensitivity of this method is <0.1 ppm, thus it fail to measure very low level of arsenic; although coprecipitation and adsorption (Talmi and Norvell 1975) and vapor-phase separation (Denyszyn et al. 1978) could somehow manage this problem.

7.3.3 Spectrophotometric Determination of As Using Molybdenum Blue Method

In this method arsenic is treated with molybdenum solution and a heteropolymolybdoarsenate complex is formed which is reduced by hydrazinium sulphate or tin (II) chloride to form blue colour soluble complex known as "molybdenum blue". Intensity of the blue color is measured at 840 nm. To remove H_2S interference, gas is passed through a tube which is loosely packed with cotton wool soaked in lead ethanoate (Dhar et al. 2004) (Table 7.3).

7.3.4 As (III) Determination by Anode Stripping Voltammetry

Anodic stripping voltammetry is a voltammetric technique used to determine particular ionic species quantitatively (Copeland and Skogerboe 1974). Samples are electroplated on the working electrode during a deposition step and oxidized

Table 7.3 Comparison among different spectrophotometric methods used in arsenic estimation

Methods	Wavelength max (nm)	Limit of detection (μg ml^{-1})	Molar absorptivity (10^4 l mol^{-1} cm^{-1})	References
Silver diethyldithiocarbamate	520	–	1.30	Kopp (1973)
Morpholine-chloroform	510	0.006	1.40	Gupta and Gupta (1986)
Iodonitrotetrazolium	620	0.03	13.1	Kolesnikova and Lazareb (1991)
Silver diethyldithiocarbamate	530	1.00	1.50	Arbab-Zavar and Hashemi (2000)
Kinetic spectrophotometric	525	0.003	–	Afkhami et al. (2001)
Methylene blue	660	0.001	–	Kundu et al. (2002)
Micro-particle Formation of ethyl violet-molybdoarsenate	612	0.004	–	Morita and Kaneko (2006)
2-(5-bromo-2-pyridylazo)-5-di-ethylaminophenol	560	0.001	24.5	Pereira et al. (2008)

from the electrode during the stripping step. The current is measured during the stripping step. The oxidized species is registered as a peak of the current signal in their own potential range which is suitable for their oxidation. Stripping steps can be linear, square waves, stairs, or pulse. The peak widths and stripping peak currents on the electrode surface (Hg or alternate) are function of the coverage, size, and distribution of the metal phase. This technique is improved over previous technologies in the areas of better sensitivity (ppt level), reproducibility and provides real-time data of in situ measurement.

7.3.5 Hydride Generation: Atomic Absorption Spectroscopy (HG-AAS)

The most widely accepted method for arsenic analysis in ppb level is based on the principle of hydride gas generation of arsenic. After generation of hydride gas, it is then thermally decomposed to give elemental arsenic for atomic detection in AAS. Combined mixture of sodium borohydrate and sodium hydroxide solution is used as reductant, and hydrochloric acid is used to acidifying the solution. Combination of these two solutions acts as carrier solution and helps to reduce the analyte in hydride form. A cathode lamp is used to produce characteristic resonance frequency of arsenic (193.7 nm) and light absorption is measured by following the Bear–Lambert rule. Little bit speciation analysis could be possible in hydride generation AAS as the process of formation of arsine follows two steps reaction. First step is the reduction of As(V) to As(III) and the second step is the formation of AsH. The levels of the redox reaction involving transferring electrons are rather slow and pH-dependent, so it could be possible to distinguish between the two species if the first phase of the response at elevated pH values is slower than the second. Therefore, species differentiation could be possible using pH-selective arsine generation technique. In this methodology, strongly acidic solutions ($pH \leq 1$) are required for the determination of As(V) and mild acidic solution (pH 5) is suitable for hydride formation of As(III) (Howard and Comber 1992). In lieu of total arsenic estimation, pretreatment of the sample is done by applying acid mixture of hydrochloric acid, potassium iodide, and ascorbic acid for at least 45 min. There are some major interferences of this method. Cu (II), Co (II), and Ni (II) form specific chemical species between As and their reduction products decompose $NaBH_4$ and this problem could be solved by applying relatively low concentration of $NaBH_4$. During the hydride atomization, interferences from flame radical absorption of resonance lines could be managed by administrating the hydride in a heated quartz tube.

7.3.6 Chromatographic Methods

In case of chromatographic methods, gas chromatography (GC) and high-performance liquid chromatography (HPLC) deliver more detailed information about arsenic estimation (Niedzielski and Siepak 2003). Gas liquid chromatography runs on the principle of partitioning. In GLC, the components of vaporized samples are fraction due to partition between gaseous and mobile phase (unreacted carrier gas, e.g., N_2, He, Ar) and liquid stationary phase (nonvolatile liquid) held in a column. The column separate of compounds according to their different strength of interaction with the stationary phase. Strong affinity between stationary and mobile phases result extended retention of mobile phase, thus more time required to migrate through the column. In high pressure liquid chromatography (HPLC) separation of a sample into its constituent parts on the basis of difference in the relative affinities of different molecules for the mobile phase and stationary phase. Various detectors used in gas liquid chromatographic method are ECD, TCD, FID, PID, UV-Vis and IR these are non selective to compounds of different metals, whereas refractive index detector, ultra-violate detector, and luminescence detectors are used in HPLC. Chromatography is very much useful in speciation analysis of arsenic, and major disadvantage of this method is it does not have sufficiently low detection limits (Table 7.4).

7.3.7 Inductively Coupled Plasma-Mass Spectroscopy

Mass spectroscopy is an analytical technique that ionized the chemical species and sorts ions based on mass to charge ratio. Actually, a mass spectrum measures the mass within the sample. In this spectroscopic technique, a sample is ionized by

Table 7.4 Application of hyphenated techniques in As speciation

Analytes	Analytical column	Mobile phase	Method	Matrix	References
As^{3+}, As^{5+}, MMA, DMA	Hamilton PRP X100	$(NH_4)_2HPO_4$, MeOH	HPLC-ICP-MS	Soil extracts	Guerin et al. (1997)
As^{3+}, As^{5+}	Dionex AS9	NaOH, Na_2CO_3 $NaHCO_3$	HPLC-SF-ICP-MS	Soils	Koellensperger et al. (2002)
As^{3+}, As^{5+},MMA, DMA, AB, AC, TMAO	Hamilton PRP X 100 Zorbax 300-SCX	Pyridine, $NH_4H_2PO_4$	HPLC-SF-MS	Sediments	Zheng et al. (2004)
As^{3+}, As^{5+}, MMA, DMA	Hamilton PRP-X100	10–200 mM $NH_4H_2PO_4$	HPLC-ICP-DRC-MS	Sediments	Orero Iserte et al. (2004)
As^{3+}, As^{5+}, DMA	G 3154A/101	EDTA, $NH_4H_2PO_4$	HPLC-ICP-MS	Soils	Liang et al. (2012)
As^{3+}, As^{5+}, MMA, DMA, AB	Hamilton PRP X100	$NH_4H_2PO_4$ NH_4HPO_4, MeOH	HPLC-ICP-MS	Soils	Sanz et al. (2007)
As^{3+}, As^{5+}, MMA, DMA	Dionex AS11, AG11	10–100 mM NaOH	HPLC-HG-AFS	Polluted soil	Yuan et al. (2007)

bombarding it with the electron. It may cause some of the sample molecules to break into charged fragments. These ions are then separated according to mass to charge ratio by accelerating and subjecting them to an electric or magnetic field. Ions having the same mass to charge ratio undergo same amount of deflection, and these ions are detected by an electron multiplier. Results are displayed as spectra or signal of the relative abundance of detected ions as a function of mass to charge ratio. The atoms or molecules in the samples can be identified by correlating known mass to identified mass. ICP-MS undeniably belongs to the most often used hyphenated techniques because of its detection limits equal to or better than AAS, ability to handle both simple and complex matrices, minimum matrix interferences, superior detection capability to ICP-AES, and its ability to obtain isotopic information. Disadvantages and weaknesses of the ICP-MS detection are due to polyatomic interferences ($^{75}As^+$ with $^{40}Ar^{35}Cl^+$).

7.3.8 Other Methods Available for As Determination

- Radiochemical methods:
  - Neutron activation analysis (Terada et al. 1978).
  - Isotopic dilution technique (Krachler et al. 2002).
- Nuclear magnetic resonance (NMR) (Faucher et al. 2014).
- Hyperspectral remote sensing (Shi et al. 2016).
- Ion selective electrodes (Kang 1974).
- Micro XRF for in situ element mapping (Voegelin et al. 2007).

But these methods are lacking in the sensitivity required for ultra-low level detection of arsenic.

7.4 Conclusion

Finally, it can be concluded that the analytical techniques existing for the estimation and speciation of arsenic are diverse in nature. Each method has its own advantages and disadvantages that must be considered with respect to the type of research conducted and laboratory facilities available. When environmental observation is used to evaluate the toxic compound exposure, it is important that instrument can able to differentiate between toxic species and non-toxic species. Suitable methods for speciation analysis are HPLC-HG-AAS, HPLC-ICP, and HPLC-ICP-MS having high sensitivity and selectivity. These instruments are expensive and not available in many laboratories. In such cases, the development of rapid arsenic testing kits with color charts is expected to play a crucial role in the estimation of this dreadful heavy metal.

References

Afkhami A, Madrakian T, Ass AA (2001) Kinetic-spectrophotometric determination of trace amounts of As (III) based on its inhibitory effect on the redox reaction between bromate and hydrochloric acid. Talanta 55(1):55–60

Arbab-Zavar MH, Hashemi M (2000) Evaluation of electrochemical hydride generation for spectrophotometric determination of As (III) by silver diethyldithiocarbamate. Talanta 52(6):1007–1014

Cherukuri J, Anjaneyulu Y (2005) Design and development of low cost, simple, rapid and safe, modified field kits for the visual detection and determination of arsenic in drinking water samples. Int J Environ Res Public Health 2(2):322–327

Copeland TR, Skogerboe RK (1974) Anodic stripping voltammetry. Anal Chem 46(14):1257–1268

Das J, Sarkar P, Panda J, Pal P (2014) Low-cost field test kits for arsenic detection in water. J Environ Sci Health A Tox Hazard Subst Environ Eng 49(1):108–115

Denyszyn RB, Grohse PM, Wagoner DE (1978) Sampling and atomic absorption spectrometric determination of arsine at the 2 mu g/m^3 level. Anal Chem 50(8):1094–1096

Deuel LE, Swoboda AR (1972) Arsenic solubility in a reduced environment 1. Soil Sci Am J 36(2):276–278

Dhar RK, Zheng Y, Rubenstone J, van Geen A (2004) A rapid colorimetric method for measuring arsenic concentrations in groundwater. Anal Chim Acta 526(2):203–209

Faucher A, Terskikh VV, Wasylishen RE (2014) Feasibility of arsenic and antimony NMR spectroscopy in solids: an investigation of some group 15 compounds. Solid State Nucl Mag 61:54–61

Guerin T, Astruc A, Astruc M, Batel A, Borsier M (1997) Chromatographic ion-exchange simultaneous separation of arsenic and selenium species with inductively coupled plasma-mass spectrometry on line detection. J Chrom Sci 35:213–220

Gupta PK, Gupta PK (1986) Microdetermination of arsenic in water, spectrophotometrically, by arsine-silver diethyldithiocarbamate-morpholine-chloroform system. Microchem J 33(2):243–251

Howard AG, Comber SDW (1992) Hydride-trapping techniques for the speciation of arsenic. Microchim Acta 109(1–4):27–33

Hudson-Edwards KA, Houghton SL, Osborn A (2004) Extraction and analysis of arsenic in soils and sediments. Trends Analyt Chem 23(10–11):745–752

Jacobs LW, Syers JK, Keeney DR (1970) Arsenic sorption by soils. Soil Sci Am J 34(5):750–754

Johnson LR, Hiltbol AE (1969) Arsenic content of soil and crops following use of methanearsonate herbicides. Soil Sci Am J 33(2):279–282

Johnston SE, Barnard WM (1979) Comparative effectiveness of fourteen solutions for extracting arsenic from four Western New York soils. Soil Sci Am J 43(2):304–308

Kang SG (1974) The development of an arsenic (V) selective electrode. Dissertations and Theses, Portland State University

Kearns J, Tyson J (2012) Improving the accuracy and precision of an arsenic field test kit: increased reaction time and digital image analysis. Anal Methods 4(6):1693–1698

Koellensperger G, Nurmi J, Hann S, Stingeder G, Fitz WJ, Wenzel WW (2002) CE–ICP-SFMS and HPIC–ICP-SFMS for arsenic speciation in soil solution and soil water extracts. J Anal Atomic Spectr 17:1042–1047

Kolesnikova AM, Lazarev AI (1991) Spectrophotometric determination of arsenic in copper- and nickel-base alloys by using iodonitrotetrazolium. Zh Anal Khim 46:169–174

Kopp JF (1973) 1-ephedrine in chloroform as a solvent for silver diethyldithiocarbamate in the determination of arsenic. Anal Chem 45(9):1786–1787

Krachler M, Mohl C, Emons H, Shotyk W (2002) Analytical procedures for the determination of selected trace elements in peat and plant samples by inductively coupled plasma mass spectrometry. Spectro Chim Acta 57(8):1277–1289

Kundu S, Ghosh SK, Mandal M, Pal T (2002) Silver and gold nanocluster catalyzed reduction of methylene blue by arsine in micellar medium. Bull Mater Sci 25(6):577–579

Liang QQ, Zhao DY, Qian TW, Freeland K, Feng YC (2012) Effects of stabilizers and water chemistry on arsenate sorption by polysaccharide-stabilized magnetite nanoparticles. Ind Eng Chem Res 51:2407–2418

Mahfuzar R (2007) Arsenic and contamination of drinking-water in Bangladesh: a public-health perspective. J Health Popul Nutr 20:193–197

Montperrus M, Bohari Y, Bueno M, Astruc A, Astruc M (2002) Comparison of extraction procedures for arsenic speciation in environmental solid reference materials by high performance liquid chromatography–hydride generation atomic fluorescence spectroscopy. Appl Organomet Chem 16(7):347–354

Morita K, Kaneko E (2006) Spectrophotometric determination of arsenic in water samples based on micro particle formation of ethyl violet-molybdoarsenate. Anal Sci 22(8):1085–1089

Niedzielski P, Siepak M (2003) Analytical methods for determining arsenic, antimony and selenium in environmental samples. Pol J Environ Stud 12(6):653–667

Orero Iserte L, Roig-Navarro AF, Hernández F (2004) Simultaneous determination of arsenic and selenium species in phosphoric acid extracts of sediment samples by HPLC-ICP-MS. Anal Chim Acta 527:97–104

Pereira SFP, Ferreira SLC, Oliveira GR, Palheta DC, Barros BC (2008) Spectrophotometric determination of arsenic in soil samples using 2-(5-bromo-2-pyridylazo)-5-di-ethylaminophenol (Br-PADAP). Eclet Quím 33(3):23–28

Reed JF, Sturgis MB (1936) Toxicity from arsenic compounds to rice on flooded soils. Agron J 28(6):432–436

Rosenfels RS, Crafts AS (1939) Arsenic fixation in relation to the sterilization of soils with sodium arsenite. Hilgardia 12(3):201–229

Sanz E, Olivas MR, Cmara C, Sengupta KM, Ahamed S (2007) Arsenic speciation in rice, straw, soil, hair and nails samples from the arsenic-affected areas of middle and lower Ganga plain. J Environ Sci Health A Tox Hazard Subst Environ Eng 42:1–11

Shi T, Liu H, Chen Y, Wang J, Wu G (2016) Estimation of arsenic in agricultural soils using hyperspectral vegetation indices of rice. J Hazard Mater 308:243–252

Talmi Y, Norvell VE (1975) Determination of arsenic and antimony in environmental samples using gas chromatography with a microwave emission spectrometric system. Anal Chem 47(9):1510–1516

Terada K, Okuda K, Maeda K, Kiba T (1978) A neutron activation analysis of selenium, arsenic and antimony in rocks and sediments by simultaneous evolution with bromide-strong phosphoric acid reagent. J Radioanal Nucl Chem 46(2):217–228

Vandecaveye SC, Horner GM, Keaton CM (1936) Unproductiveness of certain orchard soils as related to lead arsenate spray accumulations. Soil Sci 42(3):203–216

Voegelin A, Weber FA, Kretzschmar R (2007) Distribution and speciation of arsenic around roots in a contaminated riparian floodplain soil: micro-XRF element mapping and EXAFS spectroscopy. Geochim Cosmochim Acta 71(23):5804–5820

Woolson EA (1973) Arsenic phytotoxicity and uptake in six vegetable crops. Weed Sci 21(6):524–527

Woolson EA, Axley JH, Kearney PC (1971) Correlation between available soil arsenic, estimated by six methods, and response of corn (Zea mays L.). Soil Sci Soc Am J 35(1):101–105

Woolson EA, Axley JH, Kearney PC (1973) The chemistry and phytotoxicity of arsenic in soils: II. Effects of time and phosphorus. Soil Sci Soc Am J 37(2):254–259

Yamamoto M (1975) Determination of arsenic, methane arsenate and dimethyl arsenite in water and sediment extracts. Soil Sci Soc Am Proc 39:859–861

Yuan CG, He B, Gao EL, Lü JX, Jiang GB (2007) Evaluation of extraction methods for arsenic speciation in polluted soil and rotten ore by HPLC-HG-AFS analysis. Microchim Acta 159(1–2):175–182

Zheng Y, Stute M, Van Geen A, Gavrieli I, Dhar R, Simpson HJ, Ahmed KM (2004) Redox control of arsenic mobilization in Bangladesh groundwater. Appl Geochem 19(2):201–214

Approach to Study Clay-Organic Complexes 8

Nintu Mandal, Arnab Bhowmik, Pritam Ganguly, and Samar Chandra Datta

Abstract

Clay-organic complexation is the dominant mechanism for retention of carbon in soil. Conventional approaches such as density fractionation have some limitations. Now-a-days, spectroscopic and microscopic techniques are being employed for studying clay-organic complexes. Synchrotron radiation-based Fourier transform infrared (SR-FTIR) spectroscopy, synchrotron radiation-based micro-X-ray fluorescence microscopy (μ-XRF) and two-dimensional correlation spectroscopy (2DCOS) analysis, 13 C NMR spectra, synchrotron-based X-ray absorption near edge fine structure (XANES) spectroscopy, X-ray photoelectron spectroscopy (XPS), scanning electron microscopy (SEM), and transmission electron microscopy (TEM) are being used for studying in-depth the mechanisms of clay-organic complexes. The in situ SR-FTIR analysis showed that clay-OH clusters, C–H, C=C, Si–O, and Al–O, were the dominant functional groups throughout soil microaggregates and demonstrated the significantly positive correlation among these functional groups. NMR analysis confirmed the presence of alkyl C, O-alkyl C, aryl C, and carboxyl C in the soil deposits. Spatially resolved observations at the submicron scale with STXM-NEXAFS clearly showed that mineralogy influences SOM stabilization. Detailed identifi-

N. Mandal (✉) · P. Ganguly
Department of Soil Science and Agricultural Chemistry, Bihar Agricultural University, Sabour, Bihar, India

A. Bhowmik
Department of Natural Resources & Environmental Design, North Carolina A&T State University, Greensboro, NC, USA

S. C. Datta
Division of Soil Science and Agricultural Chemistry, ICAR-Indian Agricultural Research Institute, New Delhi, India

A. Rakshit et al. (eds.), *Soil Analysis: Recent Trends and Applications*,
https://doi.org/10.1007/978-981-15-2039-6_8

cation and quantification about the reactive mineral complexes need to be conducted, which are responsible for locating the key factors of regulating SOM sequestration. Fourier transform ion cyclotron resonance mass spectrometry (FT–ICR–MS), NanoSIMS, and synchrotron scanning transmission X-ray microscopy (STXM) should be integrated to in situ explore molecular structures and binding coordination of soil microaggregates.

Keywords
Clay-organic complex · SR-FTIR · 2DCOS · XANES · STXM-NEXAFS

8.1 Introduction

Organo-mineral complexes have been divided into two basic categories: (1) primary organo-mineral complex and (2) secondary organo-mineral complex. Christensen (1996) defined primary organo-mineral complexes as the "primary structure of soils as defined by the soil texture" resulting from the association of organic matter (OM) with primary mineral particles, and as complexes that are isolated "after complete dispersion of soils." Secondary organo-mineral complexes were basically made up of primary organo-mineral complexes. Approaches to study organo-mineral complexation were based dominantly on density fractionation where over-emphasis was given on primary definition of organo-mineral complexes. Primary particles and similarly sized aggregates in the silt-sized fraction had recently been quantified (Balabane and Plante 2004).

Capturing and long-term storage of organic carbon in soil was governed by several mechanisms. Organo-mineral complexation was predicated as dominant mechanism for retention of carbon in soil. Co-precipitation of iron (Fe) and soil organic matter (SOM) was often predicated as driving mechanism for carbon sequestration. Surface properties of minerals (surface charge density and electrochemical properties) and reactivity and structure of Fe hydroxides often modified owing to interaction of organic matter with Fe oxides (Giannetta et al. 2019). Fe had been predicted as key regulator of soil organic matter stability. Fe (III) was soluble under acidic conditions whereas it precipitated as Fe(III) hydroxides under alkaline conditions.

Several conventional techniques had been reported for studying the clay-organic interaction dominantly based on density fractionation. Emphasis on using spectroscopic and microscopic techniques had been given for in-depth study of clay-organic complexes. Conventional techniques as well as spectroscopic and microscopic techniques employed in the characterization of clay-organic complexes had been described hereunder.

8.2 Studying Clay Organo-Mineral Complexes: Conventional Approaches

8.2.1 Density Fractionation Scheme of Clay-Organic Complexes

Density fractionations have been used for many years to separate SOM that is more or less bound to minerals. Several authors have applied this technique to the <2 mm clay-bound OM using separation solutions of various densities, e.g., 1.7 g cm^{-3} (Wattel-Koekkoek and Buurman 2004), 1.80 g cm^{-3} (Turchenek and Oades 1979), and 2.0 g cm^{-3} (Kiem and Kögel-Knabner 2002), without a full assessment of the degree of binding between the OM and the clay minerals. Assuming a density of 2.6 g cm^{-3} for a mixed-mineralogy clay fraction, Fig. 8.1 illustrates how the density of organo-clay complexes decreases with increasing OM concentration. Assuming a density of 1.4 g cm^{-3} for OM (Mayer et al. 2004), the selection of a separation solution with a density of 1.6 g cm^{-3} would isolate a light fraction in the supernatant that consists of relatively mineral-free OM with >72.9% OM, and an organic C content >424 mg C g^{-1} fraction assuming 58% C content of OM (Mikutta et al. 1996). On the other end, the selection of a separation solution with a density of 2.2 g cm^{-3} would isolate a heavy fraction in the pellet representing relatively OM-free minerals (<21.2% OM and <123 mg C g^{-1} fraction). Hence, such fractionation techniques should enable quantification of the degree of binding of OM with minerals in clay-sized fractions.

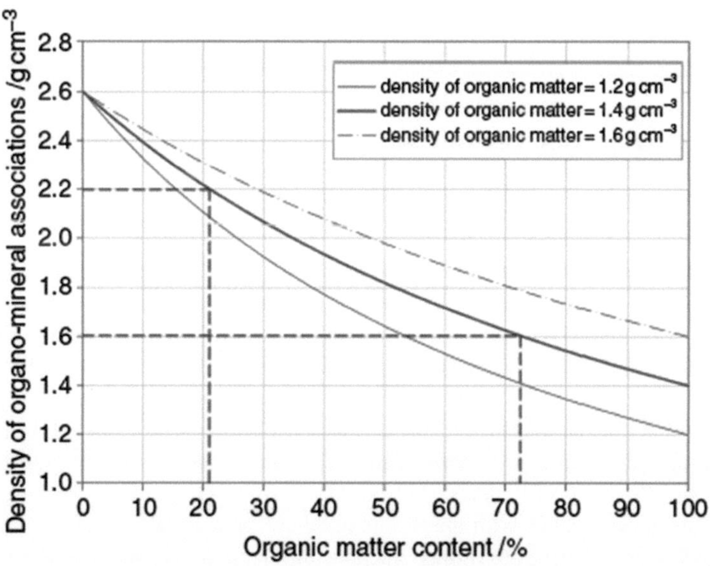

Fig. 8.1 Calculated density of organo-mineral associations assuming a density of either 1.2, 1.4, and 1.6 g cm^{-3} for the organic matter and a density of 2.6 g cm^{-3} for the mineral phase. Dashed drop-lines represent the separations performed at densities of 1.6 and 2.2 g cm^{-3} (Chenu and Plante 2006)

8.3 Spectroscopic Techniques for Studying Clay-Organic Complexes

The Fourier transform infrared (FTIR) spectroscopy is a popular facility to address the organic ligands in various environments, which can differentiate both fluorescent and non-fluorescent substances in comparison with fluorescence excitation–emission matrix (EEM) spectroscopy.

However, one-dimensional FTIR usually exhibits a variety of overlapped peaks because of the heterogeneous nature of an examined soil organic carbon (SOC). Combined with two-dimensional correlation spectroscopy analyses (2DCOS), it is able to resolve such peak overlapping problems by distributing the spectral intensity within a data set along a second dimension. This method could provide valuable information of the complexes of organic ligands with metals in soil dissolved organic matter (DOM) (Wen et al. 2014).

Recent innovations in the synchrotron-based X-ray absorption near-edge fine structure (XANES) spectroscopy, which is an element-specific technique and sensitive to both the oxidation state and the local structure of the absorber element (Prietzel et al. 2007), make it possible not only identify but also quantify the mineralogy of Fe presented in soils. Since Fe phases in soils are highly complex, this tool is helpful in comparison with the Mössbauer spectroscopy that might mask magnetically weak phases (Huang et al. 2016). In addition, the X-ray photoelectron spectroscopy (XPS) can provide valuable information on the bonding state of C framework in the surface layers of soil particles, examine the inherently stable structures of SOC, and thus the nature of SOC transformations under different fertilization regimes (Xiao et al. 2015).

8.4 Microscopic Techniques

8.4.1 Scanning Electron Microscopy (SEM)

Two sections of soil micro-aggregates (Sections 1 and 2) were randomly selected and successfully characterized by electron microprobe analysis (EMPA)μ-XRF (X-ray fluorescence microscopy) and synchrotron radiation-based Fourier transform infrared (SR-FTIR) (Fig. 8.2) (Xiao et al. 2019). The BSE images revealed that the sections were aggregated by a random and heterogeneous mixture of particles (Fig. 8.2a). Element images by EMPA showed that the spatial distribution of Al and Fe was in a similar pattern which was heterogeneously different than C (Fig. 8.2b). Also, μ-XRF maps showed that Fe was occluded in micro-regions, which was in accordance with the results obtained from EMPA (Fig. 8.2c).

2DCOS was performed to further elucidate the stable sequestration reactivity from the outside (edge) to the inside (nuclei) (4ROIs from two sections). Figures 8.3 and 8.4 showed the 1D SR-FTIR spectra and 2DCOS analysis for two representative ROIs from the Sections 1 and 2, respectively. To start with, the 1D SR-FTIR spectra in ROI1 of Section 1 had major peaks at 3695, 3619, 2922, 1613, 1465, 1030, and 915 cm^{-1}. And also, the other three ROIs also possessed similar major peaks, indicating that the

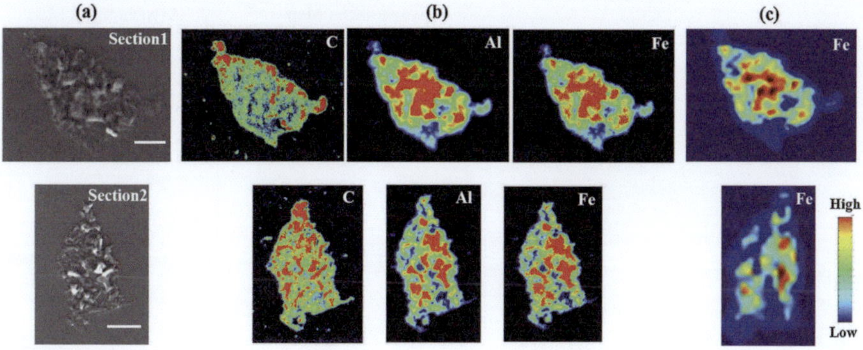

Fig. 8.2 In situ sketch of the measurement of Sections 1 and 2 from soil microaggregates. (**a**) BSE image by EMPA, section 1, Bar = 20 μm; section 2, Bar = 50 μm. (**b**) C, Al, Fe distribution images by EMPA. (**c**) Distribution of Fe by μ-XRF

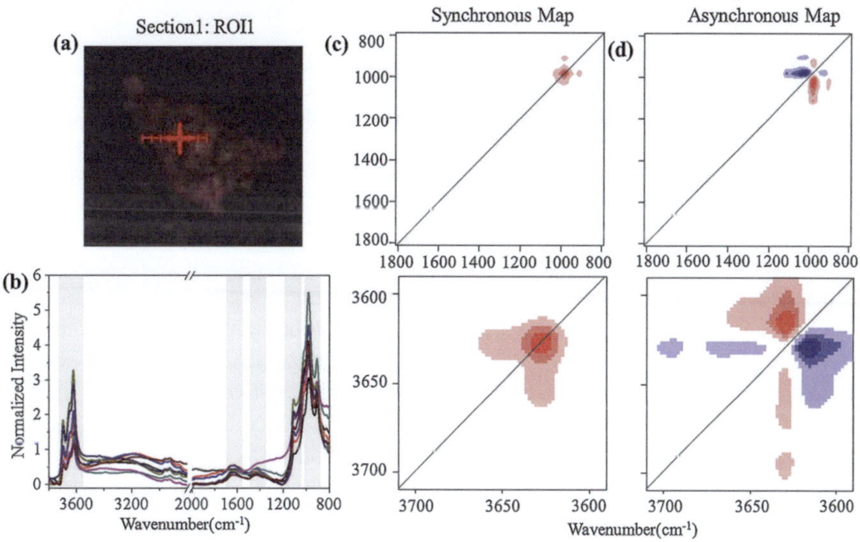

Fig. 8.3 1D SR-FTIR spectra and 2D correlation maps generated from the ROI1 of Section 1. (**a**) The red line showed ROI1, (**b**) 1D SR-FTIR spectra extracted from ROI1, (**c**) synchronous 2D correlation maps generated from 1800 to 800 cm^{-1} and 3700 to 3600 cm^{-1}, and (**d**) asynchronous 2D correlation maps generated from 1800 to 800 cm^{-1} and 3700 to 3600 cm^{-1}. The blue (red) regions are defined as negative (positive) correlation intensities; the higher color intensity indicates a stronger negative or positive correlation

same constituents' clay-OH clusters, C–H, C=C, Si–O, and Al–O, were the dominant functional groups throughout soil microaggregates. Since the main functional assignments focused on the regions of 3700–3600 cm^{-1} and 1800–800 cm^{-1}, 2DCOS analysis mostly concentrated in the two regions (Wen et al. 2014).

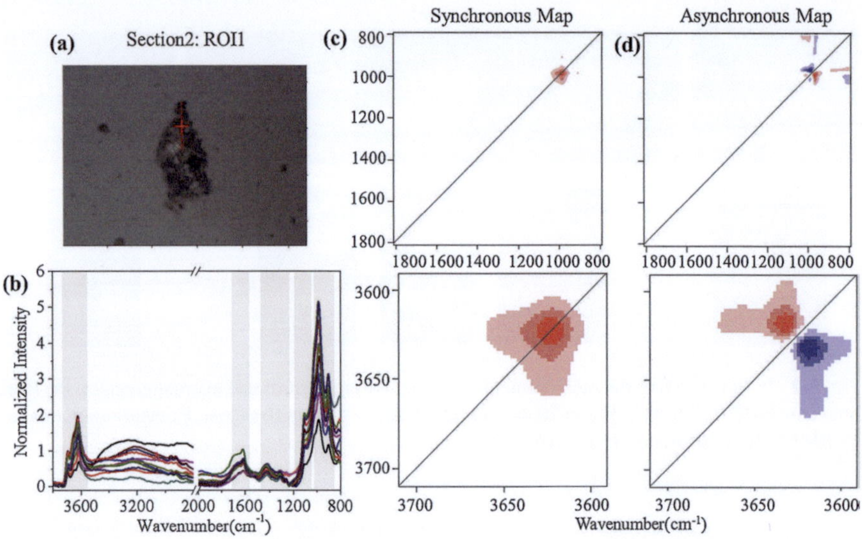

Fig. 8.4 D SR-FTIR spectra and 2D correlation maps generated from the ROI1 of Section 2. (**a**) The red line showed ROI1, (**b**) 1D SR-FTIR spectra extracted from ROI1, (**c**) synchronous 2D correlation maps generated from 1800 to 800 cm^{-1} and 3700 to 3600 cm^{-1}, (**d**) asynchronous 2D correlation maps generated from 1800 to 800 cm^{-1} and 3700 to 3600 cm^{-1}. The blue (red) regions are defined as negative (positive) correlation intensities; the higher color intensity indicates a stronger negative or positive correlation

All auto-peaks along the diagonal were positive in synchronous 2DCOS analysis, which presented the sequestration susceptibility of the corresponding assignments due to the perturbation. A positive peak off the diagonal (a cross peak) in 2DCOS spectrum indicated that the two spectral features were positively correlated, while a negative cross-peak indicated that the peaks were negatively correlated. In this study, the perturbation was Fe–OH and Si–O vibrations could keep stable sequestration of SOC at the submicron scale, which contributed to improve the aggregate hierarchy concept.

8.5 Fourier Transform Infrared (FTIR) Spectra

The broad bands at 3300–3400 cm^{-1} and at 3400–3700 cm^{-1} represent –OH groups of carboxyl and alcohols and stretching of adsorbed H_2O (or H-bonding to Si–O–Al linkage) of illite and chlorite and structural hydroxyl (Al–OH) group of illite. Absorption bands at 2922 and 2852 cm^{-1} are linked to aliphatic C–H stretching of CH_3 or CH_2 groups. Absorption bands at 1640 cm^{-1} are attributed to C=O stretching of protonated carboxyl groups and C=C stretching of aromatic compounds. The band at 1380 cm^{-1} represents aliphatic C–H stretching of methyl groups, C–H bending, O–H deformation, C=O stretching of phenolic groups, and COO– antisymmetrical stretching of carboxylates (Fig. 8.5). The band at 1250 cm^{-1}

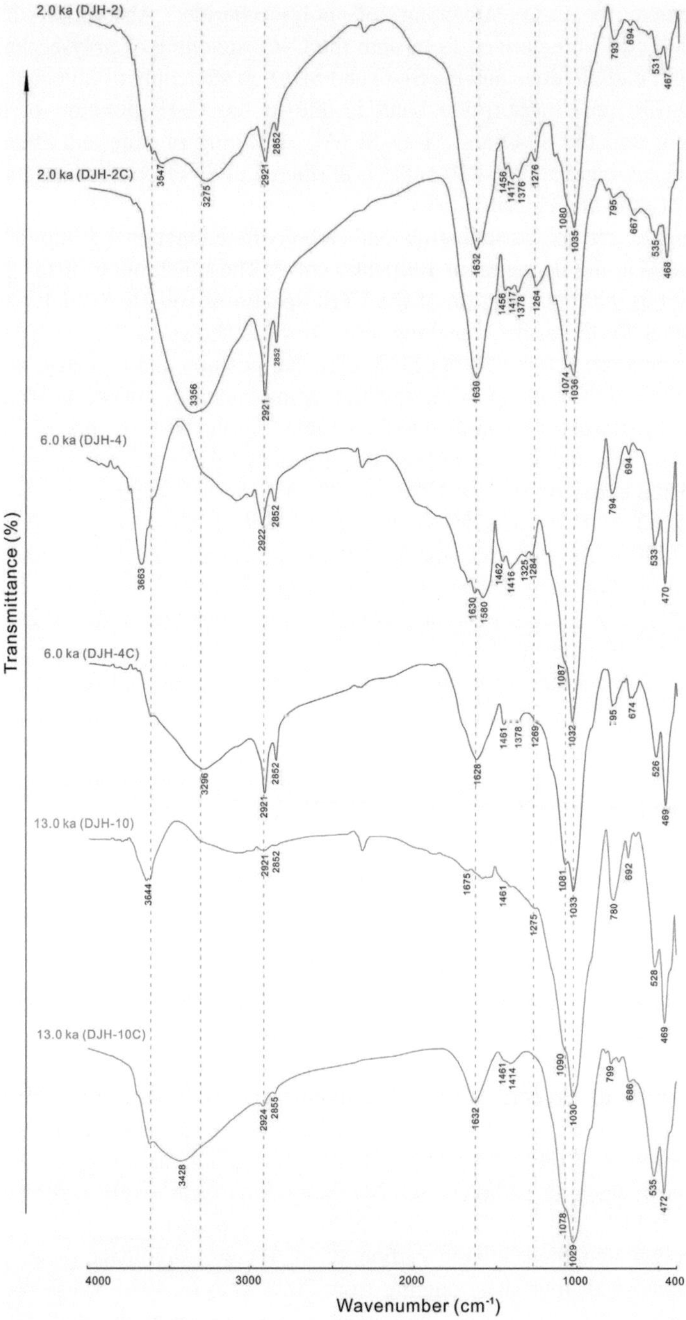

Fig. 8.5 FTIR spectra of the soil samples showing changes in intensity and shift of absorption bands of characteristic functional groups of OM in the peat sequences and their difference between coarse fractions (DJH-2, DJH-4, DJH-10) and clay fractions (DJH-2C, DJH-4C, DJH-10C)

is attributed to C–O stretching of polysaccharide. Absorption bands at 1100–1000 cm^{-1} are derived from both the C–O stretching of polysaccharide and hydroxyl or carbohydrate and the Si–O and Si–O–Si stretching of illite and chlorite. The 910–730 cm^{-1} absorption band is due to the C–H vibration of aromatic compounds and the Si–O and Si–O–Si (Al) stretching of illite and chlorite. The low frequency band at 700–420 cm^{-1} is attributed to Si–O–Al stretching and Si–O bending of illite and chlorite.

The amide, carboxylic acid, ester, and carbohydrate functional groups are mostly concentrated in the IR region of 1800–900 cm^{-1}. The synchronous maps generated from 1800 to 900 cm^{-1} region of the FTIR spectra of soil DOM under NPK and NPKM over Fe(III) at the four long-term sites are shown in Fig. 8.6 (Wen et al. 2019). In general, auto-peaks appear at diagonal position and represent the overall susceptibility of the corresponding spectral region to changes in spectral intensity as an external perturbation is applied to the system. For the NPK treatment, the spectra displayed 6, 5, 1, and 2 major auto-peaks, in which the change in the band intensity followed the order of $905 > 1270 \approx 1380 \approx 1550 \approx 1690$ in the Black soil, $905 > 1380 > 1270 > 1550$ in the Brown soil, 1380 in the Desert soil, and $1380 \approx 1100$ in the Red soil. For the NPKM treatment, only 1, 2, 1, and 1 of the major auto-peaks were observed in the Black, Brown, Desert, and Red soils, respectively. The spectral band at 1380 cm^{-1} was consistent in the Black, Brown, and Desert soils, whereas the band at 1010 cm^{-1} was the only observed peak in the Red soil. The various spectral bands were assigned as follows: the band at 1690 cm^{-1} was assigned to the C–O stretching of amide I in the protein compounds, the band at 1550 cm^{-1} to N–H deformation and C–N stretching of amide II in the protein compounds, the band at 1380 cm^{-1} to the CH deformations in the aliphatic groups, the band at 1120 cm^{-1} to the C–OH stretching of aliphatic O–H, and the band at 1010 cm^{-1} to the C–O stretching of polysaccharides, the Si–O of silicate impurities, or phosphate groups (Wen et al. 2014).

8.6 Functional Composition and Speciation of C in Water-Dispersible Soil Colloids by C 1 s XPS Techniques and Relationship with Fe Fraction

Spectral shifts in the core level C 1 s binding energy were assigned to different chemical environments of C: (1) aromatic carbon (Ar–C–C(H):284.2 eV), (2) aliphatic carbon (C–C(H): 284.8 eV), (3) ether or alcohol carbon (C–O; 286.2 eV), (4) ketonic or aldehyde carbon (C–O; 287.9 eV), and (5) carboxylic carbon (C–O–O; 289 eV) (Mikutta et al. 1996). The XPS C 1 s peak-fitting results (Fig. 8.7) demonstrated that the aliphatic carbon (C=C/H) was dominant under all three fertilizations at all four sites, ranging from 30.92% to 62.49%. These peaks were significantly higher under NPK than under NPKM, except in the Red soil. For the aromatic C (Ar–C–C(H)), the percentage ranged from 10.18% to 36.18%, with the highest under NPKM (20.77–36.18%) in each site, higher under the Control (11.81–17.65%) and least under NPK (10.18–11.87%). The ether or alcohol carbon

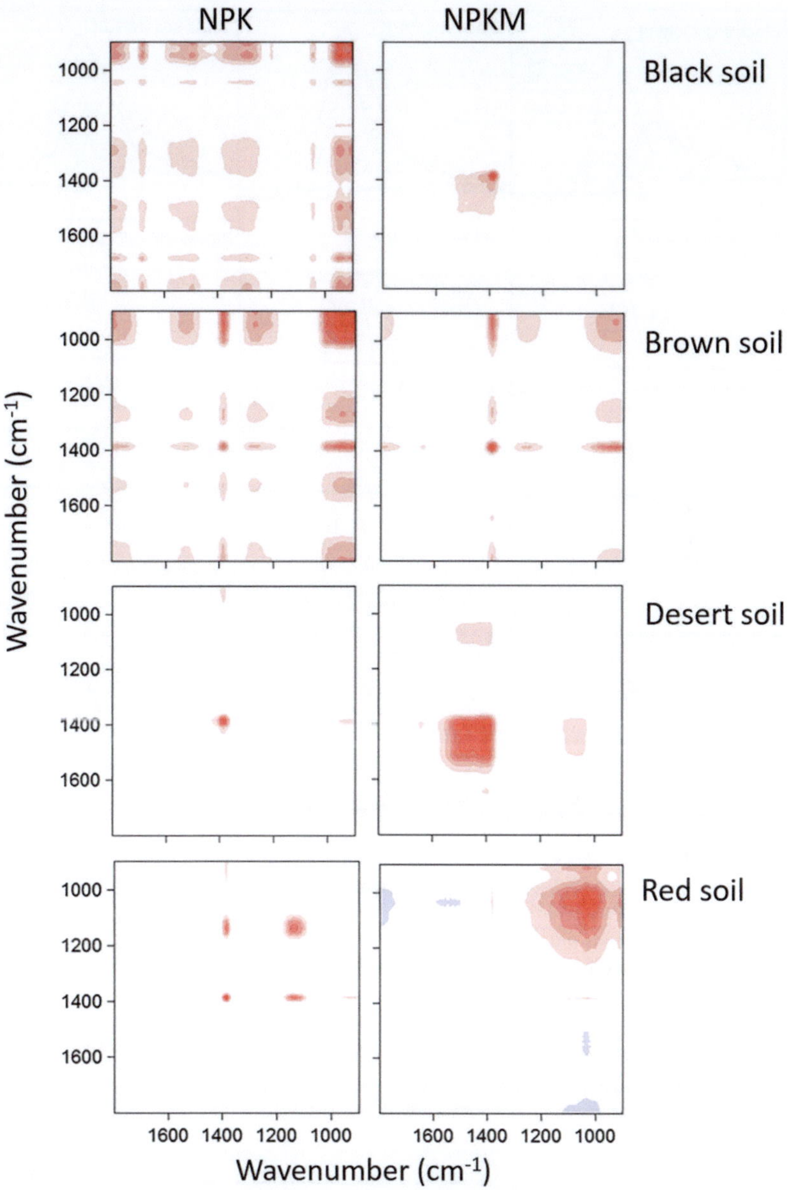

Fig. 8.6 Synchronous maps generated from 1800 to 900 cm^{-1} region of the FTIR spectra over Fe (III) of soil DOM under NPK and NPKM at the four long-term fertilization sites. The greater red or blue color intensity represents a positive or negative correlation

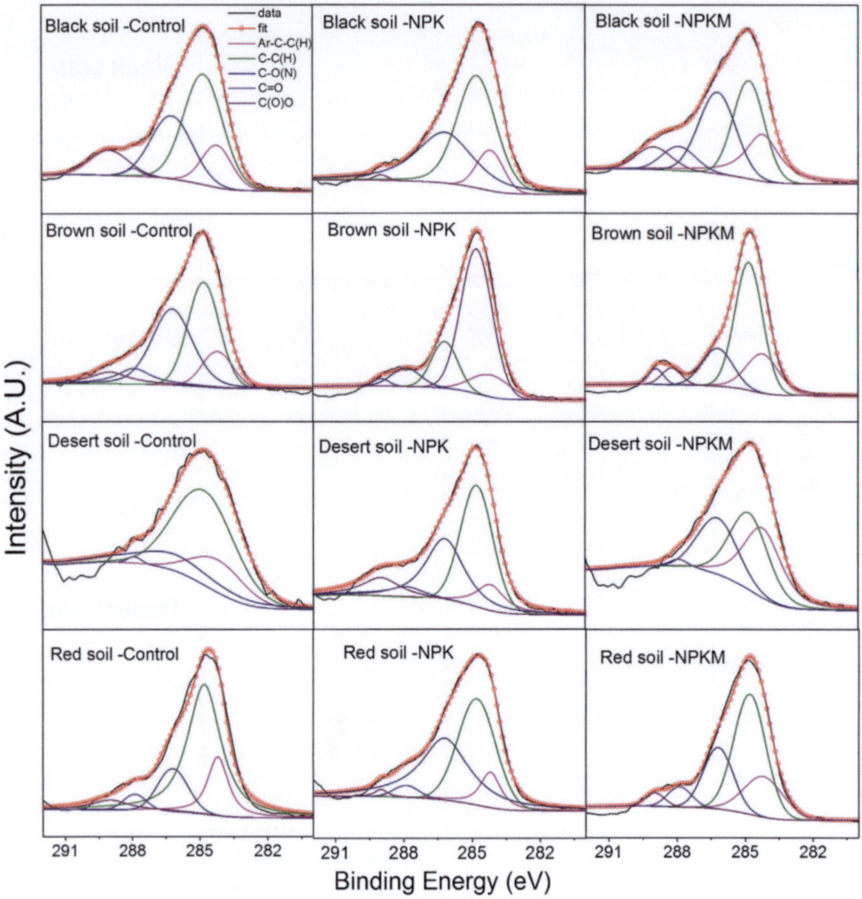

Fig. 8.7 XPS peak-fitting images recorded from water-dispersible soil colloids extracted under control, NPK, NPKM treatments at four long-term fertilization sites across China

(C–O(N)) ranged from 16.74% to 44.65% in all sites, with no significant changes under the three contrasting fertilization treatments (Wen et al. 2019).

8.7 Microscopic Observation

Clay minerals play important roles in stabilizing organic matter (OM) in soils. In order to investigate the nature of OM in association with clay minerals in peatland soils, the coarse (>0.3 mm) and clay (<2 μm) fractions of samples from a 250-cm-long core from the Dajiuhu peatland in Hubei Province (central China) were analyzed using gas chromatography–mass spectrometry (GC–MS), solid state 13C nuclear magnetic resonance (NMR) spectroscopy, and Fourier transform infrared spectrophotometry (FTIR) methods (Hong et al. 2019).

The Dajiuhu sediments exhibited a loose texture under SEM observation (Fig. 8.8). They contained dominantly plant roots and plant debris, with small amounts of clay minerals. Plant roots and debris occurred in various particle sizes with irregular outlines, which decayed intensively and decomposed into smaller grains (Fig. 8.8c). Clay aggregates occurred commonly in the sediments in close association with OM, with a size mainly of 5–30 μm. Clay mineral flakes are usually 0.2–2 μm in size, with ragged or bay-shaped edges and a poorly developed lateral dimension and small platy thickness (Fig. 8.8d), consistent with their detrital source (Hong et al. 2019).

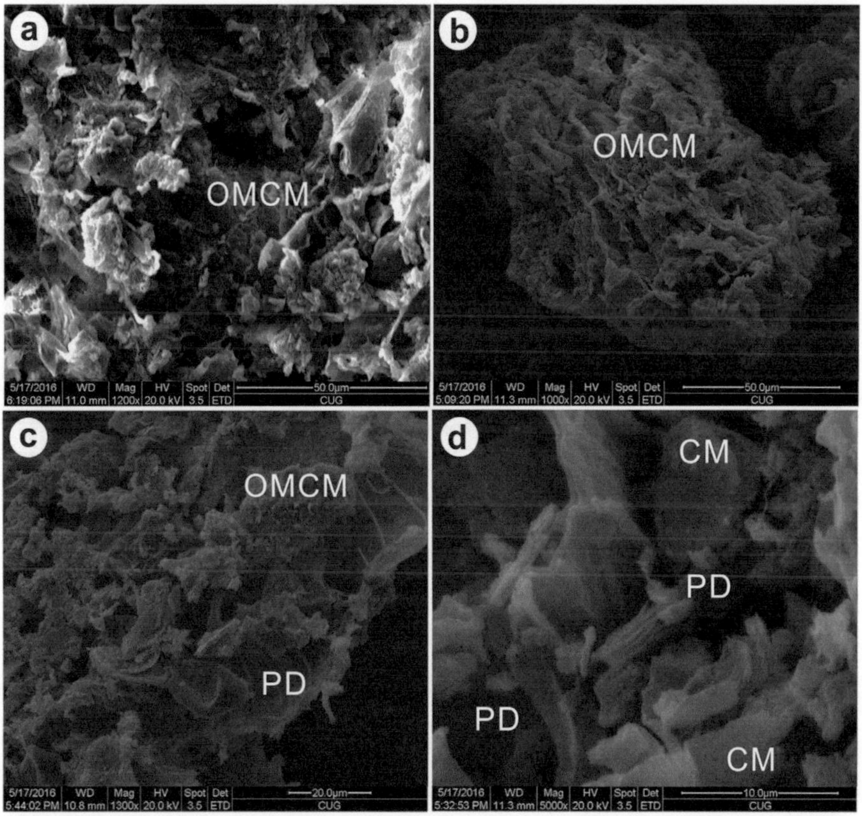

Fig. 8.8 Scanning electron photographs of the representative Dajiuhu sample (DJH-1). (**a**) Sediments consist of OM-clay mineral aggregates and exhibit a loose texture; (**b**) Clay aggregate displays typically porous structure; (**c**) Plant debris decayed to form OM-clay mineral aggregates; (**d**) Clay mineral flakes occur in association with plant debris. *OMCM* OM clay mineral aggregate, *CM* clay mineral, *PD* plant debris

8.8 Use of TEM to Study Clay-Organic Complexes

When observed with TEM, the light free organic fraction (<1.6 g cm^{-3}) consisted of free organic particles, such as microbial remains, cell wall remains, or amorphous and shapeless organic particles (Fig. 8.9) (Chenu and Plante 2006). The organic particles were often encrusted with a thin layer of small clay platelets, 0.05 0.2 mm long (Fig. 8.9a, b), which were, however, not abundant in the fraction. By contrast, the heavy organo-mineral fraction (>1.6 g cm^{-3}) consisted mainly of clay mineral particles (Fig. 8.3c, d). Organic matter was observed (Fig. 8.9c, d), and was locally identified using EDS X-ray analysis where detectable peaks of Os (the fixative) and Pb (the stain) indicated the presence of organic compounds.

8.9 Organo-Mineral Microaggregate Formation

Organo-mineral microaggregates were clearly identified in a study conducted by Chenu and Plante (2006). These consisted of several clay particles surrounded or occluded by plant debris (Fig. 8.10), bacteria (Fig. 8.10b, c), or amorphous OM, and

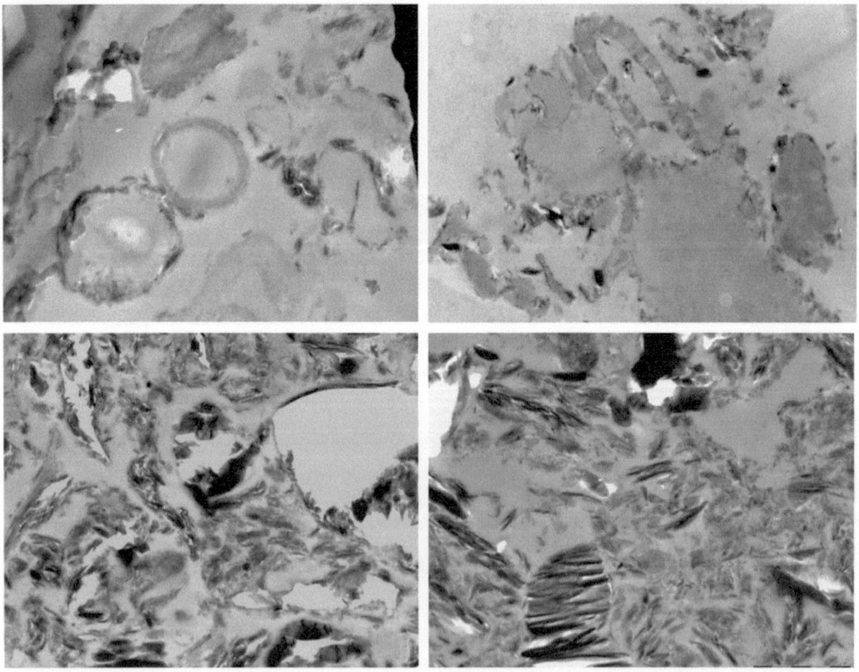

Fig. 8.9 TEM observations of the clay-sized density fractions. Free organic matter fractions (<1.6 g cm^{-3}) isolated from (**a**) forest and (**b**) cultivated soils, and organo-mineral fractions (>1.6 g cm^{-3}) isolated from (**c**) forest and (**d**) cultivated soils. *c* clay platelet, *om* organic matter, *w* cell wall, *μ* microorganism

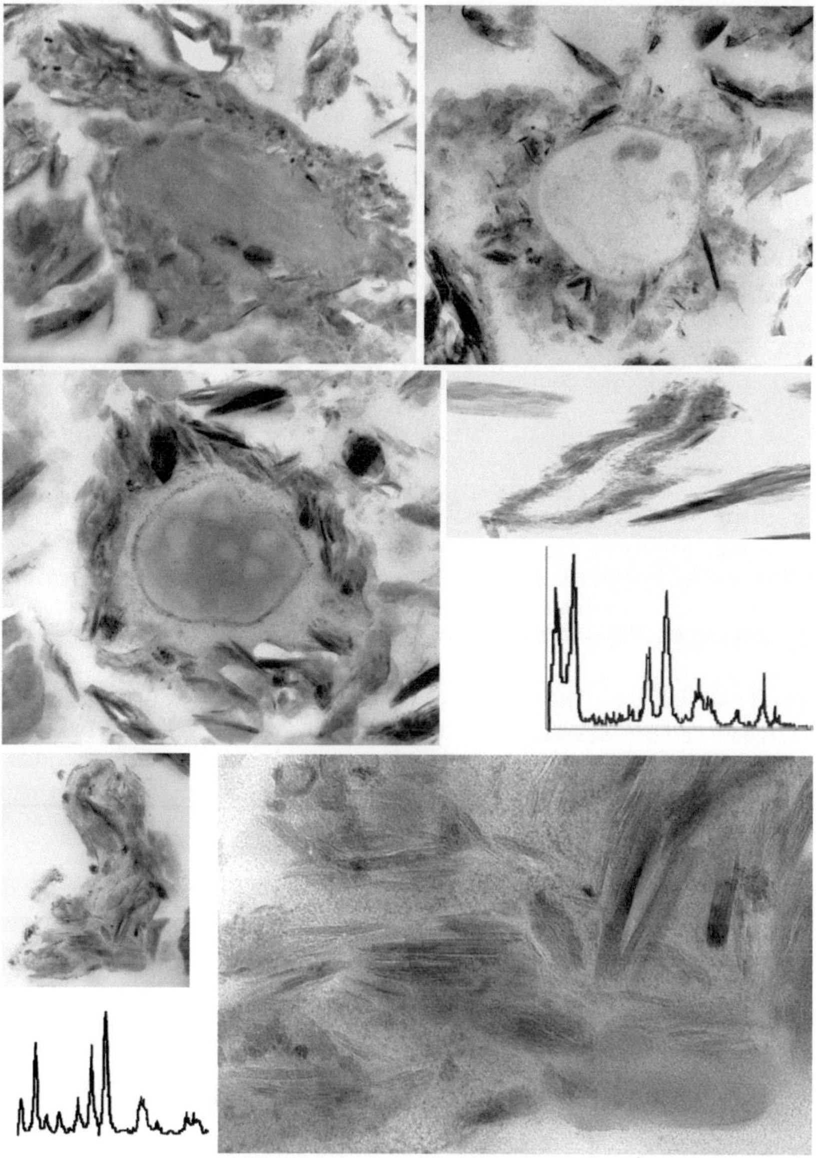

Fig. 8.10 Clay-bound organic matter observed with TEM. (**a**) Microaggregate where clay minerals encrust a plant cell wall residue (U, Pb staining). (**b**) Bacterial microaggregate (U, Pb staining). (**c**) Bacterial microaggregate (Ag staining of polysaccharides). (**d**) Organo-mineral particle. Black dots correspond to Ag grains and locate polysaccharides. (**e**) X-ray-EDS elemental analysis of outlined zone in Fig. 8.4d, showing an Ag peak. C, Cu, Au, and Cl peaks are partly or totally due to equipment (e.g., grids or resin) and not to the sample. (**f**) Microaggregate where organic matter is amorphous and located between stacks of clay platelets. (U, Pb staining). (**g**) Detail of (**f**). (**h**) X-ray-EDS elemental analysis of outlined zone, showing Pb and U peaks from the OM stains. C, Cu, and Cl peaks are partly or totally due to equipment (e.g., grids, resin) and not to the sample

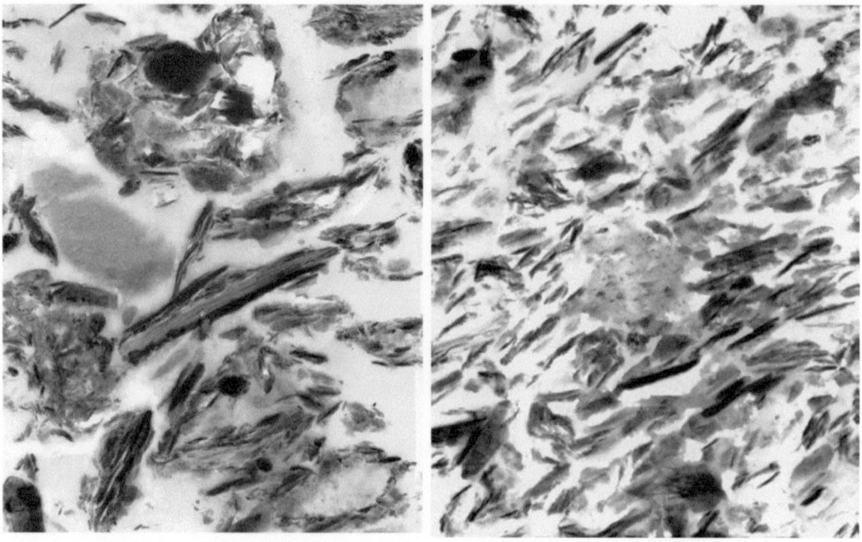

Fig. 8.11 Examples of the fabric of clay size fractions, observed with TEM. Photographs were taken at random in forest soil sample and sample from soil cultivated for 35 years. Several microaggregates are outlined with dotted lines

varied in the relative proportions of mineral particles and OM. The material appeared more microaggregated, visually, in the forest clay-sized fraction than in the soils cultivated for 7 or 35 years (Fig. 8.10). The mineral particles in organo-mineral microaggregates had, in general, the morphology and the elemental characteristics (Si/Al ratio and K content) of illite and vermiculite (Fig. 8.10). Fe coatings were sometimes detected on the surface of clay particles or within microaggregates by their morphology and using X-ray EDS elemental analysis, but no free iron oxide particles were observed. At a working magnification of 37,000, a certain proportion of objects appeared as mineral particles. Observing these particles at higher magnification and using X-ray EDS elemental analysis either confirmed the purely mineral nature of the particles or showed that OM was bound to it as thin layers on the surface of the clay or between packets of clay sheets (Fig. 8.11d–h). Organic matter was always observed between packets of several clay sheets rather than between individual clay sheets.

8.10 ^{13}C NMR Spectra of the Peat Soils

The ^{13}C CPMAS NMR spectra of samples are similar throughout the study core (Fig. 8.4). The assignments of the NMR peaks were referred to the literature. The NMR spectra are all characterized by a dominant signal at 0–50 ppm (centered at 37 ppm) of the alkyl C region, which is generally attributed to aliphatic compounds, including lipids, cut-in, and amino acids. Resonance signals at 50–100 ppm are linked to O-alkyl C compounds, which are usually assigned to carbohydrate C of polysaccharides and to amide C of proteins, including carbohydrates, cellulose, methoxyl C, and hemicellulose. Signals at 100–160 ppm and 160–200 ppm correspond to the aryl C and the carboxyl regions (Fig. 8.12). The former includes lignin, tannin, aromatic compounds, and olefins, whereas the latter includes carboxyl/amide and carbonyl C (mainly carboxylic acids, amide, aldehyde, and ketone) (Hong et al. 2019).

8.11 Spectroscopic Techniques

8.11.1 IR Spectra (ATR and FIIR)

Physical size fractionation by ultrasonic dispersion and wet sieving was performed, allowing for splitting particles into four different size fractions: coarse sand (CSa: 2000–200 μm diameter), fine sand (FSa: 200–50 μm), coarse silt (CSi: 50–20 μm), and fine silt and clay (FSi + Cl: b20 μm).

8.11.1.1 Attenuated Total Reflectance Fourier Transform Infrared Spectroscopy

The ATR-FTIR spectra of reacted and unreacted samples were recorded after freeze-drying with a Bruker Tensor 27. Samples were scanned over the 4000–600 cm^{-1} range, with a resolution of 2 cm^{-1} and 64 scans min^{-1}. Automatic baseline correction and normalization was applied to all spectra.

The ATR-FTIR spectra of the unreacted FSi + Cl fractions of GL and CF show a predominant broad band centered around 1420 cm^{-1}, which revealed the presence of symmetric COO– stretching (Giannetta et al. 2019). The dominant broad band at 1000–1100 cm^{-1}, found in all spectra, may be indicative of stretching of carbohydrate and polysaccharides-like substances (Fig. 8.13).

At the same time, however, Si–O–Fe and Al–Al–OH bonds of several (alumino) silicates may also cause absorption between 900 and 1000 cm^{-1} (Giannetta et al. 2019). The lower intensity of carbohydrate peaks suggests the comparatively scarce presence of labile organic materials in coarse fractions where they act as binding agents in stable aggregates (Giannetta et al. 2019). Attenuated total reflectance-Fourier transform infrared (ATR-FTIR) spectroscopy indicated that Fe-mediated organic C stabilization can be mainly ascribed to the formation of complexes between carbohydrate OH functional groups and Fe oxides.

8.12 Synchrotron Radiation-Based Fourier Transform Infrared (SR-FTIR) Spectroscopy

These identifiable vibrations (v) were OH clay (3619 cm^{-1} illite-OH clay), C–H (2922 cm^{-1}, aliphatic-C), C=C (1613 cm^{-1}, aromatic-C), Si–O (1030 cm^{-1}, silicates), and Al–O (915 cm^{-1}, kaolinite and smectite), respectively (Fig. 8.14).

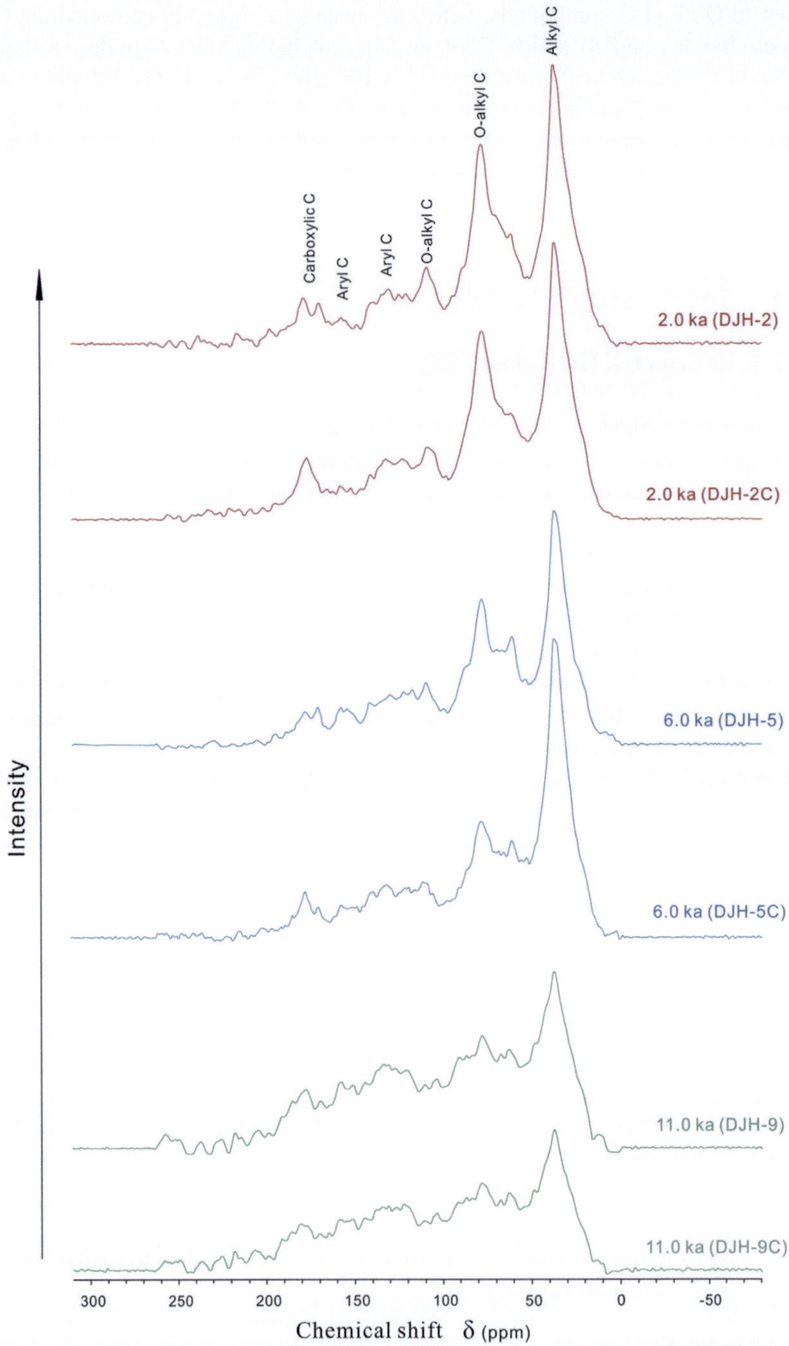

Fig. 8.12 The 13C CPMAS NMR spectra of peat soils showing the time-dependent OM components along the peat profile and the difference between coarse fractions (DJH-2, DJH-5, DJH-9) and clay fractions (DJH-C2, DJH-5C, DJH-9C)

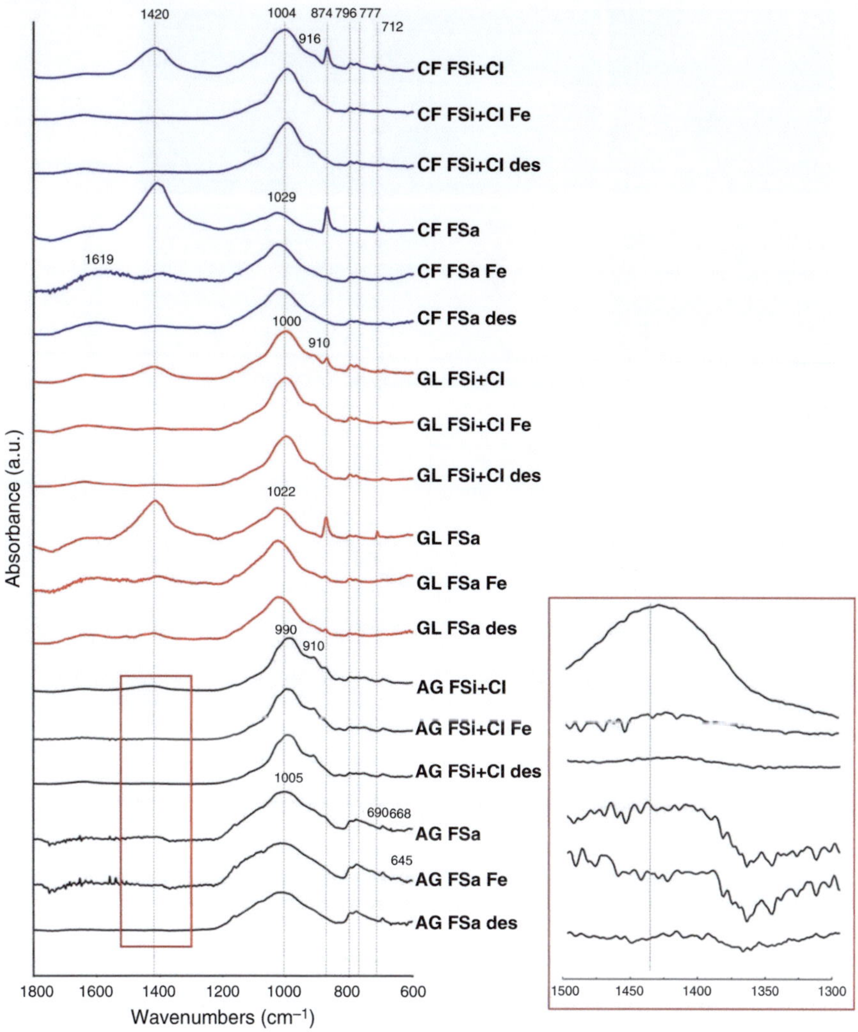

Fig. 8.13 ATR-FTIR spectra of the fine silt and clay (FSi + Cl) and fine sand (FSa) fractions before reaction with Fe, after reaction with Fe(III), and after desorption (des) reactions (*CF* coniferous forest soil, *GL* grassland soil, *AG* agricultural soil). In both FSi + Cl and FSa fractions, ATR-FTIR spectra collected after the sorption reaction show the disappearance of the band at 1420 cm^{-1}, as well as of peaks at 874 and 712 cm^{-1}; this is due to the removal of carbonates when the acidic Fe nitrate stock solution was added (pH < 2). On the bottom right corner, a zoom of the region 1500–1300 cm^{-1} is reported for AG spectra

The colors of functional groups in both Sections 1 and 2 were different, whereas the colors from red to blue corresponded to the relatively strong SR-FTIR absorbance to the relatively weak one in these images (Fig. 8.14f). The results showed that the SR-FTIR absorbance of Si–O (1030 cm^{-1}, silicates) and Al–O (915 cm^{-1}, kaolinite

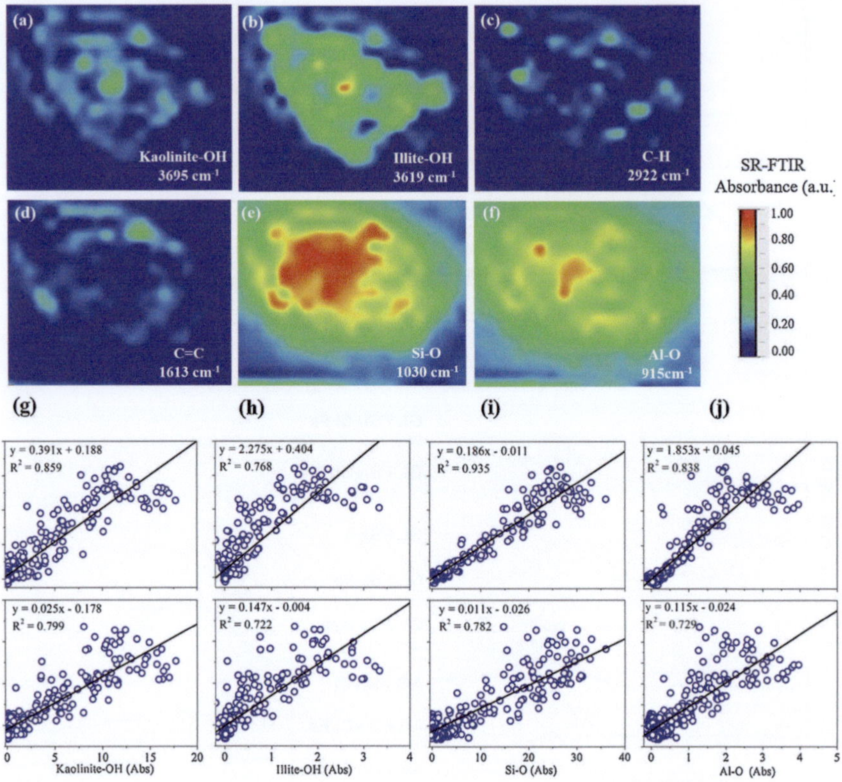

Fig. 8.14 Quantitation distribution and correlation between clay clusters and biopolymers by SR-FTIR in Section 1. (**a**–**f**) The SR-FTIR images of clay clusters and biopolymers; note that the color bar (from red-yellow to green-blue) corresponds to the relative SR-FTIR absorbance in the images from (**a**) to (**f**). (**g**–**j**) The correlation plots between the clay clusters and biopolymers ($N = 357$)

and smectite) was stronger than the other functional groups in both Sections 1 and 2, which demonstrated the spatial distribution of clay clusters and biopolymers was heterogeneous. Furthermore, according to the distribution of ratio images, it was obvious to show the spatial heterogeneity since the clay clusters (Kaolinite-OH/Illite-OH minerals, Si–O/Al–O minerals) were associated as nuclei with rims, while the biopolymers (C=C/C–H) were distributed randomly with blurry rims in both soil microaggregate sections (Fig. 8.14).

In addition, Fig. 8.14g–j plotted the spatial correlation between the clay clusters and biopolymers, whereas the absorbance spots in Section 1 ($N = 357$) and Section 2 ($N = 432$) were collected. For Section 1, the Kaolinite-OH/Illite-OH minerals correlated well with both C–H (aliphatic C, $R^2 = 0.799/R^2 = 0.722$) and C=C (aromatic C, $R^2 = 0.859/R^2 = 0.768$). Meanwhile, the same functional groups in Section 2 also showed the significant correlations (Fig. 8.2). In both Sections 1 and 2, the correlation between Si–O and C=C ($R^2 = 0.935$ and $R^2 = 0.941$) was more significant than that between Si–O and C–H ($R^2 = 0.782$ and $R^2 = 0.748$). The correlation between Al–O and C=C ($R^2 = 0.838$ and $R^2 = 0.956$) was also better than that between Al–O and C–H ($R^2 = 0.729$ and $R^2 = 0.782$).

8.13 SR-FTIR, μ-XRF, and 2DCOS Analysis

Synchrotron radiation-based Fourier transform infrared (SR-FTIR) spectroscopy, synchrotron radiation-based micro-X-ray fluorescence microscopy (μ-XRF), and two-dimensional correlation spectroscopy (2DCOS) analysis were used to in situ visualize the interiors of intact microaggregates from a typical Ferralic Cambisol in China, which had endured 25-year organic fertilization was examined by Xiao et al. (2019).

Spatial distribution and correlation between clay clusters and biopolymers were heterogeneous and significant, and also demonstrated that clay clusters were associated as nuclei with the potential of binding carbon at the submicron scale. Furthermore, the combination of SR-FTIR mapping and2DCOS analysis could explore the strategy of identifying overlapped spectra and quantifying the sequestration reactivity for the first time. Specifically, carbon retention correlated as the binding sequence orders: 3630 cm^{-1} > 3610 cm^{-1}, 985 cm^{-1} > 898 cm^{-1}, indicating that Fe/Al oxyhydroxides and phyllosilicates could regulate the organic matter sequestration without the influence of spatial perturbations.

8.14 NEXAFS Spectra of Organo-mineral Particles

Four different types of assemblages were observed in the clay subfractions (Lutfalla et al. 2019) (Fig. 8.15): (1) SOM-poor K-rich minerals, for which the abundance increased with bare fallow duration; (2) organo-mineral complexes rich in C, N, and K; (3) organo-mineral complexes rich in C, N, K, and Ca; and (4) K-poor particulate OM. Of note, pure OM without any signal from K minerals was not identified in the present samples (a possible explanation could be that the signal corresponds to the average of at least ten pixels of 40–40 nm in size, it is therefore unlikely that pure OM of pure K would be isolated at this submicron scale), which is why there is a distinguishable contribution of K in the OM-rich reference spectra. Still, these K-poor OM particles can be described as particulate OM. Similarly, pure mineral particles were not identified in the samples investigated: there was always a distinguishable contribution from C (same possible explanation).

8.14.1 Evolution of Organo-Mineral Particles in the Three Clay Subfractions with Time

Over the course of the bare fallow treatment, CC subfractions displayed particles ranging from isolated OM (particulate OM) to mineral-rich, OM-bearing particles (Fig. 8.16). After 22 years of bare fallow conditions, mineral particles exempt of OM

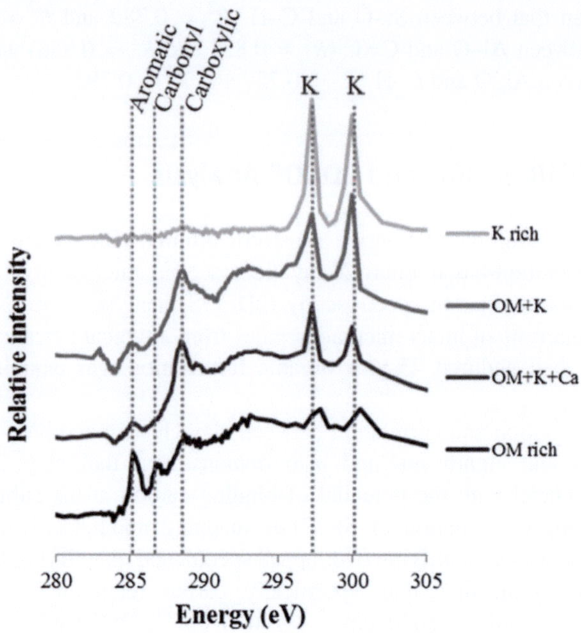

Fig. 8.15 Selected NEXAFS spectra of the four types of particles identified in the clay subfractions by STXM. Individual spectra are shifted on the intensity axis for better discrimination

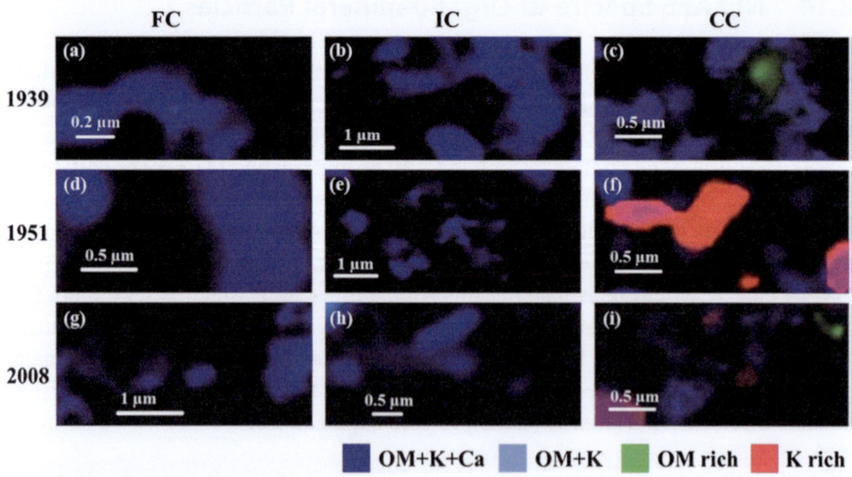

Fig. 8.16 STXM-NEXAFS compositional maps of organo-mineral particles contained in the fine clay (**a**, **d**, **g**), intermediate clay (**b**, **e**, **h**), and coarse clay (**c**, **f**, **i**) subfractions on three different dates: 1939 (**a–c**), 1951 (**d–f**), and the final sampling in 2008 (**g–i**). The scale bar is represented by a white line in each panel, it varies from 0.2 to 1 μm and depends on the sample. Dark blue represents OMCKCCa, light blue represents OMCK, green represents OM-rich, and red/pink represents K-rich

appeared in the CC subfractions. Conversely, mineral particles exempt of OM and OM-rich particles were virtually undetected in the IC and FC subfractions. These subfractions exhibited a homogeneous signal similar to the spectra of OMCKCCa organo-mineral particles, i.e., assemblages involving SOM and smectite or mixed-layer clay particles (Fig. 8.16).

8.15 Conclusions

The in situ SR-FTIR analysis showed that clay-OH clusters, C–H, C=C, Si–O, and Al–O, were the dominant functional groups throughout soil microaggregates and demonstrated the significantly positive correlation among these functional groups. In addition, the results from the chemical and 2DCOS analysis illustrated that the mineral clusters maintained the stable reactivity of SOC sequestration from the outside (edge) to the inside (nuclei) of the soil microaggregates, which Al−OH, Fe−OH, and Si–O bonds could regulate the binding microenvironments at submicron scale.

NMR analysis confirmed the presence of alkyl C, O-alkyl C, aryl C, and carboxyl C in the soil deposits. Aromatic compounds were prone to complexation with clay minerals, and their abundance increases as diagenesis proceeds. Polysaccharides were preferentially degraded by microbial activity and were mainly present in the upper soil profile due to their structural instability. OM in the clay fraction was strongly bound to clay-mineral surfaces. In particular, interactions between aromatic compounds and clay minerals influenced vibrations of silicon-basal oxygen and perpendicular silicon-apical oxygen, confirming binding on the basal surface of clay minerals. Strong chemical interactions between aromatic compounds and clay minerals were observed in older soil layers.

Long-term organic fertilization could stimulate the formation of poorly crystalline Fe minerals and aromatic C fraction involving in the mineral associated C, regardless of the soil types, which would improve the capacity of long-term SOC sequestration. These findings provide insights into the organo-mineral associations regulated by agricultural fertilization.

Spatially resolved observations at the submicron scale with STXM-NEXAFS clearly showed that mineralogy influences SOM stabilization. Indeed, several illite particles (identified by the presence of K and the absence of Ca) were devoid of OM, and the relative abundance of OM-depleted illites increased over time.

8.16 Future Aspects

Detailed identification and quantification about the reactive mineral complexes need to be conducted, which are responsible for locating the key factors of regulating SOM sequestration. Fourier transform ion cyclotron resonance mass spectrometry [FT-ICR-MS, NanoSIMS, and synchrotron scanning transmission X-ray microscopy

(STXM)] should be integrated to in situ explore molecular structures and binding coordination of soil microaggregates.

Furthermore, the development of structure modeling is also needed from macroaggregates to silt-clay sized aggregates in different soil types in order to explain the carbon system interactions, which has become embedded in studies relevant to agriculture, climate change, and the sustainability of the biosphere. Regulation of soil Fe minerals and their roles in efficient soil C storage need to be investigated further.

References

Balabane M, Plante AF (2004) Aggregation and carbon storage in silty soil using physical fractionation techniques. Eur J Soil Sci 55:415–427

Chenu C, Plante AF (2006) Clay-sized organo-mineral complexes in a cultivation chronosequence: revisiting the concept of the 'primary organo-mineral complex'. Eur J Soil Sci 57:596–607

Christensen (1996) Carbon in primary and secondary organomineral complexes. In: Carter MR, Stewart BA (eds) Structure and organic matter storage in agricultural soils. CRC Press, Boca Raton, pp 97–165

Giannetta B, Zaccone C, Plaza C, Siebecker MG, Rovira P, Vischetti C, Sparks DL (2019) The role of Fe(III) in soil organic matter stabilization in two size fractions having opposite features. Sci Total Environ 653:667–674

Hong H, Chen S, Fang Q, Algeo TJ, Zhao L (2019) Adsorption of organic matter on clay minerals in the Dajiuhu peat soil chronosequence, South China. Appl Clay Sci 178:105–125

Huang C, Liu S, Li R, Sun F, Zhou Y, Yu G (2016) Spectroscopic evidence of the improvement of reactive iron mineral content in red soil by long-term application of swine manure. PLoS One 11:e0146364

Kiem R, Kögel-Knabner I (2002) Refractory organic carbon inparticle-size fractions of arable soils. II. Organic carbon in relation to mineral surface area and iron oxides in fractions <6 micro m. Organic Geochem 33:1699–1713

Lutfalla S, Barré P, Bernard S, Guillou CL, Alléon J, Chenu C (2019) Multidecadal persistence of organic matter in soils: multiscale investigations down to the submicron scale. Biogeosciences 16:1401–1410

Mayer LM, Schick LL, Hardy KR, Wagal R, McCarthy J (2004) Organic matter in small mesopores in sediments and soils. Geochim Cosmochim Acta 68:3863–3872

Mikutta R, Schaumann GE, Gildemeister D, Bonneville S, Kramer MG, Chorover J, Nelson DW, Sommers LE (1996) Total carbon, organic carbon and organic matter. In: Sparks DL, Page AL, Helmke PA, Loeppert RH, Soltanpour PN, Tabatabai MA, Johnston CT, Sumner ME (eds) Methods of soil analysis. Part III. Chemical methods. Soil Science Society of America, Madison, pp 961–1010

Prietzel J, Thieme J, Eusterhues K, Eichert D (2007) Iron speciation in soils and soil aggregates by synchrotron-based X-ray microspectroscopy (XANES, μ-XANES). Eur J Soil Sci 58:1027–1041

Turchenek LW, Oades JM (1979) Fractionation of organomineral complexes by sedimentation and density techniques. Geoderma 21:311–343

Wattel-Koekkoek EJW, Buurman P (2004) Mean residence time of kaolinite and smectite-bound organic matter in Mozambican soils. Soil Sci Soc Am J 68:154–161

Wen YL, Li H, Xiao J, Wang C, Shen QR, Ran W, He XH, Zhou QS, Yu GH (2014) Insights into complexation of dissolved organic matter and Al(III) and nanomineral formation in soils under contrasting fertilizations using two-dimensional correlation spectroscopy and high resolution-transmission electron microscopy techniques. Chemosphere 111:441–449

Wen Y, Liu W, Deng W, He X, Yu G (2019) Impact of agricultural fertilization practices on organo-mineral associations in four long-term field experiments: implications for soil C sequestration. Sci Total Environ 651:591–600

Xiao J, Wen YL, Li H, Hao JL, Shen QR, Ran W, Mei XL, He XH, Yu GH (2015) Insitu visualisation and characterisation of the capacity of highly reactive minerals top reserve soil organic matter (SOM) in colloids at submicron scale. Chemosphere 138:225–232

Xiao J, Wen YL, Dou S, Bostick BC, He XH, Wei R, Yu GH, Shen QR (2019) A new strategy for assessing the binding microenvironments in intact soil microaggregates. Soil Tillage Res 189:123–130

Recent Trends in Soil Salinity Appraisal and Management

9

Nirmalendu Basak, Arijit Barman, Parul Sundha, and Arvind Kumar Rai

Abstract

A comprehensive understanding on soil salinity appraisal is prerequisite for development and execution of rehabilitation programs and productive agricultural use of salt-affected soil (SAS). Indian agriculture since pre-independent era has been providing top priorities to rehabilitate SAS for considering the welfare of farming communities. Broadly SAS are categorized into: saline which are loaded with soluble electrolytes; sodic soil carries disproportionately high levels of sodium relative to calcium and magnesium in soil exchange sites and soil solution with alkaline reaction; and saline-sodic soil with twin problems of sodicity and salinity; and special categories SAS having regional occurrence. Saline and sodic underground water, rainfall deficit and high evapotranspiration demand for crops are usually associated threats with SAS. Therefore, intensive and temporal survey, delineation and mapping are required for capturing the extent of salinity problem. Here, we tried to highlight classification and behaviours of SAS, threat associated with these soils, methodological appraisal for salinity assessment both conventional laboratory chemical analysis and standardized rapid and non-destructive methods like EM-38 based on electromagnetic geophysical tools, time domain reflectometry, resistivity survey, hyper-spectral remote sensing, etc. Besides, several technological options, viz., need of amendment for rehabilitation of sodic soils and agro-technologies for managing salinity are described to combat salinity and sustain crop production.

Keywords

Salinity · Sodicity · Electromagnetic tools · Reclamation technology

---

N. Basak (✉) · A. Barman · P. Sundha · A. K. Rai
ICAR-Central Soil Salinity Research Institute, Karnal, Haryana, India
e-mail: nirmalendu.basak@icar.gov.in

A. Rakshit et al. (eds.), *Soil Analysis: Recent Trends and Applications*,
https://doi.org/10.1007/978-981-15-2039-6_9

9.1 Introduction

Soil salinity is one of the major environmental threats that deleteriously affect soil properties and crop growth and yield. Countries in the arid and semi-arid regions are much distress with problem of soil salinity which limits plant growth, nutrition, resulting in poor yields and thus affecting the livelihood and socio-economic conditions of the farmers. Presently, expansion of irrigated farming and increased evapotranspiration (ET) demand of crop because of climate change associated with rise in temperature and rainfall deficit will drive the accumulation of more salt in surface soils of canal commands. Therefore, monitoring the progressive development of soil salinity and its degree of severity is imperative to quantify adverse effect on crop productivity and environmental degradation. Classification and characterization of salt-affected soils (SAS) is first step towards planned rehabilitation for sustained agricultural productivity from SAS as these categories of soils carry different nature and quantity of salts. A category of SAS affected with large deposition of soluble salts that may be neutral in reaction and/or endow with saline water table; others carrying less deposited salts but have a potentiality to give alkalinity upon water hydrolysis. Sometime prolong irrigation with sodic water (alkaline in reaction) can have provisional occurrence of sodic soil (Choudhary et al. 2011; Minhas et al. 2007). Considering the nature and stoichiometric preponderance of salts, SAS are classified into: saline, sodic and saline-sodic. The nature of adverse impact of these different categories of SAS on soil properties, crop performance, nature and extent of problem and methods of cultivation is conditional. Therefore, to cope with the adverse effect, the rehabilitation program to mitigate salinity and associated problems is varied. Presently, the soil salinity monitoring is based on traditional methods of visual observations or analytical soil analysis in laboratories. The visual assessment is unable to detect trends within the crop growing season, whereas the laboratory analytical methods are time, capital and labour intensive, which is a serious disadvantage in real-time monitoring. Additionally, monitoring and mapping of SAS in real time is a difficult taks because of dynamic nature soil salinity development in natural landscape. This chapter summarizes the comprehensive understanding about classification and behaviours of salt-affected soils, problems associated with these soils, methodologies for salinity appraisal both laboratory chemical analysis and standardized rapid and non-destructive methods. Further, technological options to combat salinity problems for crop production are also addressed.

9.2 Classification of Salt-Affected Soils

9.2.1 Saline Soil

Soluble electrolytes are abundant in typically saline soil; therefore, osmotic effects and specific ion toxicity harmfully hinder living organisms and plant root. In the presence of excess salts, crops experiences 'physiological drought' even in the range of available soil moisture. A common appearance of white encrustation on the surface is the indication of soluble salts, mainly Cl^- and SO_4^{2-} of Na^+, Ca^{2+} and Mg^{2+}. Soil chemical equilibria and nutritional imbalance when disrupted with larger concentration of soluble salts but physical soil health nearly remains unaffected. These soils generally have electrolytic conductivity of the saturated paste extract (EC_e) more than 4.0 dS m^{-1} at 25 °C, pH of saturation paste (pH_s) less than 8.2 and exchangeable sodium percentage (ESP) less than 15% (Abrol et al. 1998). The development of saline soil is because of the presence of soluble salts in soil strata, upward flux of underground saline water, long-term irrigation with saline/or waste water and impeded drainage or intrusion of saline or brackish water in sea coast area (Singh 1998).

9.2.2 Sodic Soil

Sodic soils synonymously known as 'alkali soils' contain disproportionately high levels of sodium (Na^+) relative to calcium (Ca^{2+}) and magnesium (Mg^{2+}) in the soil exchange sites and in the soil solution. These soils have an ESP of more than 15%, $pH_s > 8.2$ and variable EC_e. These soils have poor physical properties with larger clay dispersion, less pore space, restricted water and air entry (impaired hydraulic conductivity) and storage. High mobility and greater loss of organic matter are the adverse effects of sodicity. High alkaline hydrolysis and toxic appearance of Na and precipitation of Ca as $CaCO_3$ further exacerbate the Na induced toxicity and nutritional deficiency. The nitrogen deficiency in sodic soil is because of loss/and erosion of organic matter and the wasteful transformation of applied N and reduced symbiotic N fixation (Bharadwaj and Abrol 1978; Sundha et al. 2017). Because of high pH, soil organic matter gets dissolved and forms black organic-clay coatings on soil aggregates and on the surface termed 'black-alkali'. Formation of carbonates of Na and alkalization in the soil take place because of carbonation of alumino-silicate minerals.

$$NaAlSi_3O_8 + HOH + CO_2 \rightarrow HAlSI_3O_8 + NaHCO_3$$

$$2NaHCO_3 \rightarrow Na_2CO_3 + CO_2 + H_2O$$

$$Na_3CO_3 + 2HOH \rightarrow 2Na^+ + 2OH^- + H_2CO_3$$

Sodium carbonate is soluble and its hydrolysis results in high alkalinity up to pH 12.0 (Bajwa and Swarup 2009). In arid and semi-arid regions, the deposited $CaCO_3$ in the profile and constantly favours the release of OH^- ions in soil solution and maintain higher pH in calcareous alkali soils than that in non-calcareous alkali soil. Therefore, a buildup in the exchangeable sodium without large quantity of neutral soluble salts will always result in high pH. Exact value depends on the concentration of Na_2CO_3 formed or the ESP.

9.2.3 Saline-Sodic

As the EC_e of some sodic soil is variable, groups of soils having pH_s often > 8.5, ESP $>15\%$, SAR > 13 and $EC_e > 4$ dS m^{-1} at 25 °C are defined as 'saline-sodic'. These categories of soils formed because of applying irrigation waters containing high residual sodium carbonate (RSC > 2.5 me L^{-1}), and soils with shallow sodic water table.

9.3 Classical Procedure for Identification of Salt-Affected Soil

Salt-affected soils are basically classified on the basis of three soil properties: the pH of soil water saturation paste (pH_s), electrolytic conductivity of the soil water saturation paste extract (EC_e) at 25 °C and exchangeable sodium percent (ESP) or sodium adsorption ratio (SAR_e) of soil water saturation paste extract $\{Na^+/[(Ca^{2+} + Mg^{2+})/2]^{1/2}\}$. The classical protocols are described for determination of these three soil properties.

For estimation of pH_s, 250–300 g of dried and sieved soil is poured into the 500 ml capacity plastic beaker and add measured amount of distilled water (DW) with the help of burette to make a smooth saturated paste with a spatula or glass rod, till a shining top layer is obtained which may flow in one lump on tilting the beaker and keep it for 6 h or overnight. Take the pH_s of this aqueous soil water saturation paste through a pH meter after calibrating the instrument (Bhargava 2003). Calculate the saturation percentage (SP) as follows:

$$\text{Saturation percentage} = \frac{\text{Water content of saturation paste} \times 100}{\text{Oven dry weight of soil}}$$

Further, for estimation of EC_e, soil extract of the aqueous soil water saturation paste is obtained through a suction pump using a Buckner funnel with setting a Whatman filter paper 1 and the flask. Collect about 25–30 ml of the extract for soluble salt content estimation. Then record the EC_e of the soil extract using EC meter after calibrating the instrument.

For determination of ESP, cation exchange capacity (CEC) analysis is also required. CEC is the capacity of negatively charged clays and organic matter to adsorb cation by simple physical, attractive forces. With increment in CEC more cations can be retained with soil particles. A high-clay soil can hold more exchangeable cations than a low-clay soil. Whereas, exchangeable sodium is generally determined in the ammonium chloride or ammonium acetate extracts of soils and for determination of exchangeable sodium leaching with ammonium acetate is preferred. So, positively charged ions are adsorbed on the surface of clay complex can be replaced by another cation. Process by which adsorbed cation on clay complex is replaced by equivalent amount of cation is simply called as cation exchange phenomena.

$$\text{Clay} - \text{Na} + \text{NH}_4\text{OAc} \rightarrow \text{Clay} - \text{NH}_4 + \text{NaOAc}$$

For laboratory analysis, 5 g soil is placed in a centrifuge tube. Then 33 ml 1*N* NaOAc is added, and shaked for 5 min. Remove stopper and centrifuge until the supernatant liquid is clear (usually 10 min at 2000 rpm). Decant the supernatant liquid as possible and discard. Repeat the same procedure two more times, discarding the supernatant liquid each time. After the last saturation, wash the stopper and use adsorbent paper to remove any acetate drops remaining on the lip of the centrifuge tube. Add about 30 ml 60% ethanol to centrifuge tube, put in stopper, shake for 5 min, remove stopper and centrifuge until the supernatant liquid is clear. Decant and discard the supernatant liquid. Continue washing with ethanol until the electrical conductivity of supernatant liquid from the last washing is between 40 and 55 μΩ/cm (Bhargava 2003). Replace the adsorbed sodium from the sample by extracting with three 33 ml portion of 1 *N* NH_4OAc solution. Dilute to 100 ml and determine the sodium in the leachate. Sodium concentration is measured by flame photometer.

$$\text{CEC}[\text{cmol}(\text{p}+)/\text{kg}] = \frac{\text{Na from curve (me/L)} \times 10}{\text{Weight of soil (g)}}$$

Again, for determination of ESP, 5 g of soils is weighted in 100 ml of centrifuge tube and add about 30 ml 60% ethanol to centrifuge tube, put in stopper, shake for 5 min, remove stopper and centrifuge until the supernatant liquid is clear. Decant and discard the supernatant liquid. Continue washing with ethanol until the electrical conductivity of supernatant liquid from the last washing is between 40 and 55 μΩ/cm. Replace the adsorbed sodium from the sample by extracting with three portions of 33 ml 1*N* NH_4OAc solution. After the removal of free salts add 33 ml of ammonium acetate solution and collect the leachate in 100 ml volumetric flask (Bhargava 2003). Sodium concentration is measured by flame photometer.

$$\text{Exchangeable sodium(me/100 g soil)} = \frac{\text{ppm concentration(from std curve)} \times 10}{23 \times \text{weight of soil}}$$

or

$$\frac{\text{me/L Na} \times 10}{\text{Weight of soil}}$$

$$\text{Exchangeable sodium percentage} = \frac{\text{Exchangeable sodium(me/100 g soil)} \times 100}{\text{Cation exchange capacity(CEC)}}$$

9.4 Advanced Tools for Salinity Assessment

Presently, several non-destructive methods are employed for fast and real-time assessment of soil salinity at larger scale. These tools have limited effect of spatial variability. It has possibility to use under dry wet, stony, cropped and un-cropped conditions for direct measurement of soil salinity in the field.

9.4.1 Electromagnetic Induction

Electromagnetic (EM) principle based geophysical tools are becoming popular for quick diagnosis of salinity and groundwater quality. EM devices work in the low induction range where depth of investigation is controlled by separation distance between the transmitter and receiver coils, and their orientation rather than the operating frequency. The depth of investigation is also controlled by coil dipole (vertical and horizontal) orientations thus gain insight into salinity variation with depth in a farm/canal command. The EM-38 (Geonics, Canada) device is most widely used for rapid quantitative assessment of soil salinity. EM-38 transmitter coil generates a primary magnetic field (M_p) and creates eddy currents in the soil and these time-varying eddy currents induce their own magnetic field (M_i). The induced field is superimposed over the primary field and a fraction of both M_p and M_i is intercepted by the receiver coil where the signal gets amplified and formed into an output voltage, which is linearly related to apparent conductivity (EC_a) (Fig. 9.1). Na^+ and Cl^- were strongly correlated with apparent conductivity (EM_V and EM_H) measured by EM-38 as well as soil salinity (EC_e) (Narjary et al. 2017).

The EM-38 measures bulk soil EC_a of the maximum depth of 1.5 and 0.75 m in vertical and horizontal mode, respectively (Fig. 9.2a, b). The device with integrated GPS or with separate GPS integrates values of different depths EC_a and GPS coordinates with data logger. EM-38 requires a correlation for conversion of EC_a to soil salinity (EC_e) for establishing relationship between apparent to actual degrees of salinity. These depth-wise salinity data with GPS coordinates are exported to ArcGIS software for data analysis which generates spatial salinity maps of different depths using IDW/kriging methods. These maps help agency/farmers' to understand and interpret yield variation with soil salinity in order to take corrective measures for improving crop yield.

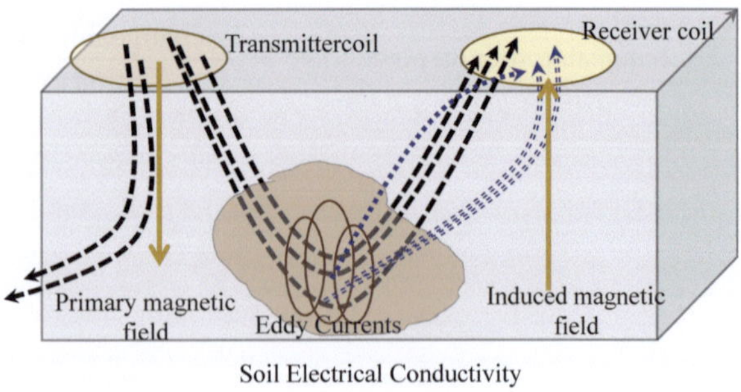

Fig. 9.1 EM-38 working principle diagram

Fig. 9.2 EM38 measurement in (**a**) vertical mode (**b**) horizontal mode

9.4.2 Time Domain Reflectometry (TDR)

Time domain reflectometry is an in-situ technique for measuring volumetric soil water content, soil temperature and bulk soil EC. When TDR probe is immersed in solutions of different electrical conductivity at 25 °C, the shape of TDR electromagnetic wave pulse changes and TDR signal gets attenuated. Changes in the wave form are used to estimate EC of the media. TDR in the soil measures EC_a in water-soil matrix. The estimation of EC_a depends on the characteristics of the soil, and hence does not relate linearly with water salinity in the soil pores, i.e., the soil solution EC_e.

9.4.3 Resistivity Survey

The use of geo-electrical resistivity survey has gained importance for assessing the soil salinity. Electrical resistivity methods introduce an electrical current through current electrodes at the soil surface and the difference in current flow potential is measured at potential electrodes placed in the vicinity. The electrode configuration is referred to as 'Wenner array' as the four electrodes being equidistantly spaced in a straight line at the soil surface. In Wenner array, two outer electrodes serve as the current or transmission electrodes and the two inner ones as the potential or receiving electrodes (Corwin and Hendrickx 2002). After the probe is inserted at the desired depth in soil, an electrical current I is induced between the two outer electrodes, and the potential drop E is measured between the two inner electrodes. The ratio $R = E/I$ is recorded as a resistance, which can be converted to soil EC_a (Fig. 9.3). Mandal et al. (2015) observed a very good correlation between bulk soil EC_a measured by resistivity EC probe and EC_e for coastal soils of West Bengal ($r^2 = 0.94$). They have concluded that the technique is reliable and easy for the spatio-temporal characterization of soil salinity.

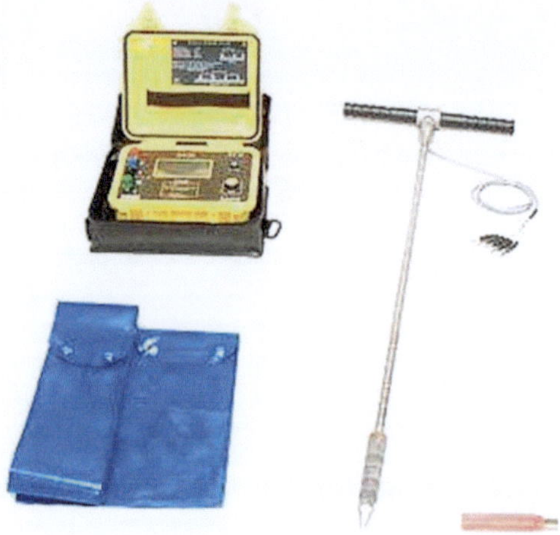

Fig. 9.3 Electrical conductivity probe (resistivity principle) of Eijkelkamp (Germany)

9.4.4 Soil Salinity Characterization Using Hyper-Spectral Remote Sensing

The complexity of salinization processes, spatial and temporal variability makes soil salinity mapping a difficult proposition. Severely salt-affected soils (SAS) can be easily detected due to high reflectance from salt crust on soil surface, whereas, detection of low and medium salinity is difficult due to intricate association of salt, soil, water and vegetation. An attempt has been made to characterize such SAS using hyper-spectral remote sensing (HRS) data. A methodology was developed at CSSRI, Karnal through integrating HRS data with limited ground truth and further quantifying through statistical modeling. The variability of salinity and sodicity attributes such as EC_e, Na^+, Cl^-, CO_3^{2-} and HCO_3^- (me L^{-1}) of the soil water aqueous saturation extract were related quantitatively ($r^2 = 90\%$) by HRS data. The spectral regions of 1400, 1900 and 2200 nm showed prominent peak due to the changes in soil salinity. At 1900 nm prominent shifting facilitated in establishing a significant correlation with salt concentration. Barman et al. (2017) developed a methodology for characterizing salt-affected soils using hyper-spectral data and was found useful for delineating SAS from the space platform (Fig. 9.4).

9.4.4.1 Case Study

An 11 ha of experimental farm, CSSRI, Nain, Panipat, Haryana was divided in regular grid (30 × 30 m) to collect surface soil samples and spectroradiometer data. The collected soils of surface layers (0–30 cm) during the post monsoon season were analysed for EC_e, pH_s and concentration of Na^+, K^+, Ca^{2+}, Mg^{2+}, CO_3^{2-}, HCO_3^-, Cl^- and SO_4^{2-} in aqueous soil paste saturation extract. Simultaneously, hyper-spectral remote sensing data (HRS) was collected in different wavelength regions using a spectroradiometer and standardized with a statistical model to find

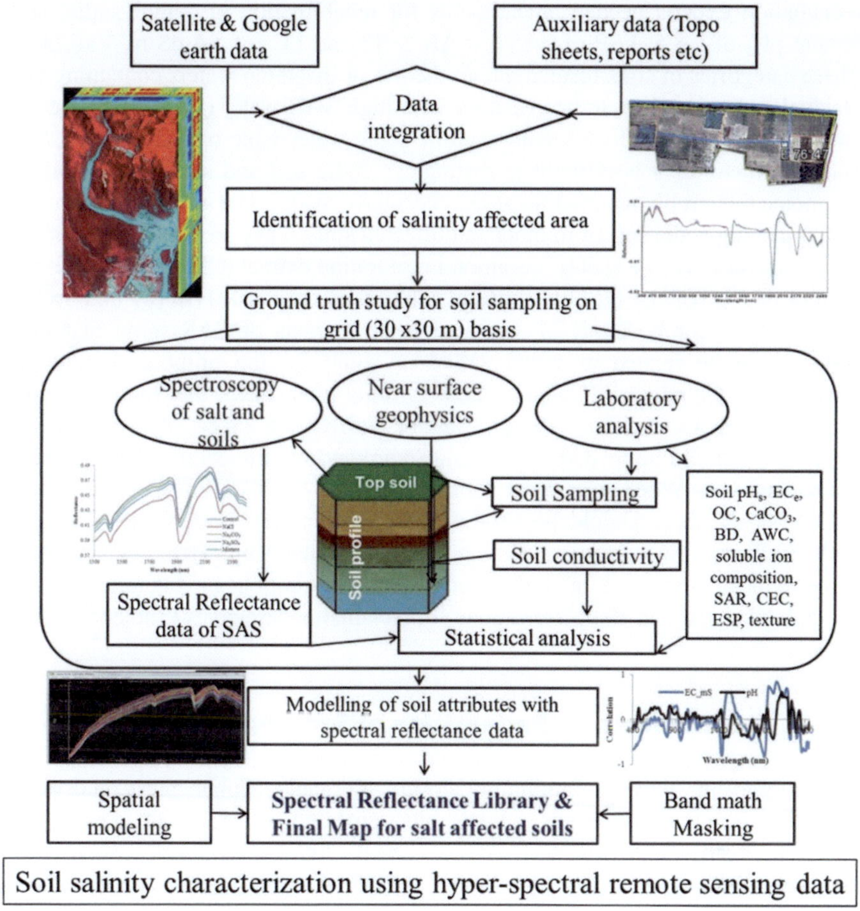

Fig. 9.1 Methodology for characterizing salt-affected soils using hyper-spectral data

prominent absorption region between 1420 and 2020 nm. Further, a salinity model was developed integrating HRS data with soil physicochemical properties by multivariate statistical data analysis, viz., principal component analysis, partial least square regression, random forest, support vector machine, etc.) and was validated using band math techniques. A spectral library thus developed is used for further mapping of salt-affected soils with limited ground truth data.

9.5 Technological Options for Management of Salt-Affected Soils

Soil electrical conductivity and nature of dissolve salts of SAS are variable (EC_e); therefore, categorization of SAS is prerequisite before application of any amendments or recommending strategies for reclamation. Very often land reclamation experts needed amendments for rehabilitation of 'saline-sodic' soils having pH_s of >8.5, ESP of >15%, SAR > 13 and EC_e of >4 dS m^{-1} at 25 °C. These categories of soils formed due to the use of irrigation waters containing high residual sodium carbonate or irrigation with high SAR water (RSC > 2.5 me L^{-1} and SAR > 10.0), and soils with shallow sodic water table or underground SAR water. Sometimes it is difficult to distinguish sodic soil and saline-sodic soil. To overcome this problem numerous researchers have made some advanced attempts at distinguishing sodic soil from other salt-affected soils. They are: dominance of Na^+ as cation and CO_3^{2-} + HCO_3^- as anion in saturation extract (Chhabra 2004); or the ratio of $[Na^+]/([Cl^-] + [2SO_4^{2-}])$ in soil solution more than 1.0 (Bajwa and Swarup 2009). These compositions of saturated paste extracts of SAS show that these soils are to be treated as either saline or sodic for the purpose of adopting reclamation techniques (Chhabra 2004). Soils that have the ratio of either $(2CO_3^{2-} + HCO_3^-)/(Cl^- + 2SO_4^{2-})$ and/or $Na^+/(Cl^- + 2SO_4^{2-}) > 1$, expressed in me L^{-1}, should be treated as 'natric' (Sodic) and require chemical amendments for reclamation. When soils have both these ratios <1, then, irrespective of their pH and SAR, these should be treated as 'salic' (Saline). Under such situations, both SAR and EC_e decrease simultaneously on leaching. Here, we compiled some typical SAS soil in India with ionic distribution in saturation extract for classifications and characterizations for reclamation strategies (Table 9.1).

9.5.1 Amendments for Reclamation of Sodic and Saline-Sodic Soils

Mineral gypsum, pyrites, aluminium chloride, inorganic sulphur, press mud, acids, acid-formers, phosphogypsum, fly ash, bioaugmented material with gypsum, etc. have long history to reclaim soil sodicity (CSSRI 2006). Additionally, alkali water and scheduling of sodic waters for safe agricultural use are prioritized in sodic or calcareous sodic soils in semi-arid regions of North-Western India and waterlogged *Vertisols* of southern India. The reclamation of soil sodicity is largely a onetime investment for sustaining production if irrigation water quality safe for both soil and crop (Rai et al. 2019). It is expected that productivity for rice-wheat cycle reaches its potential level nearly in 3 year after application of amendments. But, gypsum or other amendment needs a recurring application to overcome incipient sodicity when water quality is alkali (sodic) which is unsafe for both soil and crop. The gypsum recommendation is advocated when RSC of irrigation water exceeded 2.5 me L^{-1}. Other amendments such as one time pyrite application before the sowing of wheat are better for improving irrigation water than split application (Sharma and Swarup 1997). Organic matter was shown to enhance dispersion of soil due to greater inter-particle interactive forces at high pH. The response to farmyard manure, however, decreased with increase in RSC of irrigation water. Incorporation of organic materials decreased the precipitation of Ca and carbonates, increased the removal of Na in drainage waters, decreased soil pH and ESP and improved crop yield. The effectiveness followed the order: paddy straw > green

Table 9.1 The chemical behaviours of SAS determining reclamation strategies (Rai et al. personal communication)

Sites/ Soil attributes	Nain, Panipat, Haryana	Trichy, Tamilnadu	Haibatpur, Karnal Haryana	Dharamgarh, Panipat, Haryana	Shivri, Lucknow, Uttar Pradesh	Barwah, Khandawa, Madhya Pradesh
pH_s	7.93	8.22	9.89	10.06	10.87	9.67
EC_e	10.04	3.81	1.34	1.81	20.57	2.76
Na^+	65.7	38.2	9.6	17.5	170.9	21.3
Ca^{2+}	19.7	7.5	0.25	0.41	0.67	0.25
Mg^{2+}	12.6		0.75	0.67	1.08	0.91
K^+	1.0	0.06	0.09	0.31	0.25	0.05
CO_3^{2-}	2.8	Nd	0.86	2.11	110.0	1.6
HCO_3^-	1.5	10.0	4.8	7.1	23.6	3.2
Cl^-	44.7	32.0	1.23	3.33	6.0	9.9
SO_4^{2-}	48.0	2.8	4.3	11.6	96.2	9.0
SAR_e	9.8	13.9	13.8	23.8	184.1	28.1
CBCS	0.05	0.27	0.66	0.43	1.23	0.23
SCSR	0.5	1.01	1.0	0.7	0.9	0.8
Reclamation strategies	Saline, leaching with best/good quality water based on underground water table	Sodic soil and produce alkaline hydrolysis, application of Ca (agricultural grade mineral gypsum) source to reduce ESP and SAR_e followed by leaching with best quality available water				

pH_s pH soil water saturation, *EC_e* electrical conductivity of soil water saturation paste extract, dS m^{-1}, Na^+, Ca^{2+}, Mg^{2+}, K^+, CO_3^{2-}, HCO_3^-, Cl^- and SO_4^{2-}: concentration of all electrolytes in soil water saturation paste extract in me L^{-1}, SAR_e $Na^+/[(Ca^{2+} + Mg^{2+})/2]^{1/2}$ sodium adsorption ratio of soil water saturation paste extract in $mmol^{1/2}\ L^{-1/2}$, *CBCS*: $(2CO_3^{2-} + HCO_3^-)/(Cl^- + 2SO_4^{2-})$, *SCSR* $Na^+/(Cl^- + 2SO_4^{2-})$

manure > farmyard manure (Minhas and Bajwa 2001). The choice of an amendment at any place will depend upon its relative effectiveness as judged from improvement of soil properties and crop growth and the relative costs involved (Table 9.2).

9.5.2 Reclamation and Management of Saline Soils

Leaching of soluble salts retained in some soil layer by ponding of good quality water received from rain, canal or groundwater is the most effective strategy for reclamation of these soils (Rao and Visvanatha 1998). The quality of salts leached from soils depends on the quality of irrigation water and texture of soils. Gypsum application or irrigation with low SAR water is favoured for improvement in soil physical structure and desirable leaching (Rai et al. 2014). Sometimes flushing with water is practiced to remove surface deposited salts. This procedure is advocated in low permeable soil and soil susceptible to hard crust formation. Scraping of deposited salts is prescribed to manage small land holding affected with salinity; however, a frequent removal of salts is required to achieve desired and productive plant growth and production (Chhabra 1996). Different categories of saline soils are rehabilitated with specific management options:

9.5.2.1 Inland Saline Soil with Shallow Water Table with Poor Quality Water

Sub-surface or surface drainage is a long-term solution for lowering water table and leaching of salts and to provide a favourable salt balance in root zone (Rao and Visvanatha 1998). Perforated corrugated PVC pipe covered with synthetic filter are mechanically installed in proper plan below the rhizosphere depth to lower down poor quality water table and leach excess salt and water (Chinchmalatpure et al. 2015). Bio-physical characteristics of salinity affected sites like soil texture, geology, hydrology, rainfall, potential evapotranspiration, growing degree day (GDD), concentration and nature of salt present and predominant cropping systems are the factors that determine the spacing and depth of drainage lines. Several countries like USA, Egypt and Gulf countries use this technology to manage a sizeable area of saline soil. In India, ~40,000 ha waterlogged saline areas have been reclaimed using this technology (Chinchmalatpure et al. 2015). Appreciable yield is achieved in fields having a sub-surface drainage system than in fields with a deep water table and the differences were larger at applied water salinities of more than 10 dS m^{-1} as horizontal sub-surface drainage improves aeration in the rhizosphere by lowering water table and lowering salts concentration. Around INR 0.6 and 0.75 lakh is the implementation cost for this technology for managing salinity in alluvial Gangetic saline land and heavy texture vertisols of southern states, respectively.

9.5.2.2 Costal and Deltaic Saline Soil

Preventing the ingress of brackish saline water and seawater tides is possible by constructing high and well-designed earthen dykes, and these embankments prevent

Table 9.2 Commonly used amendment for reclamation of sodic soil

Amendments	Nature and mechanism to neutralize soil sodicity	Experimental evidence/ comments
Gypsum ($CaSO_4.2H_2O$)	Sparingly soluble in water and widespread in nature as soil component and advantages with gypsum is relatively faster reclamation Na-clay-Na + $CaSO_4 \rightarrow$ Ca-clay + Na_2SO_4	8 years (1994–2004) field experiment with rice-wheat with application of gypsum (5 Mg ha^{-1}) sustain the productivity having sodic groundwater (irrigation water pH 9.0 and RSC 8.5 me L^{-1}, and SAR 8 used) for irrigation in sodicity affected soils (Soil pH_e 8.6 and SAR 29) (Yaduvanshi and Swarup 2005)
Calcium chloride ($CaCl_2.2H_2O$)	Highly soluble and rarely deposited in nature Na-clay-Na + $CaCl_2 \rightarrow$ Ca-clay + 2NaCl	Control laboratory experiment showed saline water with low SAR (<5) dominant with $CaCl_2.2H_2O$ is efficient to reclaim sodic soils ($pH_{1:2}$ 10.7 and $EC_{1:2}$ 12.2 dS m^{-1}) (Rai et al. 2014)
Ground limestone ($CaCO_3$), native or industrial wastes (pressmud)	Low solubility of limestone; therefore, other corrective measures like vegetative or crop planting, supply external electrolytes is advocated to further increase native CaCO3 dissolution and to improve the reclamation efficiency by intermittent leaching in medium to low sodicity affected soils (Li and Keren 2009)	
Sulphuric acid (H_2SO_4)	Handling problem; however, this acid once used extensively in some parts of western United States and USSR to combat sodicity	
Pyrite (FeS_2); iron sulphate ($FeSO_4$, $7H_2O$) Elemental sulphur (S); lime sulphur (CaS_5)	Acids forming amendments; The oxidation of elemental S/pyrite is mediated by *Thiobacillus thiooxidans,* which require a warm, well aerated and moist soil with low pH condition $2S + 3O_2 = 2SO_3$ (microbiological oxidation) $SO_3 + H_2O = H_2SO_4$ $NaHCO_3 + H_2SO_4 = Na_2SO_4$ (leachable) + $H_2O + CO_2$ $Na_2CO_3 + H_2SO_4 = Na_2SO_4$ (leachable) + $H_2O + CO_2$ Na^+-[soil]-Na^+ + $H_2SO_4 = H^+$-[soil]-H^+ + Na_2SO_4 (leachable)	Pyrites declined soil $pH_{1:2}$ 9.2 and ESP 29% than the unamended soil $pH_{1:2}$ 10.4 and ESP 87% in rice-wheat rotation after 1 year of application (Sharma and Swarup 1997) Sulphur is effective in the reclamation of sodic soils, especially if large amount of $CaCO_3$ is present (Brady 1988). Elemental Sulphur has also been used successfully to reclaim poor quality soils in Egypt (El-Mowafy 1982). In another lab scale study application of the 0.221 milligram equivalent elemental Sulphur in 100 g soil resulted in 32% transformation of sulphate in 120 days with resultant decrease in soil pH from 10.1 to 8.4 (Kubenkulov et al. 2013). However, both S and pyrite must first be oxidized to sulphuric acid by soil

(continued)

Table 9.2 (continued)

Amendments	Nature and mechanism to neutralize soil sodicity	Experimental evidence/ comments
		microorganisms before they are available for reaction, the amendments are relatively slow acting. Further, conjunctive application of sulphur and gypsum in 50:50 ratio significantly reduced the exchangeable sodium (Stamford et al. 2007). Research and development are advocated for improvement on elemental sulphur properties for its improvement
Farmyard manure (FYM) and green manuring (GM)	Organic amendments carry organic acids which solubilizes Ca from inherent and precipitated $CaCO_3$ in calcareous soils and consumption of mineral gypsum decline for sodicity reclamation in sodic water irrigation for achieving sustainable yields	The 50 GR (laboratory gypsum requirement) and FYH or GM (20 Mg ha^{-1}) and wheat straw (6 Mg ha^{-1}) declined soil pH (pH 9.3 and ESP 38.8%), (pH 9.4 and ESP 39.5%) and (pH 9.6 and ESP 40.3%) compared to control (pH 10.0 and ESP 58.6%), respectively (Choudhary et al. 2011)
Compost	Similarly, GR25 and 20 Mg ha^{-1} city compost are recommended for reducing alkalinity and salinity stress of soil under use of poor quality water	GR25 gypsum + Delhi compost declined soil $pH_{1:2}$ 8.7; EC_e 1.1 dS m^{-1}; GR25 gypsum + Karnal compost declined soil $pH_{1:2}$ 8.9; EC_e 1.0 dS m^{-1} compared to sodicity ($pH_{1:2}$ 10.7; $EC_{1:2}$ 12.2 dS m^{-1}; ESP 70%) (Sundha et al. 2018)

the back flow of this water into rivers and estuaries. In these regions ponds and other land shaping techniques are adopted to capture monsoon rains for irrigating *rabi* crops and leaching salts from rhizospheric zones in dry seasons (Chinchmalatpure et al. 2015). 'Land shaping techniques' is an advanced practice of modifying land surface by developing raised and sunken bed by alternately digging soil from one strip and putting it on the other. This minimizes the capillary rise to avoid salt deposition in root zone. For ease in using available farm machinery, minimum width of raised bed is taken as 2.0 m and the height of sunken bed is 1.0 m above ground surface. The average depth of sunken bed is 0.5 m below ground surface and side slope is 1:1. Vegetables and forages are grown in raised and deep water paddy in sunken beds.

9.5.2.3 Bio-drainage

Physiological transpiration of tree is used to remove excess soil water to manage shallow water table. It is an effective option to prevent the development of water-logged and saline soils with congestion of drainage problem. This eco-friendly low-cost technology is easily adopted by farmers and with additional benefit of sale of wood and promotion of social forestry. This technology is also effective in controlling seepage from higher elevation, industrial and wastewater disposal, and waterlogged canal commands.

Field scale salinity management needs proper soil, water and crop management strategies to sustain cultivation in saline soil and mitigation of increasing risk of soil salinization and sustaining soil and environment quality. Properly levelled crop field, conservation tillage (minimum tillage), mulching, conjunctive use of saline water, cycling and mixing mode of irrigation, frequent application of saline irrigation for reduction of accumulation of soluble salts in root zone, irrigating with best available water at germination and seedling emerging stages, pre sowing irrigation for *kharif* crop, and improving water use efficiency practice by pressurized sprinkler irrigation facilitate in reducing root zone salinity and sustaining crop production in salt-affected lands and use of saline water (Minhas 1998; Rai et al. 2017) (Table 9.3). Mulching prevents crust formation and stores rainwater for the longer period and checks the growth of fungus. The use of dry grass mulch at 5 t ha^{-1} increases seedling emergence and forage productivity of pearl millet, maize and sorghum in calcareous red soils.

9.5.3 Crop Management

Salinity and sodicity tolerance can be exploited for screening purpose based upon satisfactory higher yield under given levels of root zone salinity/sodicity (Minhas and Gupta 1992). Less water requiring crops like oilseed crops can tolerate higher levels of irrigation water salinity over the salinity sensitive pulses and vegetables. Usually, mono cropping is advocated for maintaining salt-balance in arid and semi-arid zone having rainfall <400 mm. Semi-tolerant to tolerant (mustards, wheat, cotton) crops is a recommendation for successful use of saline

Table 9.3 Effect mulching and consumptive use of saline water on reduction of soil salinity (dS m^{-1}) (Basak et al. 2017)

Soil and crop management strategies	EC_e (dS m^{-1})
Fallow	10.04^{a}
100% best available water	4.15^{c}
100% water requirement with 8.0 dS m^{-1}	7.74^{abc}
100% water requirement with 8.0 dS m^{-1}—mulch 5 t ha^{-1}	8.78^{ab}
60% water requirement with 8.0 dS m^{-1}	5.71^{bc}
60% water requirement with 8.0 dS m^{-1}—mulch 5 t ha^{-1}	9.59^{a}

Numbers followed by different letters are significantly different at $P < 0.05$ by Tukey's test

Table 9.4 Salt-tolerant varieties

Crop	Tolerant varieties	Abiotic stressors		
		Sodic	Saline	Coastal saline
		$pH_{1:2}$	EC_e ($dS\ m^{-1}$)	EC_e ($dS\ m^{-1}$)
Rice	CSR 10[a], CSR 11, CSR 12, CSR 13[a]	9.8–10.2	6–11	–
	CSR19, CSR23[a], CSR27[a], CSR30[a], CSR36[a]	9.4–9.8	6.11	–
	CSR1–3, CSR4[a], CST7–1[a], SR26B, Sumati[a]	–	6–9	4
Wheat	KRL 1–4[a], WH157	<9.3	6–10	–
	Raj3077, KRL19[a]	<9.3	6–10	
Indian	Pusa bold, Varuna	8.8–9.2	6–8	–
Mustard	Kranti, CS52[a], CSTR330–1,	8.8–9.3	6–9	–
(Raya)	CST609-B 10, CS54[a]	8.8–9.3	6–9	–
Gram	Karnal Chana 1	<9.0	<6.0	–
Sugarbeet	Ramonskaaya 06, MariboResistapoly	9.5–10	<6.5	–
Sugarcane	Co453, Co1341	<9.0	EC_e–10	–

[a]CSSRI varieties released by Central Varietal Release Committee

waters (Minhas et al. 2004), whereas crops like rice, sugarcane and forages are normally avoided. However, in coastal ecosystem, rice is advised due to receiving heavy downpour in *kharif* and facilitates leaching of soluble salt. Tolerance limits to the use of saline waters and salt accumulation in soil can also be regulated by inherent soil properties (soil texture, water table depth, soil depth, etc.), seasonal rainfall and its distribution, ionic constituents of saline water. Concentrated saline water with EC more than 12 $dS\ m^{-1}$ can be useful for salt-tolerant and semi-tolerant crops in light textured soils, contrarily, diluted saline water only EC upto 2 $dS\ m^{-1}$ can build up salinity in heavy textured soils (Sharma and Minhas 2005). Further, frequent irrigations with sodic water are not advisable for high water requiring crops like rice and sugarcane. Besides ageing, crop cultivars and the presence of other toxic constituents along with salinity/sodicity also can change the tolerance of crops to osmotic and abiotic stress (Sharma et al. 2018). A list of all salt-tolerant varieties along with their level of tolerance to soil salinity is given in Table 9.4.

9.6 Conclusions

Salts are indispensable in irrigated agriculture and salt-affected soils. Therefore, inventory, assessment and regular monitoring of soil, land and groundwater is prerequisite for planning agricultural practices to achieve good yield and strengthening livelihood of farming community. Conventional analytical assessment for soil salinity is widely accepted for identification of soil salinity problem.

Further, advanced soil chemical analytical procedure is useful to discriminate sodicity from salinity and is very vital for land manager/crop planner in decision making of amendment needs. Besides gypsum, elemental S/pyrites, organic manure and city generated composts are the potential amendment to reclaim sodic soils and increasing productivity of these unproductive lands. Further, recent analytical tools EM-38 based on electromagnetic geophysical tools, time domain reflectometry, resistivity survey, hyper-spectral remote sensing, etc., are effective for delineating the extension of soil salinity in temporal basis and generate knowledge about regional salinity. Sub-surface drainage is a useful technology to restore saline soil-affected soil with shallow water table/and saline underground water; storing rainwater in *kharif* season and use of salt-tolerant varieties are recommended for coastal areas; effective management of waterlogged salinity with bio-drainage is useful option to combat salinity and ensuring social forestry. Therefore, a comprehensive knowledge of soil salinity can guide soil and crop managers to adopt the need based rehabilitation program of specific categories of soils.

References

Abrol IP, Yadav JSP, Massoud FI (1998) Salt-Affected Soils and their Management. FAO Soils Bulletin 39. Soil Resources, Management and Conservation Service, FAO Land and Water Development Division, Rome

Bajwa MS, Swarup A (2009) Soil salinity and alkalinity. In: Goswami et al (eds) Fundamentals of soil science. Indian Society of Soil Science, New Delhi

Barman A, Mandal AK, Srivastava R, Yadav RK, Sharma PC (2017) Soil salinity characterization using hyper-spectral remote sensing data. ICAR News 23(4):11–12

Basak N, Gobinath R, Rai AK, Sundha P, Yadav RK, Narjary B, Gajender Y, Kumar S, Sharma D K (2017) Deficit saline water irrigation and mulch affect root zone salinity and micronutrient availability in salt-affected soil. In: National Seminar on "nutrients and pollutants in soil-plant-animal-human continuum for sustaining soil, food and nutritional security—way forward" June 9–10, 2017, Lake Hall, Bidhan Chandra Krishi Viswavidyalaya, Kalyani

Bharadwaj KKR, Abrol IP (1978) Nitrogen management in alkali soils. In: Sen SP, Abrol IP, Sinha SK (eds) Nitrogen assimilation and crop productivity. Associated Publishing Company, New Delhi

Bhargava GP (2003) Training manual for undertaking studies on genesis of sodic/alkali soils. Central Soil Salinity Research Institute, Haryana

Brady NC (1988) The nature and properties of soils, 9th edn. MacMillan Publishing Co. Inc., London

Chhabra R (1996) Soil salinity and water quality. Oxford & IBH Publications, New Delhi

Chhabra R (2004) Classification of salt-affected soils. Arid Land Res Manag 19(1):61–79

Chinchmalatpure AR, Ali S, Kulshrestra N, Singh RK, Bundela DS, Kumar P, Sharma DK (2015) Intellectual property management and commercialization of ICAR-CSSRI technologies for management of salt-affected and waterlogged soils of India. ICAR-Central Soil Salinity Research Institute, Haryana

Choudhary OP, Ghuman BS, Bijay-Singh TN, Buresh RJ (2011) Effects of long-term use of sodic water irrigation, amendments and crop residues on soil properties and crop yields in rice–wheat cropping system in a calcareous soil. Field Crops Res 121:363–372

Corwin D, Hendrickx JMH (2002) Solute content and concentration—indirect measurement of solute concentration—electrical resistivity: Wenner array. In: Dane JH, Topp GC (eds)

Methods of soil analysis, part 4—physical methods, Soil Science Society of America book series 5. Soil Science Society of America, Madison
CSSRI (2006) CSSRI: a journey to excellence (1969–2006). Central Soil Salinity Research Institute, Haryana
El-Mowafy AM (1982) Proc Int Sulphur Conf 1982:489–496
Kubenkulov K, Naushabayev A, Hopkins D (2013) Reclamation efficiency of elemental Sulphur on the soda saline soil. World Appl Sci J 23(9):1245–1252
Li F, Keren R (2009) Calcareous sodic soil reclamation as affected by corn stalk application and incubation: a laboratory study. Pedosphere 9:465–475
Mandal UK, Burman D, Mahanta KK, Sarangi SK, Raut S, Mandal S, Maji B, Bandyopadhyay BK (2015) Bulk soil electrical conductivity for coastal salt affected soils of West Bengal. J Indian Soc Coastal Agric Res 33(1):11–18
Minhas PS (1998) Crop production in saline soils. In: Tyagi NK, Minhas PS (eds) Agricultural salinity management in India. Central Soil Salinity Research Institute, Haryana
Minhas PS, Bajwa MS (2001) Use and management of poor quality waters for the rice–wheat based production system. J Crop Prod 4:273–305
Minhas PS, Gupta RK (1992) Quality of irrigation water assessment and management. ICAR Publication, New Delhi, p 123
Minhas PS, Sharma DR, Chauhan CPS (2004) Management of saline and alkali waters for irrigation. In: advances in Sodic land reclamation, international conference on the sustainable management of sodic lands, Lucknow, India, 9–14 February 2004, pp. 121–162
Minhas PS, Dubey SK, Sharma DR (2007) Comparative affects of blending, intera/inter-seasonal cyclic uses of alkali and good quality waters on soil properties and yields of paddy and wheat. Agric Water Manag 87(8):3–90
Narjary B, Jangra P, Abhishek R, Kumar N, Raju R, Thimappa K, Meena RL, Kumar S, Kumar P, Chichmatalpure AR, Kamra SK (2017) Quantitative assessment of soil salinity using electromagnetic induction technique and geostatistical approach. J Soil Salin Water Qual 9(2):156–166
Rai P, Bardhan G, Chaudhari SK (2014) Water retention, hydraulic conductivity and soil-water diffusivity of three biodraining Sodic soils as influenced by concentration and composition of water. J Indian Soc Soil Sci 62(3):197–208
Rai AK, Basak N, Kumar K, Sundha P, Dixit AK, Yadav RK (2017) Forage production in problematic soil: Indian perspectives. In: Ghosh PK, Mahanta SK, Singh JB, Vijay D, Kumar RV, Yadav VK, Kumar S (eds) Approaches towards fodder security in India. Studera Press, New Delhi
Rai AK, Basak N, Sundha P (2019) Alternate reclamation sources of sodic soils. In: Minhas PS (ed) Salinity management in India. Central Soil Salinity Research Institute, Haryana
Rao KVGK, Visvanatha NA (1998) In: Gupta SK, Sharma SK, Tyagi NK (eds) Salinity control through subsurface drainage. Central Soil Salinity Research Institute, Haryana
Sharma BR, Minhas PS (2005) Strategies for managing saline/alkali waters for sustainable agricultural production in South Asia. Agric Water Manag 78:136–151
Sharma P, Swarup A (1997) Comparison of pyrites varying in water-soluble sulfur with gypsum for the reclamation of alkali soil under a rice-wheat rotation. Biol Fertil Soils 24:96–101
Sharma SK, Kulshreshtha N, Kumar A, Yaduvanshi NPS, Singh M, Prasad KRK, Basak N (2018) Waterlogging effects on elemental composition of wheat genotypes in sodic soils. J Plant Nutr 41(10):1252–1262. https://doi.org/10.1080/01904167.2018.1434541
Singh OP (1998) Salinity and waterlogging problems in irrigation commands. In: Tyagi NK, Minhas PS (eds) Agricultural salinity management in India. Central Soil Salinity Research Institute, Haryana
Stamford NP, Ribeiro MR, Cunha KPV, Freitas ADS, Santos CERS, Dias SHL (2007) Effectiveness of sulphur with *Acidithio bacillus* and gypsum in chemical attributes of a Brazilian sodic soil. World J Microbiol Biotechnol 23:1433–1439

Sundha P, Basak N, Rai AK, Yadav RK, Sharma DK, Sharma PC (2017) N and P release pattern in saline-sodic soil amended with gypsum and municipal solid waste compost. J Soil Salin Water Qual 9(2):145–155

Sundha P, Basak N, Rai AK, Yadav RK, Sharma DK (2018) Utilization of municipal solid waste compost in reclamation of saline-sodic soil irrigated with poor quality water. J Indian Soc Soil Sci 66(1):28–39. https://doi.org/10.5958/0974-0228.2018.00004.X

Yaduvanshi NPS, Swarup A (2005) Effect of continuous use of sodic irrigation water with and without gypsum, farmyard manure, pressmud and fertilizer on soil properties and yields of rice and wheat in a long term experiment. Nutr Cycl Agroecosyst 73:111–118. https://doi.org/10.1007/s10705-005-3361-1

Modern Sample Preparation Techniques for Pesticide Residues Analysis in Soil

10

Prithusayak Mondal

Abstract

Pesticides are one of the mandatory defense weapons in this modern world to win over the vast populations of plant pests attacking crops during and after cultivation. But injudicious application of these chemicals creates nuisance to the environment leading to residues, resistance, and pest resurgence problem. These residues bind to the environment and revolve in the food chain resulting in bioaccumulation and biomagnification. As the presence of trace amounts of both pesticide residues and their degradation products could be potential health hazards, the International organizations like FAO, WHO have already raised concerns regarding presence of these toxic chemicals in soil, food, and feed samples. Codex Alimentarius Commission after years of trial determined a value called maximum residue limit (MRL) with the aim to establish restrictive measures to protect the environment against pollution. Due to intensive use of pesticides, their residues have become an intrinsic part of the environment including soil, and they are often detected in various samples and therefore their monitoring has been frequently performed throughout the world. Considering low concentration levels of pesticide residues in soil matrices and the determination of these residues often requires extensive sample extraction and clean-up prior to the analysis. This article describes the different sample preparation techniques including their extraction and clean-up that are widely applied for soil sample analysis for pesticide residues.

Keywords

Pesticide · Soil residues · MRL · Sample preparation · Extraction · Clean-up

P. Mondal (✉)
Regional Research Station (Terai Zone), Uttar Banga Krishi Viswavidyalaya, Cooch Behar, West Bengal, India

A. Rakshit et al. (eds.), *Soil Analysis: Recent Trends and Applications*,
https://doi.org/10.1007/978-981-15-2039-6_10

10.1 Introduction

Since time immemorial, we have an inseparable relationship with the soil. Soil, being a natural resource, has considerably affected our ability to cultivate crops and influenced the development of civilizations. This relationship between humans, the earth, and food sources affirms soil as the foundation and one of the critical parts of successful agriculture. To enhance production and productivity, the application of pesticide is compulsory and unavoidable. Pesticides are basically a heterogeneous group of compounds including insecticides, herbicides, fungicides, nematicides, etc. having different physicochemical and biological properties. After introduction into the soil, pesticides undergo various movements and transformation processes which ultimately produce their derivatives or metabolites, degradation products, reaction products, and other impurities having toxicological significance, those are collectively called as pesticide residues (Đurović 2011). Owing to the misuse and overuse of these pesticides, their residues are continuously increasing and they have become an unavoidable portion of the pedosphere. Considering the persistence of residues and their deleterious effects, it seems that soil contamination over a long period of time is the biggest threat in terms of food safety as these compounds are mobile and capable of bioaccumulation (Damalas 2009). Exposure to contaminated soil samples may be detrimental to the health of not only humans but also of all other living organisms (Odukkathil and Vasudevan 2013). Therefore, the concentration levels of pesticides and their derivatives in the soil must be frequently monitored. Maximal residue limits (MRLs) for pesticides have been established by the United Nation's Food and Agriculture Organization (FAO) and the World Health Organization (WHO) for this purpose (Codex 2019) and any quantity above this MRL value is a concern to human health. The MRL is the maximum level of a pesticide residue (expressed in mg kg^{-1}) which is legally permitted in or on food or animal feed (EU, MRL 2019). In recent years, some MRLs have been significantly lowered from ppm to ppb levels to meet the expectations for securing human health at the top level. Therefore, to detect these pesticides meticulously, reliable methods that can analyze dilute mixtures of parent substances and their metabolites are required. So, there is also increasing interest in industrial and government sectors to develop more precise, sophisticated, and cost-effective methods to generate large amounts of residue data on new and existing products.

The current trend in pesticide residues analysis is developing multi-residual methods that not only provide a simultaneous determination of a large number of pesticides but also can be applied to large numbers of samples of different origin. The entire chemical analysis involves several important stages like sample preparation, analyte separation (i.e., quantification and data analysis) of which sample preparation step is considered as the most critical one. Conventional sample preparation techniques (solvent extraction, sonication assisted extraction, etc.) are laborious, expensive, time consuming, and require large amounts of organic solvents and usually involve many steps, leading to loss of some analyte quantity. Additionally, consequences of use of hydrocarbon solvents, such as depletion of ozone layer and generation of considerable carcinogenic waste, lead to a reduction of not only their use but their manufacture also. As a result, modern sample preparation procedures, such as solid phase extraction (SPE), solid phase micro-extraction (SPME), supercritical fluid extraction (SFE), microwave assisted extraction (MAE), microwave-assisted micellar extraction (MAME), accelerated solvent extraction (ASE), matrix solid phase dispersion (MSPD) extraction, and QuEChERS (quick, easy, cheap, effective, rugged and safe) method, have been developed to overcome the limitations of the conventional approaches. SFE, ASE, and MAE are instrumental techniques, and often use SPE (for purification of obtained extracts) and SPME (for purification and concentration of obtained extracts) for desired results.

Most residue analysis procedures fall within the scheme shown in Fig. 10.1. Design of experiment deals with strategic planning for evaluation of several factors such as selection of site, plot size, replications of sample, time element, maximum residue limit, formulation, type or variety of crops, etc. Sampling is the process to obtain a representative quantity from the large consignment, so that the selected representative quantity can be handled conveniently. Sample preparation for laboratory analysis is considered the most crucial step as the success of entire experiment depends on the proficiency at this level. It is done by selecting the components of interest, thereafter mixing, subdividing, and systematically reducing the sample size. Once a valid, representative sub-sample has been selected for residue analysis, it is processed for isolation of pesticide or its metabolites having toxicological significance from the surrounding biological environment. Extraction must be adequate to remove the toxicant in sufficient quantities from sample into a suitable solvent. The method of extraction and the type of solvent or solvent combinations will be dependent on the physical and chemical properties of the pesticide to be extracted, the type of substrate from which it will be quantitatively removed, and the final method of analysis. While extracting the pesticide with solvents from the plant materials, proteins, tannins, lipids, fat, waxes, chlorophyll, and terpenoids, they are co-extracted from matrix of substances (Erwin et al. 1955). These co-extractives can prevent the reaction of pesticides with chromogenic reagents, colored extracts directly interferes in the colorimetric analysis and can also contaminate the columns and detectors in the analysis. To achieve necessary sensitivity, the interfering substances have to be removed from pesticide, and this step is known as clean-up. It usually begins with some form of extraction technique and the degree of clean-up required is dependent on the scope of analysis, the complexity of sample, and the

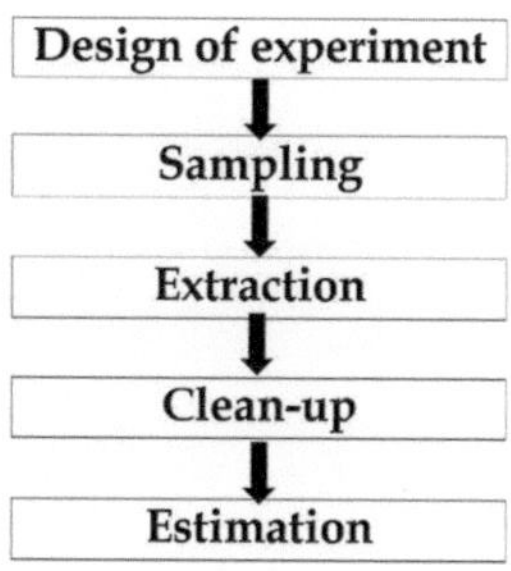

Fig. 10.1 Steps of pesticide residue analysis

sensitivity and selectivity of detection methods available for the contaminant sought (Handa et al. 1999; Singh 2000). Estimation step, covering both detection and quantification of target compound, wind up the story of residue analysis. It is always desirable that chosen analytical procedure allows the simultaneous determination of large number of pesticides.

This article describes the basic principles of sample preparation techniques, especially soil sampling, extraction cum clean-up techniques, both conventional and modern approaches, comparing their advantages and disadvantages, and their ability and applicability for pesticide residues determination, with special emphasis on soil samples.

10.2 Soil Sampling Methods

The sampling of soil is typically done to detect pesticide residues or to routinely monitor environmental samples (Sharma, 2007). Soil samples should be taken from growing fields in the grid pattern uniformly distributed so that each area of the field is sampled. A 3 × 3 grid with nine total sample proportions is suggested for smaller fields, with 4 × 4 (16 sample portions) for the medium-sized fields, and 5 × 5 and even larger grids are used for very large fields. Each sample site represents one portion of the total sample, and at each site, two soil plugs about 15 cm deep and 3–5 cm in diameter are to be taken. The two plugs, when combined, become sample portion of that sample site. Another common soil sampling method for a field or other area is to take "5" portions in a "Z" pattern. An example of a 3 × 3 sampling grid is designed by X-pattern sampling (Fig. 10.2).

Sampling tools include soil augers. Place each portion of the soil sample into a separate glass jar covered with aluminum foil. It is recommended to chill soil samples to 4 °C for transport to the laboratory. The glass jars for collecting soil samples should be rinsed thoroughly with acetone or methanol and dried.

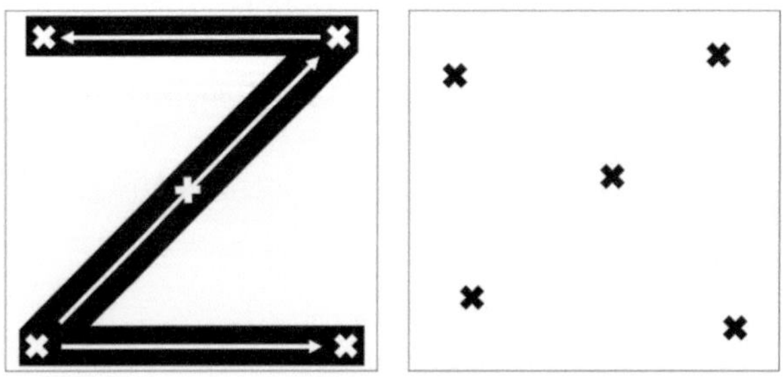

Fig. 10.2 Soil sampling patterns: Z-pattern and X-pattern

10.3 Conventional to Modern Approaches for Extraction and Clean-up: A Paradigm Shift

Traditional sample preparation methods (liquid–liquid extraction, Soxhlet extraction, sonication assisted extraction, etc.) are laborious, time consuming, expensive, require large amounts of organic solvents and usually involve many steps, leading to loss of some analyte quantity. Additionally, consequences of hydrocarbon solvents use, such as ozone depletion and generation of considerable cancer waste, lead to reduction of not only their use but also their manufacture. As a result, modern sample preparation procedures, such as accelerated solvent extraction (ASE), supercritical fluid extraction (SFE), microwave assisted extraction (MAE), solid phase extraction (SPE), solid phase microextraction (SPME), matrix solid phase dispersion (MSPD) extraction and QuEChERS (quick, easy, cheap, effective, rugged and safe), have been developed to overcome the drawbacks of the traditional approaches. It should be noted that some (SFE, ASE and MAE) are instrumental techniques, and often use SPE (for purification of obtained extracts) and SPME (for both purification and concentration of obtained extracts) for desired purpose.

10.3.1 Solvent Extraction

For extraction of toxicants, either any suitable solvent or mixture of solvents is used. Soil samples were extracted by shaking with suitable solvent or solvent mixture in a mechanical shaker for definite period. The mixture was filtered, washed, and stored for further action. Liquid–liquid extraction (LLE) is one of the earliest and most commonly used extraction techniques employed for pesticide residue analysis in complex media (Dean 1998). The principle of LLE is that the sample is distributed or partitioned between two immiscible solvents in which the analyte and matrix have different solubility or it is based on the low value of the partition coefficient for most organic compounds between different solvents. The main advantage of this technique is the wide availability of pure solvents and use of low cost apparatus. Khan et al. (2011) employed ethyl acetate and hexane for the LLE of pentachloronitrobenzene and hexachlorobenzene and its metabolites prior to their HPLC determination. Another method that can be applied for dry materials like soil is Soxhlet extraction. Although the method is very efficient, sometimes formation of fine capillary tubes in the sample mass obstructs complete extraction.

10.3.2 Sonication Assisted Extraction (SAE)

Sonication provides a more efficient contact between the solid and solvent than Soxhlet method, usually resulting in a greater recovery of analyte (Poole et al. 1990). The extraction procedure should be optimized with regard to the solvent amount, the duration of sonication, and the number of extraction steps. The ultrasonic solvent extraction is more rapid than conventional shake-flask or Soxhlet extraction methods, and the solvent consumption is significantly lower. Additionally, the extracts from sonication can be chromatographed without subsequent clean-up step, and the analysis time is considerably reduced. Sonication assisted extraction has been used by Sánchez-Brunete et al. (2003) for carbamate pesticides.

10.3.3 Liquid Solid Extraction (LSE)/Solid Phase Extraction (SPE)

Solid phase extraction is one of the sorbent techniques which is used very often for pesticide residue analysis. It, being less laborious, produces low background interferences and also significantly reduces the use of organic contaminants. This method is based on the exclusion of extracts containing target analytes through a column (cartridge) filled with the appropriate solid phase called sorbent (which was previously conditioned by an appropriate solvent or solvent mixture), or passing of an appropriate solvent through the SPE column to which a suitable amount of sample was previously added (Moors et al. 1994). So, SPE basically separates compounds of interest from impurities in three distinct ways: selective extraction (the compounds of interest retained by the packing material and the impurities are eluted out), selective washing (the column is washed with strong solutions to remove impurities but the solution should not be so strong that it carries away the compound), and selective elution (the compound of interest is eluted in a solvent but the impurities are retained in the column). Method of operation can be divided into five steps: wetting the sorbent, conditioning of the sorbent, loading of the sample, rinsing or washing the sorbent to elute extraneous material, and elution of the analyte of interest. Each step is characterized by the nature and type of solvent used which in turn is dependent upon the characteristics of the sorbent and the sample (Dean 1998). Using selective solvents, first the co-extractants from the SPE column can be successfully eluted, and then the target analytes (Fig. 10.3, A), or the elution of analytes can be direct, where undesirable co-extractants derived from the sample matrix remain in the SPE column (Fig. 10.3, B).

The SPE sorbents used frequently in pesticide residues determination include reverse phase octadecyl (C18), normal-phase aminopropyl (–NH_2) and primary-secondary amine (PSA), anion-exchanger three-methyl ammonium (SAX)

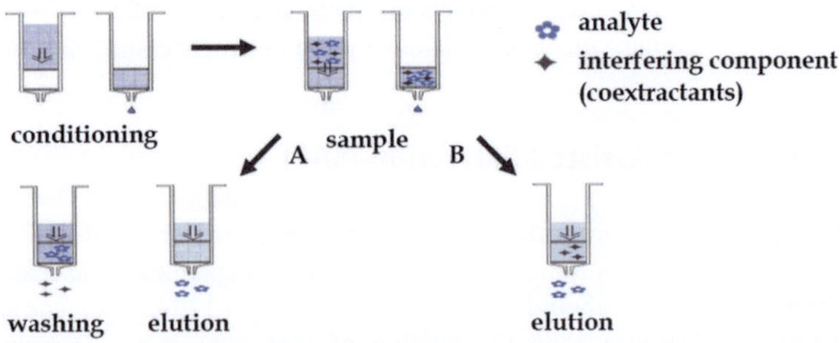

Fig. 10.3 Steps of solid phase extraction technique

and adsorbents such as graphitized carbon black (GCB). Normal-phase sorbents such as florisil ($MgSiO_3$), aluminum oxide (Al_2O_3), and silica (SiO_2) are usually used in combination with the previously mentioned sorbents. The SPE cartridge should be chosen depending on the physicochemical properties of pesticides that are searched for in a particular sample, and the nature of the sample matrix (Đurović and Đorđević 2011). C18 cartridges have been found a good choice for determination of carbamates in soil (Santalad et al. 2010) and silica gel has proven effective in determination of OCPs in soil samples (Lehnik-Habrink et al. 2010).

10.3.4 Solid Phase Micro-Extraction (SPME)

Solid phase micro-extraction, one of the latest extraction techniques, is widely used in the pesticide residues analysis because of the fact that purification and concentration of the sample extract (analytes of interest) run simultaneously here. SPME syringe is the main part of the SPME system that visually resembles on the chromatographic system; however, it also contains a 1 cm long fiber located within a syringe needle, which is made of an appropriate polymer deposited on the holder of fused silica. Micro-extraction process is based on the redistribution of analytes between micro-extraction fiber and sample matrix, i.e., on the selective sorption of target analytes in the active layer of the fiber and direct desorption in the chromatograph injector (thermal in the case of the gas chromatography or, i.e., by solvent elution in the case of liquid chromatography). The basic micro-extraction procedure of analytes from the solution is shown in Fig. 10.4 (Đurović and Đorđević 2011).

Before the analysis, the fiber is drawn into a metal tube of the SPME syringe. After breaking through the vial septum in which a certain sample amount was previously inserted, the fiber is pulled out from the syringe, i.e., it is exposed to the sample by lowering the syringe plunger. After specific time, the fiber with the sorbed analytes is drawn into the needle, which is then pulled out from the vial.

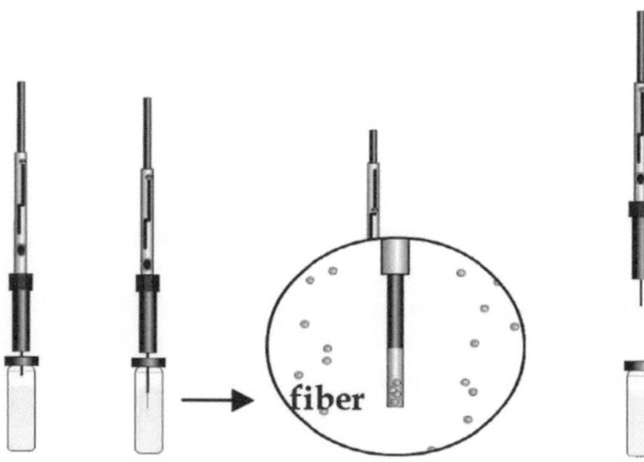

Fig. 10.4 Procedure for micro-extraction of analytes from solution

Analytes desorption from the fiber is performed by introducing a SPME syringe needle into the injector of the chromatographic system.

SPME is an equilibrium technique, where analytes are distributed between the three phases: sample, gas phase, and fiber. The fiber does not extract all analytes present in the sample, but by the proper calibration, this technique can be used for successful quantification (Đurović et al. 2007a; Pawliszyn 1997). The amount of analytes that would be adsorbed on the fiber will depend on the thickness and polarity of the active fiber layer, sampling mode (direct sampling—micro-extraction from solution, "DM-SPME" and headspace sampling—micro-extraction from gas phase, "HS-SPME"), the nature of the sample and the analyte (analyte polarity, its molecular weight, pH value, nature of matrix), the mode and speed of the sample mixing, the SPME duration, the temperature at which it is performed, and so on.

Today, about 30 different fiber types are in use (different types of polymers and their thickness), so when selecting the fiber it is necessary to take into consideration several factors: molecular weight, structure and polarity of the analyte molecules, the polarity of fibers, the mechanism of extraction (used sampling mode), the detection limit and range of linearity that is desired to be achieved. In order for a fiber to extract specific compounds from a given matrix, it must have a much higher affinity for the given analytes than a matrix, where the general rule applies: non-polar analytes are more efficiently extracted by non-polar active fiber layer, i.e., polar by polar. The research in the field of pesticide residues has indicated that, in the most of the cases, fibers with extremely non-polar polydimethylsiloxane (PDMS) and highly polar polyacrylate (PA) active layers are most effective in the analysis of samples of different origin (Doong and Liao 2001; Sakamoto and Tsutsumi 2004; Đurović et al. 2007b, c, 2010b; Fernandez-Alvarez et al. 2008). After fiber selection, it is necessary to determine optimal conditions for analytes transfer in the chromatographic system. Adsorbed analytes are desorbed from the fiber by introducing the SPME syringe needle into the injector. Defining the parameters of desorption involves determination of the optimal injector temperature, flow of the carrier gas, and desorption time in the case of GC, i.e., proper choice of elution solvent, its flow rate and desorption time, in the case of HPLC.

Although the maximum of SPME sensitivity is achieved at equilibrium times, for practical reasons, extraction time can be shortened (Đurović et al. 2007a, 2010a, b; Pawliszyn 1997). The most effective ways to overcome the kinetics restrictions are heating and efficient sample mixing. The temperature has two opposite effects. On the one side, its increase increases the analytes transfer from the sample to the fiber, while on the other side, due to the simultaneous heating of the fiber during extraction, there is enhanced desorption of analytes from it. Therefore, the necessary step in method development is optimization of the extraction temperature. The speed of extraction is also determined by the sample stirring efficiency. Intensive stirring increases the analytes mobility, and therefore reduces the equilibrium time and increases the analytes amount adsorbed on the fiber. However, in method developing it should be noted that the sample stirring leads to its warming, which may also have non-preferred effects, especially in the case of direct mode.

The matrix nature greatly influences the SPME efficiency, too. Since the analytes distribution coefficients are partially determined by analytes/matrix interaction, appropriate matrix modification can increase the analytes partition coefficients. Thus, for example, the presence of chloride and sulfate ion increases the ionic strength of the solution, which makes a large number of compounds less soluble. In this way, by weakening the matrix/analyte interaction, distribution coefficients can be significantly increased (Arthur et al. 1992). Considering the fact that SPME is a single-stage method that does not require additional purification and concentration of the sample, the problems related to the matrix occur in the analysis of samples with complex matrices. The researches have shown that the negative effect of the matrix could be significantly reduced by adequate dilution of the sample with the distilled water (Simplício and Boas 1999; Đurović and Marković 2005; Đurović et al. 2007c, 2008).

The research results indicate that the most often used SPME fibers in the pesticide residues analysis (PDMS and PA) are a good choice for determination of: OCPs in soils (Zhao et al. 2006; Herbert et al. 2006); pesticides belonging to different chemical groups in soil (Đurović et al. 2010a, b), i.e., in samples of potato, tomato, onion, cabbage, and pepper (Marković et al. 2010). Considering that in the SPME analysis only 1 cm of fiber is exposed to the sample, not only the nature, but also the size of the active surface layer will significantly affect the micro-extraction efficiency. Thus, by adding an additional material into the active layer of the fiber, its outer surface may increase, and therefore often the SPME efficiency, too. On the other side, the added material can significantly change the polarity of the fiber (similar to the GC stationary phase). Thus, for example, by using mixed PDMS/DVB (polydimethylsiloxane/divinylbenzene) fiber, Vega Moreno et al. (2006) provided satisfactory analytical parameters for SPME determination of OCPs in soil.

10.3.5 Supercritical Fluid Extraction (SFE)

Supercritical fluid extraction is the process of separating organic compounds (extractants) from solid matrices using supercritical fluids (CO_2, NO_2, SO_2, NH_3, etc.). A substance exists as a supercritical fluid (SCF) when system temperature and pressure are above a critical point (Fig. 10.5). The principle of SFE is based on the solvent power of SCF which is highly dependent on the density of SCF, which in turn depends on the pressure and temperature. Modification of little temperature and pressure changes the property of SCF which is very useful for extraction purpose. Because of low viscosity and higher diffusivity as compared to liquids, SCFs diffuse more rapidly and even penetrate solid samples.

CO_2 is the most commonly used SCF for this purpose, as it has relatively low critical temperature (31 °C) and low critical pressure (73 kPa) (Atkins and de Paula 2002). It is non-reactive and non-toxic also, available in a high degree of purity at low cost and shows absence of contamination of final products as CO_2 volatilizes off. Changes in temperature and pressure at which the supercritical CO_2 is held will increase or decrease the strength of solvent that ensures selective extraction

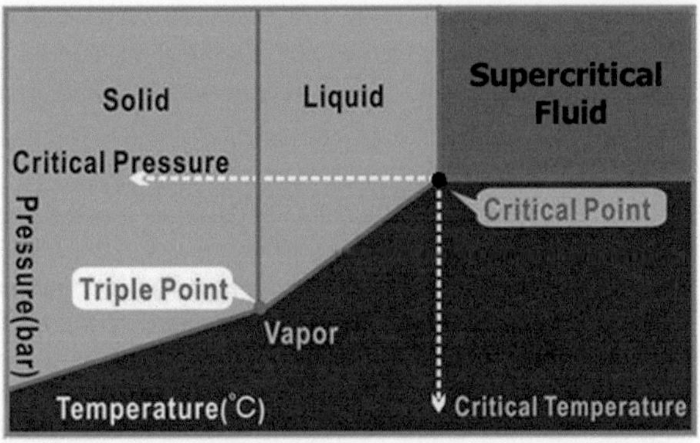

Fig. 10.5 Phase diagram of supercritical fluid

of the target compound. At constant temperature beyond critical temperature, the supercritical CO_2 will be able to extract analytes of high polarity at high pressure, and low polarity analytes at low pressure. SFE with CO_2 is usually performed at pressures that are not high enough to achieve efficient extraction of polar compounds. In such conditions, the supercritical CO_2 is a good extraction medium for non-polar compounds and moderately polar ones, such as polycyclic aromatic hydrocarbons (PAHs), polychlorinated biphenyls (PCBs), organochlorine (OC), and organophosphorus (OP) pesticides, etc. Supercritical CO_2, being non-polar, sometimes requires small amounts of polar co-solvents as modifiers, whose major role is to interact with the sample matrix to promote desorption into the fluid (Langenfeld et al. 1994). Some of the common solvents such as acetone (Kaihara et al. 2002; Ono et al. 2006) and methanol (Rissato et al. 2005a, b) are now mostly used as modifiers.

In general, extraction procedure is completed within 2 h, and further analysis can be accomplished in various ways. According to one, supercritical fluid with analytes is passed through a capillary that is immersed in an appropriate solvent. While in the capillary, it remains as supercritical fluid, but after leaving the capillary it becomes a gas (the pressure falls below the critical pressure). The largest part of this gas passes through the solvent, while the extracted analytes are retained in the solvent (the degree of retention depends on the solvent, i.e., the solubility of the analyte in it). The flow of SF can be directed to a solid sorbent, which will then bind analytes, and its elution by an appropriate solvent, analysts translate into a solution suitable for further analysis (Fig. 10.6). Also, the flow of SF could be directed directly to the capillary column of the gas chromatograph (GC), thus obtaining the on-line SFE. This approach enables the analytical scheme with the highest sensitivity for a limited amount of sample available for analysis. The recent studies have shown that SFE methods, followed by additional purification of the obtained extracts, meet the strict criteria of the pesticide residues analysis. The same sorbent was shown to be the best choice for determination of 32 pesticides in soil using SFE sample preparation (Rissato et al. 2005b).

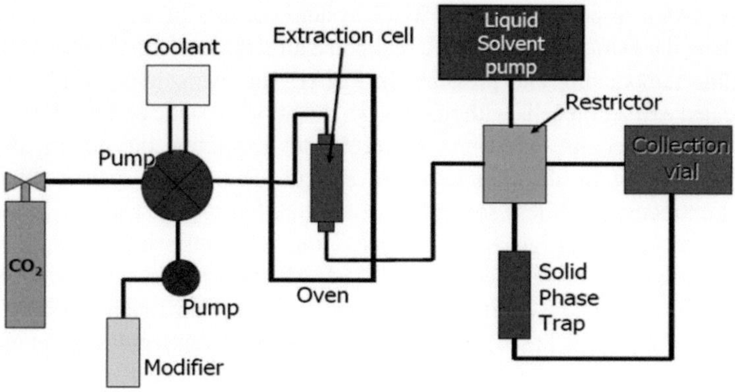

Fig. 10.6 Supercritical fluid extraction technique

10.3.6 Microwave-Assisted Extraction (MAE)

Microwave-assisted extraction (MAE) is a technique utilizing the microwave energy, and where target compounds can be extracted more selectively and rapidly, with similar or better recovery compared to traditional extraction processes. Microwave energy is a non-ionizing radiation (frequency 300–300,000 MHz), which can penetrate into certain materials and interact with the polar components to generate heat. The MAE causes a direct migration of the desired components out of the matrix, as a result of selective energy application into the matrix. Greater efficiency of extraction method effects in the matrix macrostructure destruction (Lambropoulou and Albanis 2007). During the MAE of solid material, microwave rays travel freely through the solvent and interact selectively with the free matrix water causing localized heating resulting in non-uniform temperature rise with more pronounced effects where the free water is in larger proportions which ultimately leads to a volume expansion within the systems. The walls of these systems cannot countenance the high internal pressures and rupture spontaneously, allowing the organic contents to flow freely toward the relatively cool surrounding solvent that solubilizes them rapidly (Ranz et al. 2008). For method optimization, several variables, such as solvent composition and amount, extraction temperature and time, are usually studied. In order to heat a solvent, part of it must be polar with high dielectric constant to absorb microwave energy efficiently. Non-polar solvents with low dielectric constants can be also used, by adding certain amount of polar solvent that absorbs the microwave radiation and passes it on to other molecules (Caddick 1995). For example, hexane and toluene can be modulated by the addition of small amounts of acetone or methanol (Ericsson and Colmsjö 2000).

Generally, MAE devices comprise a closed extraction vessel under controlled pressure and temperature or a focused microwave oven at atmospheric pressure. These two technologies are commonly named pressurized MAE (PMAE) or focused

MAE (FMAE), respectively. The PMAE system consists of a magnetron tube, an oven where the extraction vessels are set upon a turntable and monitoring devices for controlling temperature and pressure. In PMAE, the extractions are performed in some sealed extraction cells with microwave radiation, in static and batch mode. The increase in temperature and pressure accelerates extraction due to the ability of extraction solvent to absorb microwave energy. The closed system offers fast, efficient extraction with less solvent consumption, but it is susceptible to losses of volatile compounds and generally expensive due to its resistance to high pressure and its air-tightness (Zhang et al. 2011). FMAE involves an open MAE system developed to counter the shortcomings of the closed system, such as safety issues. The extractor design is based on the principles of a conventional Soxhlet extractor modified to facilitate accommodation of the sample cartridge compartment in the irradiation zone of a microwave oven. Solvent distillation in the FMAE extractor could be achieved by electrical heating, which is independent of extractant polarity (Luque-García and Luque de Castro 2003, 2004). It is considered more suitable for extracting thermo-labile compounds due to only part of the extraction cell being directly exposed to microwave radiation. Since the upper part of the extraction cell is connected to a reflux unit to condense vaporized solvent, sample throughput is limited (Fig. 10.7).

From economical and practical aspects, MAE is a strong competitor to other recent sample preparation techniques. The main MAE advantages are the complete automation, low temperature requirement, high extraction efficiency, and the possibility of extracting different samples at the same time without interference. The main disadvantage of MAE seems to be the lack of selectivity resulting in the co-extraction of significant amounts of interfering compounds. Additional clean-up is therefore needed before chromatographic analysis. Apart from that, the poor efficiency of microwaves when either the target compounds or the solvents are non-polar, or when they are volatile, can be regarded as another disadvantage. Besides, it is important to notice that the application of microwave energy to flammable organic compounds, such as solvents, can pose serious hazards in inexperienced hands, thus an extraordinary level of safety and attention to details when planning and performing experiments must be used by all personnel

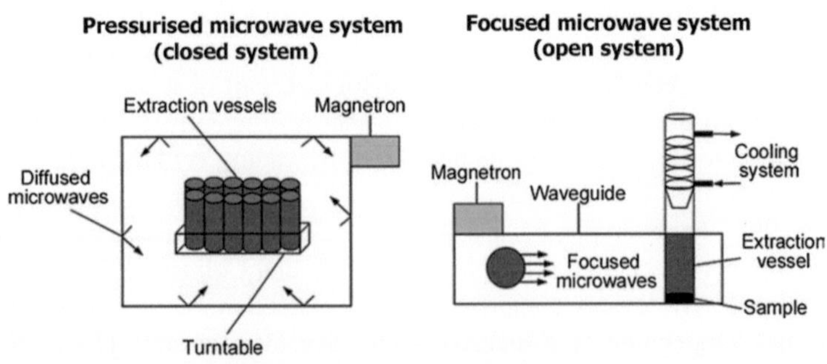

Fig. 10.7 Microwave-assisted micellar extraction (MAME) procedure

dealing with microwaves. The first use of MAE technique for pesticide residues determination (parathion and bromophos in soil) was reported by Ganzler et al. (1986). In 1994, 20 OCPs were extracted from six marine sediments and soils (Lopez-Avila et al. 1994). Investigations on MAE extractions of OCPs and OPPs from soil, optimization and comparison of method, was performed by numerous authors (Fuentes et al. 2007; Wang et al. 2007). MAE determination of triazines in soils was reported by Hoogerbrugge et al. (1997) work. MAE determination of imidazolinone herbicides have been reported by Stout et al. (1997). The investigated fungicides were hexaconazole (Frost et al. 1997), and dimethomorph (Stout et al. 1998), both extracted from soils. De Andréa et al. (2001) applied MAE for determination of methyl parathion, *p,p′*-DDE, HCB, simazine, and paraquat dichloride in soil, Sun and Lee (2003) for carbamates in soil.

10.3.7 Microwave-Assisted Micellar Extraction (MAME)

Microwave assisted micellar extraction, which uses a micellar (surfactant-rich) system to substitute organic solvent as extractant in MAE, has been applied lately to the extraction of different compounds from solid samples including soil (Wang et al. 2016). In order to escalate both extraction rate and efficiency, microwave energy is used while maintaining the sample at a suitable temperature. At this point, micelles of surfactant are formed, with analytes isolated and enriched in them. Figure 10.8 shows the three key steps for the operations of MAME;

1. Introduction of surfactant to the sample,
2. Microwave-assisted micellar extraction for definite time period, and
3. Suitable treatment of the extract.

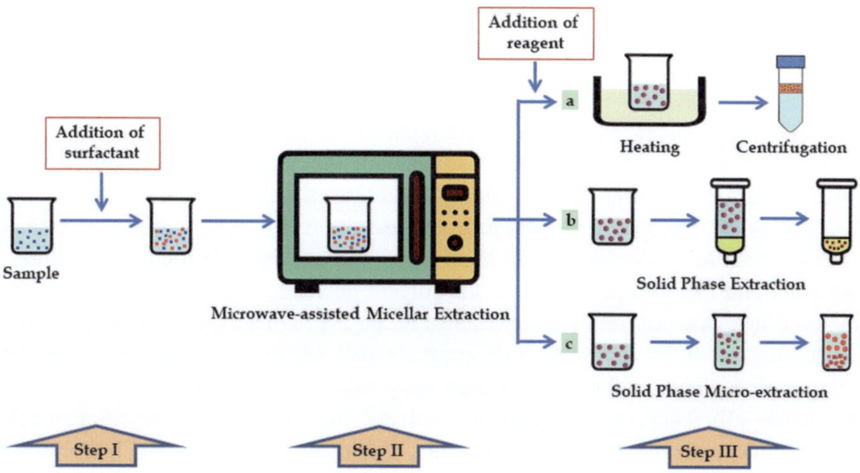

Fig. 10.8 Steps of microwave-assisted micellar extraction

Before injecting the extract into the analytical instrument, the MAME extract obtained should be suitably prepared (Step III, Fig. 10.8). Separation of two phases requires appropriate experimental conditions depending on the nature of the surfactant. Sometimes, analytes in the MAME extract were concentrated with the help of centrifugation after equilibrium at high temperature and adding salt reagent (Step III a, Fig. 10.8) (Chen et al. 2010). The analytes in the micelle-rich phase could be directly injected into HPLC for subsequent separation and detection. As the micelle-rich phase is viscous and cannot be injected directly into some analysis apparatus (e.g., LC-MS/MS), then additional clean-up and concentration of the MAME extract should be employed, such as solid phase extraction (SPE) (Cueva-Mestanza et al. 2008) or solid-phase micro-extraction (SPME) (Pino et al. 2007) (Step III b and c, Fig. 10.8). For SPE, MAME extracts went through the SPE cartridge, and the retained analytes were eluted and analyzed. For SPME, SPME fibers were directly immersed into the MAME extract under optimized conditions, and thereafter the analytes were desorbed from the fiber by a suitable solvent.

10.3.8 Accelerated Solvent Extraction (ASE)/Pressurized Fluid Extraction (PFE)

Accelerated solvent extraction (ASE™, a Dionex trademark), also known as pressurized liquid extraction (PLE) or enhanced solvent extraction (ESE), is a relatively new extraction technique which is partly derived from SFE (Camel 2001; Richter et al. 1996). It is a solid–liquid extraction process using organic solvents at an elevated temperature (usually between 50 and 200 °C) and applying higher pressure (between 10 and 15 MPa) for short periods (12–18 min) to extract samples in an extraction cell. Extractions are carried out under pressure in order to maintain the solvent in its liquid state, even at temperatures above boiling point. Moreover, pressure allows the extraction cell to be filled more quickly and helps to force the solvent into the matrix pores. Thus, the efficiency of the extraction process is improved. Extraction at elevated temperatures increases solubility, diffusion rate, and mass transfer, along with the ability of the solvent to disrupt the analyte-matrix interactions. PLE thus allows fast extraction due to increased solubility, better desorption, and enhanced diffusion, and rapid extraction.

In practice, the extraction cell is filled with the sample to be examined and placed in a furnace controllable. After the addition of a suitable solvent, the cell is brought to an elevated temperature and pressure (Fig. 10.9). Later, the extract is transferred to a collection vial for clean-up and analysis. At high temperatures, viscosity and the surface tension of the solvent decrease, resulting in a substantial increase in extraction rate (Anastassiades et al. 2003). The solvent is kept below its boiling point by applying high pressure that forces its penetration into the sample pores. The combination of high temperature and pressure results in better extraction efficiency, thus minimizing solvent use. The extraction efficiency is almost independent of sample mass, i.e., is mainly dependent on temperature (Richter et al. 1996; Smith 2002). Often a sample undergoes several extraction

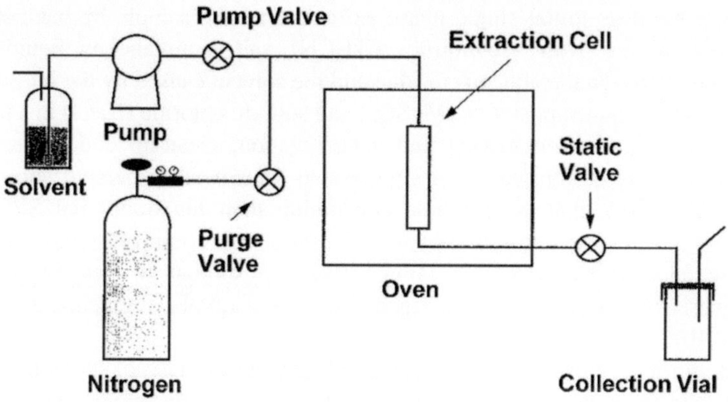

Fig. 10.9 Accelerated solvent extraction technique

cycles. Finally, the extraction cell is flushed with solvent, open the purge valve and the cell, as well as all the lines purged with nitrogen and the apparatus prepared for further extraction. Besides the type of the solvent used, the main parameters which influencing ASE efficiency are extraction temperature and time (Luo et al. 2010). Although high temperatures increase the efficiency, it may lead to degradation of thermo-labile compounds, and to the co-extraction of interfering species. Hence, a compromise between the extraction efficiency and minimization of interfering compounds must be performed carefully, and in addition, usually a further clean-up step involves.

ASE is advantageous over conventional techniques as it requires much lesser solvent and shorter extraction times. Using elevated pressure and temperatures with organic solvents, an enhanced analytes extraction can be achieved. Moreover, ASE can reduce waste levels and analysts exposure to harmful solvents. However, samples with high moisture contents are subjected to desiccation before the extraction step (Cervera et al. 2010). ASE was carried out for determination of DDT and its metabolites (Tao et al. 2004), i.e., abamectin in soil samples (Brewer et al. 2004). ASE methods for soil samples were reported for OCPs (Wang et al. 2007), for bromacil and diuron (Pinto and Lanças 2009), and dichlorvos, dimethoate, parathion, malathion, and chlorpyrifos determination (Zhang et al. 2010).

10.3.9 QuEChERS Method

"QuEChERS" is a portmanteau word derived from "Quick, Easy, Cheap, Effective, Rugged, and Safe." It is a novel multi-residue method for determining pesticide residues in different matrices and appeared to overcome the loopholes of conventional solvent extraction methods (Anastassiades et al. 2003). It is undoubtedly one of the most streamlined sample preparation approaches with excellent results for a wide range of pesticides in different soil samples. The original procedure involves initial single phase extraction of the sample by hand-shaking or vortex mixing with acetonitrile (CH_3CN) and simultaneous liquid–liquid partitioning between the aqueous residue and the solvent caused by the introduction of anhydrous magnesium sulfate ($MgSO_4$) and sodium chloride (NaCl) in a suitable ratio (4:1). After vortex mixing and centrifugation, clean-up and exclusion of residual water is performed via a simple step known as dispersive solid phase extraction (d-SPE) that is less time consuming than the traditional SPE. This procedure involves addition of anhydrous $MgSO_4$ with aliquot to remove residual moisture and primary-secondary amine (PSA) adsorbent to get rid of many polar matrix components, such as organic acids, some polar pigments, and sugars (Fig. 10.10).

Acetonitrile is selected as the QuEChERS solvent because of its high polarity, well miscibility with water, and sufficient dispersive (hydrophobic) properties to extract effectively both polar and non-polar pesticides. The original QuEChERS method was subjected to certain necessary modifications to ensure efficient extraction of pH-dependent compounds (e.g., phenoxyalkanoic acids) and to minimize degradation of susceptible compounds (e.g., base and acid labile pesticides). Anastassiades et al. (2007) realized that buffering at pH 5.0 during extraction could give the optimum balance to achieve acceptably recoveries (>70%) for pH-dependent pesticides, independent of the matrix. On the other hand, Lehotay (2007) modified the method to apply for stronger acetate-buffering conditions. Both of these versions of methods went through extensive laboratory trials and successfully met statistical criteria for acceptability by independent scientific

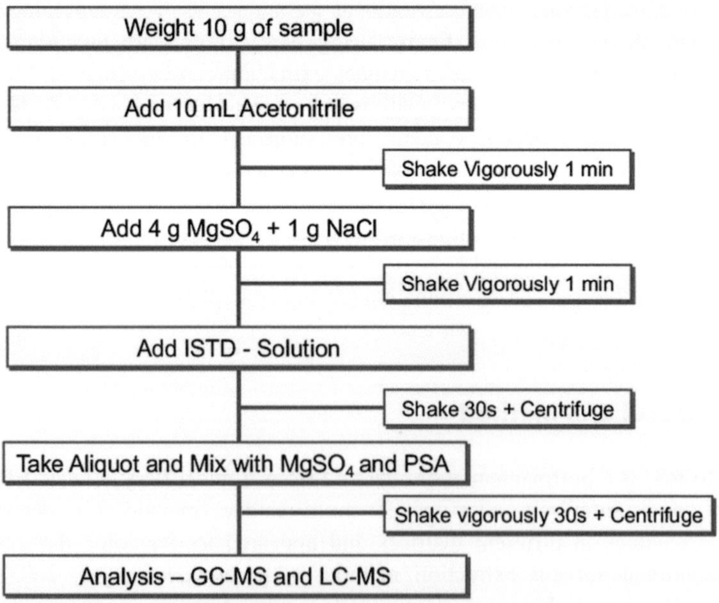

Fig. 10.10 Steps of QuEChERS method for pesticide residue analysis

standards organizations. So the acetate-buffering version was labeled as AOAC Official Method 2007.01 (Lehotay 2007) and the citrate-buffering version being entitled as Standard EN 15662 method (www.cen.eu). The QuEChERS advantages are high recovery (>85%), very accurate results (an internal standard is used), low solvent and glassware usage, high sample throughput (10–20 samples analysis in about 30–40 min), less skill, labor and bench space, lower reagent costs, and ruggedness. The main drawback of this method is that the final extract must be concentrated to furnish the necessary sensitivity, i.e., to achieve the desired limits of quantification (LOQ). QuEChERS has been successfully used for determination of metaflumizone (Dong et al. 2009), oxadiargyl (Shi et al. 2010), and 38 pesticides (Yang et al. 2010) in soil samples. As a modified version, it was applied for OCPs (Rashid et al. 2010) and OPPs determination in soil samples (Asensio-Ramos et al. 2010).

10.3.10 Matrix Solid Phase Dispersion (MSPD)

Matrix solid phase dispersion (MSPD) is a new SPE-based extraction and clean-up technique developed for pesticide multi-residue analysis (Kristenson et al. 2006). The MSPD method is based on the homogenization of a viscous, solid, or semi-solid sample with an abrasive solid support material in a glass mortar, in order to perform the complete disruption and dispersal of the sample. After blending, the sample is transferred into a column and analytes are eluted with appropriate solvent. Complete disruption of the sample and its dispersion over the support surface greatly enhance surface area for the sample extraction. Furthermore, interferences are retained on the adsorbent and in that way, extraction and clean-up are performed simultaneously, reducing the analysis time and the amount of solvent used (Barker 2000; Kristenson et al. 2006).

Reversed-phase materials such as C8 and C18-bonded silica are the most commonly used adsorbents, because their lipophilic properties enable good disruption, dispersion, and retention of lipophilic species (Lambropoulou and Albanis 2007). Basically, the adsorbent choice depends on analyte polarity and interferences which could be co-extracted from sample matrix (Fig. 10.11). Also, the nature of the elution solvent is crucial for efficient pesticides elution from the adsorbent (Albero et al. 2003; Blasco et al. 2002a, b; Bogialli et al. 2004). The original MSPD can be modified or combined with other extraction methodologies to improve the extraction yields or simplify the MSPD procedures. The schematic procedure of the original and representative modification of MSPD is shown in Fig. 10.11 (Tu and Chen, 2018).

In comparison to traditional extraction methods, MSPD approach has several advantages, including simplified and faster sample-treatment, reduced use of toxic solvents, eliminated emulsion formation, and increased selectivity and sensitivity. In MSPD, the sample extraction and clean-up are performed in the same step by use of small amounts of adsorbent and solvent, thus reducing the cost and analysis time. As a drawback, a number of applications still use large volumes of solvents

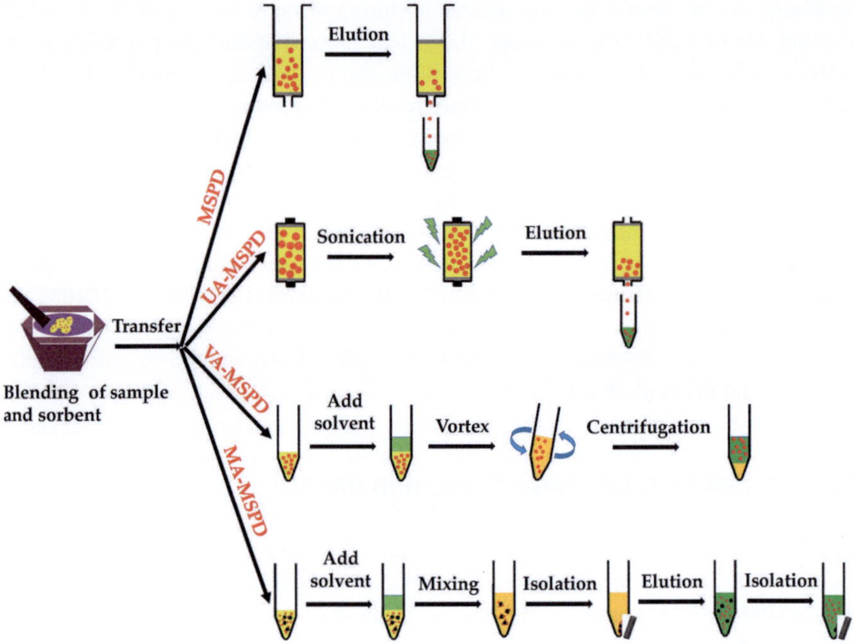

Fig. 10.11 Schematic procedure of original matrix solid phase dispersion (MSPD), ultrasonic-assisted MSPD (UA-MSPD), vortex-assisted MSPD (VA-MSPD), and magnetically-assisted MSPD (MA-MSPD)

for extraction and clean-up, which requires solvent evaporation. There is a very reason to believe that solving of this problem will make MSPD more useful in the near future. It has been successfully applied for phenthoate (Li et al. 2002), OCPs (Shen et al. 2005, 2006), and five pesticides in soil (Shen et al. 2007).

10.4 Conclusion

The sample extraction step, the most time determining step, is still the weakest link in the whole analytical procedure and also the prime cause of experimental errors and disparity between laboratories. However, in the recent times, upgradation in the existing techniques and also development of new techniques have unfolded new horizons in the sample preparation techniques in terms of saving time and reducing use of chemicals and thus undoubtedly improved the overall performance of analytical process. As a result of advancement of modern science and technology, several rapid, low cost, environmentally friendly, and readily automated methods of extraction are now available. Besides, because of the complexity of the matrices, extraction is usually followed by very specific clean-up procedures to achieve accurate sample quantification, so the new methods are modified in order to achieve a compromise between cost, selectivity, and sensitivity. Reduced solvent methods, including supercritical fluid extraction (SFE), solid phase extraction (SPE), solid phase micro-extraction (SPME), microwave assisted extraction (MAE), accelerated solvent extraction (ASE), QuEChERS and matrix solid phase dispersion (MSPD) have grown in their maturity, which increased application of these techniques in pesticide analysis of soil matrices. Although the composition of soil matrix varies from place to place, which requires application of different approaches and strategies, the development of a uniform procedure is highly encouraged. Future developments in all areas of analytical sample preparation are expected to continue to be application-driven in a quest for improved recovery, higher sample throughput, and reduced consumption of organic solvent with capability to provide accurate results.

References

Albero B, Sánchez-Brunete C, Tadeo JL (2003) Determination of endosulfan isomers and endosulfan sulfate in tomato juice by matrix solid-phase dispersion and gas chromatography. J Chromatogr A 1007(1–2):137–143

Anastassiades M, Lehotay SJ, Štajnbaher D, Schenck FJ (2003) Fast and easy multiresidue method employing acetonitrile extraction/partitioning and "dispersive solid-phase extraction" for the determination of pesticide residues in produce. J AOAC Int 86(2):412–431

Anastassiades M, Scherbaum E, Tasdelen B, Štajnbaher D (2007) Recent developments in QuEChERS methodology for pesticide multiresidue analysis. In: Ohkawa H, Miyagawa H, Lee PW (eds) Pesticide chemistry, crop protection, public health. Wiley-VCH, Weinheim, pp 439–458

Arthur CL, Killam LM, Buchholz KD, Pawliszyn J, Berg JR (1992) Automation and optimization of solid-phase microextraction. Anal Chem 64(17):1960–1966

Asensio-Ramos M, Hernández-Borges J, Ravelo-Pérez LM, Rodríguez-Delgado MA (2010) Evaluation of a modified QuEChERS method for the extraction of pesticides from agricultural, ornamental and Forestal soils. Anal Bioanal Chem 396(6):2307–2319

Atkins P, de Paula J (2002) Atkins` physical chemistry, 7th edn. Oxford University Press, New York, p 18

Barker SA (2000) Applications of matrix solid-phase dispersion in food analysis. J Chromatogr A 880(1–2):63–68

Blasco C, Picó Y, Mañes J, Font G (2002a) Determination of fungicide residues in fruits and vegetables by liquid chromatography–atmospheric pressure chemical ionization mass spectrometry. J Chromatogr A 947(1–2):227–235

Blasco C, Font G, Picó Y (2002b) Comparison of microextraction procedures to determine pesticides in oranges by liquid chromatography–mass spectrometry. J Chromatogr A 970(1–2):201–212

Bogialli S, Curini R, Di Corcia A, Nazzari M, Tamburro D (2004) A simple and rapid assay for analyzing residues of carbamate insecticides in vegetables and fruits: hot water extraction followed by liquid chromatography-mass spectrometry. J Agric Food Chem 52(4):665–671

Brewer BN, Armbrust KL, Mead KT, Holmes WE (2004) Determination of abamectin in soil samples using high-performance liquid chromatography with tandem mass spectrometry. Rapid Commun Mass Spectrom 18(15):1693–1696

Caddick S (1995) Microwave assisted organic reactions. Tetrahedron 51(38):10403–10432

Camel V (2001) Recent extraction techniques for solid matrices-supercritical fluid extraction, pressurized fluid extraction and microwave-assisted extraction: their potential and pitfalls. Analyst 126:1182–1193

Cervera MI, Medina C, Portolés T, Pitarch E, Beltrán J, Serrahima E, Pineda L, Muñoz G, Centrich F, Hernández F (2010) Multi-residue determination of 130 multiclass pesticides in fruits and vegetables by gas chromatography coupled to triple quadrupole tandem mass spectrometry. Anal Bioanal Chem 397(7):2873–2891

Chen L, Zhao Q, Xu Y, Sun L, Zeng Q, Xu H et al (2010) A green method using micellar system for determination of sulfonamides in soil. Talanta 82:1186–1192

Codex (2019) Pesticide residues in food and feeds. Codex pesticides residues in food online database. Available online: http://www.fao.org/fao-who-codexalimentarius/codex-texts/list-standards/en

Cueva-Mestanza R, Sosa-Ferrera Z, Torres-Padrón ME, Santana-Rodríguez JJ (2008) Preconcentration of pharmaceuticals residues in sediment samples using microwave assisted micellar extraction coupled with solid phase extraction and their determination by HPLC-UV. J Chromatogr B 863:150–157

Damalas CA (2009) Understanding benefits and risks of pesticide use. Sci Res Essays 4:945–949

De Andréa MM, Papini S, Nakagawa LE (2001) Optimizing microwave-assisted solvent extraction (MASE) of pesticides from soil. J Environ Sci Health B 36(1):87–93

Dean JR (1998) Extraction methods for environmental analysis. John Wiley & Sons, West Sussex

Dong F, Liu X, Cheng L, Chen W, Li L, Qin D, Zheng Y (2009) Determination of Metaflumizone residues in cabbage and soil using ultra-performance liquid chromatography/ESI-MS/MS. J Sep Sci 32(21):3692–3697

Doong RA, Liao PL (2001) Determination of organochlorine pesticides and their metabolites in soil samples using headspace solid-phase microextraction. J Chromatogr A 918(1):177–188

Đurović R (2011) Procesi koji određuju sudbinu pesticida u zemljištu. Pesticidi i fitomedicina 26(1):9–22

Đurović R, Đorđević T (2011) Modern extraction techniques for pesticide residues determination in plant and soil samples. In: Pesticides in the modern world-trends in pesticides analysis, Margarita Stoytcheva. Intech Open. https://doi.org/10.5772/17312

Đurović R, Marković M (2005) Solid phase microextraction in the analysis of Vinclozolin and Procymidone residues in Strawberries. Pesticidi i fitomedicina 20(3):163–169

Đurović R, Marković M, Marković D (2007a) Headspace solid phase microextraction in the analysis of pesticide residues—kinetics and quantification prior to the attainment of partition equilibrium. J Serb Chem Soc 72(8–9):879–887

Đurović R, Milinović J, Cupać S, Marković M (2007b). The book of abstracts, Euroanalysis XIV, p. 75, Antwerp, Belgium, September 9–14, 2007

Đurović R, Milinović J, Marković M, Marković D (2007c) Headspace solid phase Microextraction in pesticide residues analysis: 2. Apple samples. Pesticidi i fitomedicina 22(2):173–176

Đurović R, Gajić Umiljendić J, Đorđević T (2008) Determination of atrazine, acetochlor, clomazone, pendimethalin and oxyfluorfen in soil by a solid phase microextraction method. Pesticidi i fitomedicina 23(4):265–271

Đurović RD, Đorđević TM, Šantrić LR, Gašić SM, Ignjatović LM (2010a) Headspace solid phase microextraction method for determination of triazine and organophosphorus pesticides in soil. J Environ Sci Health B 45(7):626–632

Đurović RD, Gajić Umiljendić JS, Cupać SB, Ignjatović LM (2010b) Solid phase microextraction as an efficient method for characterization of the interaction of pesticides with different soil types. J Brazil Chem Soc 21(6):985–994

Ericsson M, Colmsjö A (2000) Dynamic microwave-assisted extraction. J Chromatogr A 877(1–2):141–151

Erwin WR, Schiller D, Hoskins WM (1955) Pesticide residue analysis, preassay purification of tissue extracts by wax column. J Agric Food Chem 3(8):676–679

EU, MRL (2019) European Commission (EU) legislation on maximum residue levels. Available online: https://ec.europa.eu/food/plant/pesticides/max_residue_levels/eu_rules/mrls_2019_en

Fernandez-Alvarez M, Llompart M, Pablo Lamas J, Lores M, Garsia-Jares C, Cela R, Dagnac T (2008) Simultaneous determination of traces of pyrethroids, organochlorines and other main plant protection agents in agricultural soils by headspace solid-phase microextraction–gas chromatography. J Chromatogr A 1188(2):154–163

Frost SP, Dean JR, Evans KP, Harradine K, Cary C, Comber MHI (1997) Extraction of hexaconazole from weathered soils: a comparison between soxhlet extraction, microwave-assisted extraction, supercritical fluid extraction and accelerated solvent extraction. Analyst 122(9):895–898

Fuentes E, Báez ME, Labra R (2007) Parameters affecting microwave-assisted extraction of organophosphorus pesticides from agricultural soil. J Chromatogr A 1169(1–2):40–46

Ganzler K, Salgó A, Valkó K (1986) Microwave extraction: a novel sample preparation method for chromatography. J Chromatogr 371:299–306

Handa SK, Agnihotri NP, Kulshreshtha G (1999) Multiresidue methods for estimation of pesticide residues pesticide residues: significance, management and analysis. Research Periodicals and Book Publishing House, New Delhi, pp 136–145

Herbert P, Morais S, Paíga P, Alves A, Santos L (2006) Development and validation of a novel method for the analysis of chlorinated pesticides in soils using microwave-assisted extraction–headspace solid phase microextraction and gas chromatography-tandem mass spectrometry. Anal Bioanal Chem 384(3):810–816

Hoogerbrugge R, Molins C, Baumann RA (1997) Effects of parameters on microwave assisted extraction of Triazines from soil: evaluation of an optimisation trajectory. Anal Chim Acta 348(1–3):247–253

Kaihara A, Yoshii K, Tsumura Y, Ishimitsu S, Tonogai Y (2002) Multi-residue analysis of 18 pesticides in fresh fruits, vegetables and rice by supercritical fluid extraction and liquid chromatography-electrospray ionization mass spectrometry. J Health Sci 48(2):173–178

Khan F, Prakash D, Jain RK (2011) Development of an HPLC method for determination of pentachloronitrobenzene, hexachlorobenzene and their possible metabolites. BMC Chem Biol 11(1):2

Kristenson EM, Brinkman UAT, Ramos L (2006) Recent advances in matrix solid-phase dispersion. Trends Anal Chem 25(2):96–111

Lambropoulou DA, Albanis TA (2007) Methods of sample preparation for determination of pesticide residues in food matrices by chromatography-mass spectrometry-based techniques: a review. Anal Bioanal Chem 389(6):1663–1683

Langenfeld JJ, Hawthorne SB, Miller DJ, Pawliszyn J (1994) Role of modifiers for analytical-scale supercritical fluid extraction of environmental samples. Anal Chem 66(4):909–916

Lehnik-Habrink P, Hein S, Win T, Bremser W, Nehls I (2010) Multi-residue analysis of PAH, PCB, and OCP optimized for organic matter of forest soil. J Soils Sediments 10(8):487–1498

Lehotay SJ (2007) Determination of pesticide residues in foods by acetonitrile extraction and partitioning with magnesium sulfate: collaborative study. J AOAC Int 90(2):485–520

Li ZY, Zhang ZC, Zhou QL, Gao RY, Wang QS (2002) Fast and precise determination of phenthoate and its enantiomeric ratio in soil by the matrix solid-phase dispersion method and liquid chromatography. J Chromatogr A 977(1):17–25

Lopez-Avila V, Young R, Beckert WF (1994) Microwave-assisted extraction of organic compounds from standard reference soils and sediments. Anal Chem 66(7):1097–1106

Luo L, Shao B, Zhang J (2010) Pressurized liquid extraction and cleanup procedure for the determination of Pyrethroids in soils using gas chromatography/tandem mass spectrometry. Anal Sci 26(4):461–465

Luque-García JL, Luque de Castro MD (2003) Where is microwave-based analytical equipment for solid sample pre-treatment going. Trends Anal Chem 22:90–98

Luque-García JL, Luque de Castro MD (2004) Focused microwave-assisted Soxhlet extraction: devices and applications. Talanta 64:571–577

Marković M, Cupać S, Đurović R, Milinović J, Kljajić P (2010) Assessment of heavy metal and pesticide levels in soil and plant products from agricultural area of Belgrade, Serbia. Arch Environ Contam Toxicol 58(2):341–351

Moors M, Massart DL, Mcdowall RD (1994) Analyte isolation by solid phase extraction (SPE) on silica-bonded phases: classification and recommended practices. Pure Appl Chem 66:277

Odukkathil G, Vasudevan N (2013) Toxicity and bioremediation of pesticides in agricultural soil. Rev Environ Sci Biotechnol 12:421–444

Ono Y, Yamagami T, Nishina T, Tobino T (2006) Pesticide multiresidue analysis of 303 compounds using supercritical fluid extraction. Anal Sci 22(11):1473–1476

Pawliszyn J (1997) Solid phase microextraction – theory and practice. Wiley, New York

Pino V, Ayala JH, González V, Afonso AM (2007) Focused microwave-assisted micellar extraction combined with solid-phase microextraction gas chromatography/mass spectrometry to determine chlorophenols in wood samples. Anal Chim Acta 582:10–18

Pinto JSS, Lanças FM (2009) Design, construction and evaluation of a simple pressurized solvent extraction system. J Brazil Chem Soc 20(5):913–917

Poole SK, Dean TA, Oudsema JW, Poole CF (1990) Sample preparation for chromatographic separations: an overview. Anal Chim Acta 236:3–42

Ranz A, Maier E, Motter H, Lankmayr E (2008) Extraction and derivatization of polar herbicides for GC-MS analyses. J Sep Sci 31(16–17):3021–3029

Rashid A, Nawaz S, Barker H, Ahmad I, Ashraf M (2010) Development of a simple extraction and clean-up procedure for determination of organochlorine pesticides in soil using gas chromatography-tandem mass spectrometry. J Chromatogr A 1217(17):2933–2939

Richter BE, Jones BA, Ezzell JL, Porter NL, Avdalovic N, Pohl C (1996) Accelerated solvent extraction: a technique for sample preparation. Anal Chem 68(6):1033–1039

Rissato SR, Galhiane MS, Apon B, Arruda M (2005a) Multiresidue analysis of pesticides in soil by supercritical fluid extraction/gas chromatography with electron-capture detection and confirmation by gas chromatography−mass spectrometry. J Agric Food Chem 53(1):62–69

Rissato SR, Galhiane MS, De Souza AG, Apon BM (2005b) Development of a supercritical fluid extraction method for simultaneous determination of organophosphorus, organohalogen, organonitrogen and pyrethroids pesticides in fruit and vegetables and its comparison with a conventional method by GC-ECD and GC-MS. J Brazil Chem Soc 16(5):1038–1047

Sakamoto M, Tsutsumi T (2004) Applicability of headspace solid-phase microextraction to the determination of multi-class pesticides in waters. J ChromatogrA 1028(1):63–74

Sánchez-Brunete C, Rodriguez A, Tadeo JL (2003) Multiresidue analysis of carbamate pesticides in soil by sonication-assisted extraction in small columns and liquid chromatography. J Chromatogr A 1007:85

Santalad A, Zhou L, Shang F, Fitzpatrick D, Burakham R, Srijaranai S, Glennon JD, Luong JHT (2010) Micellar electrokinetic chromatography with amperometric detection and off-line solid-phase extraction for analysis of carbamate insecticides. J Chromatogr A 1217(32):5288–5297

Sharma KK (2007) Pesticide residue analysis. In: Pesticide residue analysis manual, 1st edn, pp 8–10

Shen Z, Cai J, Gao Y, Zhu X, Su Q (2005) A new matrix solid phase dispersion-accelerate solvent extraction-gas chromatographic method for determination of organochlorine pesticides residues in soil. Chin J Anal Chem 33(9):1318–1320

Shen X, Cai J, Gao Y, Su Q (2006) Determination of organophosphorus pesticides in soil by MMSPD-GC-NPD and confirmation by GC-MS. Chromatographia 64(1–2):71–77

Shen X, Su Q, Zhu X, Gao Y (2007) Determination of pesticide residues in soil by modified matrix solid-phase dispersion and gas chromatography. Ann Chim 97(8):647–654

Shi C, Gui W, Chen J, Zhu G (2010) Determination of oxadiargyl residues in environmental samples and rice samples. Bull Environ Contam Toxicol 84(2):236–239

Simplício A, Boas L (1999) Validation of a solid-phase microextraction method for the determination of organophosphorus pesticides in fruits and fruit juice. J Chromatogr A 833(1):35–42

Singh B (2000) Significance of clean-up in residue analysis. In: Yadav PR, Kathpal TS, Rohilla HR (eds) Pesticide residue analysis. Department of Entomology, CCSHAU, Hisar, pp 22–33

Smith RM (2002) Extractions with superheated water. J Chromatogr A 975(1):31–46

Stout SJ, Da Cunha AR, Safarpour MM (1997) Simplified determination of Imidazolinone herbicides in soil at parts-per billion level by liquid chromatography electrospray ionization tandem mass spectrometry. J AOAC Int 80(2):426–432

Stout SJ, Babbitt BW, Da Cunha AR, Safarpour MM (1998) Microwave-assisted extraction coupled with gas chromatography with nitrogen-phosphorus detection or electron capture negative chemical ionization mass spectrometry for determination of dimethomorph residues in soil. J AOAC Int 81(5):1054–1059

Sun L, Lee HK (2003) Optimization of microwave-assisted extraction and supercritical fluid extraction of carbamate pesticides in soil by experimental design methodology. J Chromatogr A 1014(1–2):165–177

Tao S, Guo LQ, Wang XJ, Liu WX, Ju TZ, Dawson R, Cao J, Xu FL, Li BG (2004) Use of sequential ASE extraction to evaluate the bioavailability of DDT and its metabolites to wheat roots in soils with various organic carbon contents. Sci Total Environ 320(1):1–9

Tu X, Chen W (2018) A review on the recent progress in matrix solid phase dispersion. Molecules 23(11):2767

Vega Moreno D, Sosa Ferrera Z, Santana Rodriguez JJ (2006) Sample extraction method combining micellar extraction–SPME and HPLC for the determination of organochlorine pesticides in agricultural soils. J Agric Food Chem 54(20):7747–7752

Wang W, Meng B, Lu X, Liu Y, Tao S (2007) Extraction of polycyclic aromatic hydrocarbons and organochlorine pesticides from soils: a comparison between Soxhlet extraction, microwave-assisted extraction and accelerated solvent extraction techniques. Anal Chim Acta 602(2):211–222

Wang H, Ding J, Ren N (2016) Recent advances in microwave-assisted extraction of trace organic pollutants from food and environmental samples. Trends Anal Chem 75:197–208

Yang XB, Ying GG, Kookana RS (2010) Rapid multiresidue determination for currently used pesticides in agricultural drainage waters and soils using gas chromatography-mass spectrometry. J Environ Sci Health B 45(2):152–161

Zhang JS, Pan FD, Cheng H (2010) Determination of organophosphorus pesticide in soil by accelerated solvent extraction-gas chromagraphy/mass spectrometry 2010 4th International conference on bioinformatics and biomedical engineering (iCBBE 2010), art. no. 5517391

Zhang HF, Yang XH, Wang Y (2011) Microwave assisted extraction of secondary metabolites from plants: current status and future directions. Trends Food Sci Tech 22:672–688

Zhao R, Wang X, Fu S, Yuan J, Jiang T, Xu X (2006) A novel headspace solid-phase microextraction method for the exact determination of Organochlorine pesticides in environmental soil samples. Anal Bioanal Chem 384(7–8):1584–1589

Characterization of Nanomaterials Using Different Techniques

11

Ritika Joshi, Ashish Khandelwal, Manoj Shrivastava, and S. D. Singh

Abstract

The reduction in size through top-down and bottom-up approaches causes difference in the physical, chemical, and biological properties as compared to original particle.

Nanomaterial can be identified through several instrument techniques such as dynamic light scattering (DLS), scanning electron microscopy (SEM), transmission electron microscopy (TEM), atomic force microscopy (AFM), X-ray diffraction (XRD), thermo gravimetric analysis (TGA), synchrotron radiation (SR) based techniques. The dispersion of nanoparticles in solution either in pure form or in agglomeration state, purity of nanoparticles, shape and size, speciation, and localization of nanoparticles in plants or tissue part is an important characteristic to determine the fate of applied nanoparticles in soil-plant-atmosphere continuum and living system. Hence, identification and characterization of morphology and property of nanoparticle through several techniques is an important feature to utilize nanomaterials in the material science.

Keywords

Nanomaterial · DLS · SEM · TEM · AFM · XRD · TGA · SR

R. Joshi
Centre for Environment Science and Climate Resilient Agriculture, ICAR-Indian Agricultural Research Institute, New Delhi, India

Department of Soil Science, Punjab Agricultural University, Ludhiana, India

A. Khandelwal (✉) · M. Shrivastava · S. D. Singh
Centre for Environment Science and Climate Resilient Agriculture, ICAR-Indian Agricultural Research Institute, New Delhi, India

A. Rakshit et al. (eds.), *Soil Analysis: Recent Trends and Applications*,
https://doi.org/10.1007/978-981-15-2039-6_11

11.1 Introduction

The materials having size of 1–100 nm in at least one dimension are termed as nanoparticles. Development of nanoparticles is widening its application in different fields such as agriculture, health care, and several industries product. Its application is increasing due to its high reactivity, low dose requirement for application, higher efficacy due to reduction in size, and large surface to mass ratio. Development of several nanoparticles and their application enhances utility of nanomaterials in different areas. Development of nanoparticles of specific size and shape is an essential feature particularly in the field of agriculture. In most cases, particle size <20 nm, polydispersity index <1, zeta potential value from +30 mV to −30 mV, and cube-shaped nanoparticles is appropriate to enter through the plant pores. Nanomaterials can be synthesized through top-down or bottom-up approach. Several metal nanoparticles, metal oxide nanoparticles, nanocomposites, nano-ranged amphiphilic polymer, nano gel, and nano whiskers have been synthesized due to their catalytic, optical, electronic, magnetic, antimicrobial, pesticidal, and higher absorption ability (Hong et al. 2017; Khan et al. 2017; Mourdikoudis et al. 2018; Sun et al. 2014; Adak et al. 2012; Dubes et al. 2003). Several instrumental techniques for different parameters can be utilized (Table 11.1).

This chapter discusses about brief introduction, method of sample preparation, observation/characterization/interpretation of nanoparticles through commonly used instrumental techniques such as dynamic light scattering, scanning electron microscopy, transmission electron microscopy, atomic force microscope, X-ray diffraction, thermo gravimetric analysis, and synchrotron based techniques.

Table 11.1 Identification of different parameters of nanoparticles through instrumental techniques

S. no.	Parameters	Techniques
1.	Size	SEM, TEM, XRD, DLS, HRTEM, AFM,
2.	Shape	TEM, HRTEM, AFM, 3D-tomography
3.	Elemental-chemical composition	XRD, XPS, SEM-EDX, NMR
4.	Crystal structure	XRD, HRTEM, STEM
5.	Size distribution	DCS, DLS, SEM
6.	Ligand binding, surface composition	XPS, FTIR, TGA
7.	Surface area, specific surface area	BET
8.	Surface charge	Zeta potential
9.	Agglomeration state	Zeta potential, DLS, SEM, TEM
10.	Dispersion of NP in matrices/supports	SEM, TEM, AFM
11.	Optical properties	UV-Vis-NIR, PL
12.	Magnetic properties	SQUID

11.2 Instrumentation Techniques

11.2.1 Dynamic Light Scattering (DLS)

It is a noninvasive, non-destructive, low cost technique, and its operation is relatively simple and rapid. DLS measurement is based on the Brownian motion of particles. The particles are constantly colliding with solvent molecules and certain amount of energy transferred to the particles, which induces particle movement. It is used for determination of particle size of solid particles, polymers, emulsions, proteins in colloidal suspensions, and state of aggregation in suspension. The relation between the speed of the particles and the particle size is given by the Stokes–Einstein equation, which can be used for determination of hydrodynamic diameter of the particle.

$$D = k_B T / 6\pi\eta R_H$$

where D = Translational diffusion coefficient [m^2/s]"speed of the particles," k_B = Boltzmann constant [m^2 kg/Ks^2], T = Temperature [K], η = Viscosity [Pa s], R_H = Hydrodynamic radius [m].

Sample Preparation

1. A known quantity of sample was dissolved in distilled water (200 mg L^{-1} solution).
2. A certain amount (5 mL) of solution was taken into a glass vial.
3. Minimum quantity (50 μL) of chloroform was added to the polymer solution.
4. The vial was sonicated for 5 min.
5. DLS measurements were performed at 25 °C and light scattering was detected at a fixed angle.

Result/Observation The observation of different nanoparticle/nanopolymers is presented in Table 11.2.

Table 11.2 DLS observation of different nanoparticle/nanopolymer

S. no.	Nanoparticles	Size (nm)	Reference
1.	Imidacloprid	127.5–354	Adak et al. (2012)
2.	Thiamethoxam	51.6–206.7	Sarkar et al. (2014)
3.	Nanotized curcumin	20–50	Ghosh et al. (2014)
4.	Potassium grafted chitosan-poly(methacrylic acid)	368.1	Plofino et al. (2019)
5.	Rock phosphate	<100 nm	Bhattacharjya et al. (2019)
6.	ZnO	621	Rossi et al. (2019)

11.2.2 Scanning Electron Microscopy (SEM)

It is based on electron scanning principle. It is used to determine shape, morphology, and dispersion of nanoparticles in the bulk or matrix.

Sample Preparation
1. The dried powder was placed on carbon tape.
2. It is coated with gold palladium.
3. The images were performed at certain voltage and pressure after palladium coating at different magnifications.

Result/Observation The observation and their interpretation from different nanomaterials are presented in Table 11.3.

11.2.3 Transmission Electron Microscopy (TEM)

It is a microscopy technique, where image is formed after passing beam of an electron through ultra-thin specimen. It mainly works on the electron transmission principle. It can be used to determine size, shape, separation of nanoparticles by equal distance, presence of secondary material capping, and information about two or more layer materials and penetration of nanoparticles into slime layer can also be predicted.

Table 11.3 SEM observation of different nanomaterials

S. no.	Nanoparticles	Observation/interpretation	Reference
1.	ZnO	1. Spherical, spongy, and irregular shapes 2. Particle size of green-synthesized method is comparatively smaller than the chemical-synthesized methods	Hassan et al. (2018)
2.	Organic acid loaded nano clay polymer composites	1. Irregular and rough surface morphology 2. Presence of pores at higher magnification	Roy et al. (2016)
3.	Zincated nanoclay polymer composites	1. Fracture morphology with specific topography 2. Surface roughness increased with increased clay content 3. Absence of nonhomogeneous scattering from clay aggregates confirms exfoliation nature of composites	Mandal et al. (2018)
4.	Bacitracin A	Nano Bacitracin A applied on *E. coli* showed significant rupture of cell membrane followed by viscidity of protoplasm, whereas untreated and treatment with bacitracin A showed smooth surface with no significant cell membrane damage	Hong et al. (2017)

Sample Preparation

1. The samples were collected from control and treated samples.
2. The drop from the aqueous solution of sample was casted on a carbon-coated copper grid and allowed to be air-dried.
3. In general, samples were fixed in paraformaldehyde (2%), glutaraldehyde (2%) sodium cacodylate buffer (100 mM, pH = 7.35).
4. Each sample was centrifuged and the resulting pellets were resuspended in gel.
5. Fixed pellets were then washed with 100 mM of sodium cacodylate buffer (100 mM, pH = 7.35) and of sucrose (130 mM).
6. Secondary fixation was performed using osmium tetroxide (1%) in 2-ME buffer
7. Then, specimens were incubated (4 °C, 1 h), rinsed with cacodylate buffer, and further washed with distilled water.
8. Staining performed using 1% aqueous uranyl acetate (incubation at 4 °C overnight), then rinsing with distilled water.
9. The graded dehydration series was performed using ethanol, transitioned into acetone, and dehydrated specimens were then infiltrated with Epon resin (250 W for 3 min) and polymerized at 60 °C overnight.
10. Sections were cut (~85 nm thickness) and TEM images were observed at certain voltage and certain magnifications.
11. TEM can also be equipped with EDS detector to confirm the elemental composition of the materials.

Note: The above procedure is not universal and it can be varied according to sample type and instrument conditions.

Result/Observation The images observed from TEM can be used to interpret the following information and presented in Table 11.4.

11.2.4 Atomic Force Microscopy (AFM)

It is a microscopy technique, developed by Gerd Binnig and Heinrich Rohrer at IBM in 1986. It is based on measuring interacting forces between fine probe and the samples. When AFM scans the samples, cantilever gets reflected and bending is quantified by laser beam that reflects on the cantilever backside. AFM can scan sample in contact, non-contact, and oscillating mode, which depends upon proximity between the probe and samples. It is mainly used for analyzing several properties of nanoparticles such as topography, elasticity, adhesion, friction, electrical properties, and magnetism and is also useful for determining particle size distributions as well as image complex arrays of nanoparticles.

Sample Preparation

1. The aqueous dispersion of the nanoparticles was put on a glass coverslip.
2. The coverslip was air-dried at room temperature.

Table 11.4 TEM observation of different nanomaterials

S. no.	Nanoparticle	Shape/observation	Size (nm)	Reference
1.	ZnO	Spherical	12–48	Hassan et al. (2018)
2.	Bacitracin A	Nano Bacitracin A solution induced a significant decrease in the integrity of membrane of *E. coli* as compared to Bacitracin A solution (smooth surface and dense internal structure)	88.9–122.3	Hong et al. (2017)
3.	Gold nanowires	Rigid network structure	50–60	He et al. (2008)
4.	HSMV 1	Bullet-shaped magnetosomes	40–113	Lefevre et al. (2010)
5.	Au	Spherical	75	Liu et al. (2015)
6.	Nano imidacloprid encapsulated amphiphilic polymer	Encapsulation of imidacloprid into spherical-shaped polymer	127.5–354	Adak et al. (2012)
7.	Nano thiamethoxam encapsulated amphiphilic polymer	Spherical shape of the polymeric micellar system and size of micelle increased due to encapsulation of thiamethoxam in nano-ranged amphiphilic PEG and diacid based block polymers	51.6–206.7	Sarkar et al. (2012)
8.	Nano clay polymer composites	Platy morphology, broader dimensions and thickness of clay composites	5–10	Sarkar et al. (2014)
9.	Nano phoshphatic fertilizer	Particle size and spherical alveoli shape of different particles	–	Sarkar et al. (2018)
10.	Zincated nanoclay polymer composites	Fraction of nano bentonite ($<$100 nm)	–	Mandal et al. (2015)

3. After drying, samples were analyzed using the Nanoscope III scanning probe microscope.

Note: In this technique, surface modification or coating prior to imaging is not required. Hence, topographical analysis of small NPs ($\leq$6 nm) is possible.

Result/Observation The observation of nanoparticles through AFM is presented in Table 11.5.

Table 11.5 AFM observation of different nanoparticles

S. no.	Nanoparticle	Observation/interpretation	Reference
1.	Polystyrene	93.5–104.7 nm	Hoo et al. (2008)
2.	Solid lipid nanoparticle	Circular in shape	Dubes et al. (2003)

Table 11.6 XRD observation/Interpretation of different nanomaterials

S. no.	Nanoparticles	Observation/interpretation	Reference
1.	ZnO	1. Crystalline wurtzite ZnO (hexagonal structure) 2. Nanoparticles prepared from green method (7.1–13.96 nm) are smaller in size as compared to chemical methods (19.6 nm).	Hassan et al. (2018)
2.	Copper telluride	XRD peaks vary according to different shapes cubes, plates, and rods	Li et al. (2013)
3.	Nanoclay–polymer composites	Size of nano clay varied from 6.9 to 16.3 nm and dominated by kaolinite ($2\theta = 12.4°$), mica ($2\theta = 8.8°$), and smectite ($2\theta = 5°$)	Sarkar et al. (2014)
4.	Zincated nanoclay polymer composites	Absence of montmorillonite peak $2\theta = 6°$ in polymer composites indicates that bentonite clearly dispersed in polymer matrixes which showed it as an exfoliated type of composites	Mandal et al. (2018)

11.2.5 X-Ray Diffraction (XRD)

It is most extensively used technique used for characterization of nanoparticles. It provides information about crystalline structure, nature of the phase (single and multiphase), lattice parameters, and particle size (through Debye Scherrer formula). It can be used for solid as well as liquid samples. The compositions of the particles can be determined by comparing the positions and intensity of the peaks with the reference patterns.

The mean crystalline size (D) of the particles was determined from the XRD line broadening measurement using Scherrer equation:

$$D = 0.89\lambda/(\beta \text{Cos}\theta)$$

where D = Mean size of crystallites (nm), K = Scherrer constant crystallite shape factor (0.9), λ = Wavelength of the X-ray source (Cu Kα = 0.1541 nm), B = Full width at half the maximum of the diffraction peak (FWHM) in radians of the X-ray diffraction peak, θ = Bragg (diffraction) angle.

Sample Preparation XRD instrument operated with Cu Kα radiation and patterns were generated at xkV and ymA with a fix scan rate different 2θ values.

Observation/Interpretation The result of XRD of different nanomaterials is presented in Table 11.6.

11.2.6 Thermo Gravimetric Analysis (TGA)

It provides information about mass and composition of materials. It is also used to determine type, purity, and quantity of organic ligand on nanomaterial.

Sample Preparation

1. Sample is heated in range of temperatures in thermo gravimetric analyzer.
2. Compound at different temperature degraded and vaporized, and a change in mass is recorded.

Result The observation and interpretation of different nanomaterials are presented in Table 11.7.

Table 11.7 TGA observation of different nanomaterials

S. no.	Nanoparticle	Observation	Interpretation	Reference
1.	ZnO	First degradation step: 25–256 °C	Evaporation of surface adsorbed water	Hassan et al. (2018)
		Second degradation step: 256–599 °C	Decomposition of the condensation dehydration of the hydroxyls	
	Result	Average sample weight loss <9.1% of the material weight		
2.	Calcinated green ZnO	First degradation step: 19–200 °C	Removal of surface waste adsorbed onto zinc oxide	
		Second degradation step: 200–598 °C	Evaporation of surface adsorbed water and dehydration of the hydroxyls	
	Result	Average sample weight loss <2.2% of the material weight (extreme purity)		
3.	Non-calcinated green ZnO paste	First degradation step: 38–93 °C	Removal of surface waste adsorbed onto zinc oxide	
		Second degradation step: 93–177 °C	Evaporation of surface adsorbed water	
		Third degradation step: 177–282 °C	Decomposition of the condensation dehydration of the hydroxyls	
		Fourth degradation step: 282–491 °C	Existence of organic material, in small amounts	
	Result	Average sample weight loss <51.6% of the material weight may be due to polyphenol or natural pigments		
4.	Urea-kaolinite Nanocomposite	First degradation step (48–425 °C) describes about moisture evaporation, decomposition of the gum Arabic biopolymer, and the elimination of interlayer moisture and water coordinated in the exchangeable cations and second degradation step (475–550 °C) describes about dehydroxylation of kaolinite		Sempeho et al. (2015)

11.2.7 Synchrotron Radiation (SR) Technique

It is a non-destructive technique with high sensitivity and spatial resolution. The synchrotron radiation-based techniques are mainly synchrotron micro-X-ray fluorescence (μ-XRF) and synchrotron X-ray absorption spectroscopy (XAS). It is widely used to understand the translocation, speciation, and localization as well as redox state of nanoparticles in plant system. The multi-elemental composition along with movement of nanoparticles such as Ti, Ce, Zn, Ag, Au, etc. within the tissue of agricultural crops can be determined (Shrivastava et al. 2019).

Sample Preparation

1. Hydrated/fresh plant samples
   a. The plant samples can be analyzed in their native state or in frozen hydrated state.
   b. Plant samples washed with deionized water to eliminate any surface contaminants.
   c. Then, samples are transversally cut and frozen in liquid nitrogen for 30 min.
   d. Samples are fixed with Tissue Tek and sectioned with a cryomicrotome at −20 °C to a thickness of 30 μm.
   e. Subsequently, samples are mounted onto Kapton tape and freeze-dried for 1 h in a freeze-dryer with operating conditions of −53 °C and 0.140 mBar pressure.
2. Dehydrated plant sample preparation,
   a. Plant samples were immersed in liquid nitrogen for 45 min and lyophilized in a freeze-dryer at −53 °C and 0.140 mBar pressure for 3 days.
   b. Samples were mortar homogenized, loaded in aluminum sample holders, and covered with Mylar Tape.
3. Soil Sample Preparation.
   a. Surface soil (0–20 cm) samples are homogenized, air-dried, and sieved through a stainless steel sieve with a mesh size of 250 μm.
   b. One gram soil sample treated four times with 20 mL, 0.7 M NaOCl (pH ~8.5) for 2 h at 90 °C.
   c. The soil samples are used for μ-XRF and μ-XANES analysis
   d. In addition, thin section of air-dried soil, embedded in epoxy resin can also be used for synchrotron μ-XRF and μ-XAS analysis.

Result/Observation The observation of different nanomaterials through SR techniques is presented in Table 11.8.

11.3 Conclusion

Nanomaterials have utmost importance due to their high reactivity, higher efficacy, and low dose requirement. Their versatile application enhances tremendous process for developing nano-based material in plant, animal, and human health sector.

Table 11.8 Observation of different nanomaterials through synchrotron radiation based techniques

S. no.	Nanoparticle	Observation	Reference
1.	ZnO	Vascular region of *Prosopis julifora*	Hernandez-Viezcas et al. (2011)
2	Ag	Roots of *Lolium multiflorum*	Yin et al. (2011)
3	TiO_2	In wheat crop, TiO_2 is localized inside the parenchyma and vasuclar tissues of root portion	Laure et al. (2011)
4	TiO_2	Preferential translocation in the leaf of cucumber plants and localized in root tissue	Servin et al. (2012)
5	CeO_2	Mainly absorbed in the cortex and vascular region of root in the form of CeO_2	Zhao et al. (2012)

Nanoparticles developed by different processes can be distinguished on the basis of size, shape, crystal structure, surface area, zeta potential, surface charge, agglomeration state, and optical as well as magnetic properties and their physical, chemical, and biological process were affected due to variation in their size and shape. Hence, it is very important to distinguish developed nanomaterials through characterization techniques. The characterized nanomaterials based on their properties can be used in different sectors and can be acted as boon for growth and development in suitable manner.

References

Adak T, Kumar J, Shakil NA, Walia S (2012) Development of controlled release formulations of imidacloprid employing novel nano-ranged amphiphilic polymers. J Environ Sci Health B 47(3):217–225

Bhattacharjya S, Adhikari T, Kundu S, Sahu A, Patra AK (2019) Evaluation of microbial solubilisation of Nano rock phosphate. Int J Curr Microbiol App Sci 8(1):1055–1069

Dubes AH, Parrot-Lopez P, Shahgaldian AW (2003) Coleman, interfacial interactions between amphiphilic cyclodextrins and physiologically relevant cations. Eur J Pharm Biopharm 55:279–282

Ghosh A, Banerjee T, Bhandary S, Surolia A (2014) Formulation of nanotized curcumin and demonstration of its antimalarial efficacy. Int J Nanomedicine 9:5373–5387

Hassan HS, Elkady MF, El-Sayed EM, Hamed AM, Hussein AM, Mahmoud IM (2018) Synthesis and characterization of zinc oxide nanoparticles using green and chemical synthesis techniques for phenol decontamination. Int J Nanoelectron Mater 11(2):179–194

He S, Zhang Y, Guo Z, Gu N (2008) Biological synthesis of gold nanowires using extract of *Rhodopseudomonas capsulata*. Biotechnol Prog Am Inst Chem Eng 24:476–480

Hernandez-Viezcas J, Castillo-Michel H, Servin A, Peralta-Videa J, Gardea-Torresdey J (2011) Spectroscopic verification of zinc absorption and distribution in the desert plant *Prosopis juliflora*-velutina (velvet mesquite) treated with ZnO nano-particles. Chem Eng J 170:346–352

Hong W, Gao X, Qiu P, Yang J, Qiao M, Hong S, Zhang D, Tian C, Niu S andliu M (2017) Synthesis, construction, and evaluation of self-assembled nano-bacitracin a as an efficient antibacterial agent in vitro and in vivo. Int J Nanomedicine 12:4691–4708

Hoo CM, Starostin N, West P, Mecartney ML (2008) A comparison of atomic force microscopy (AFM) and dynamic light scattering (DLS) methods to characterize nanoparticle size distributions. J Nanopart Res 10:89–96. https://doi.org/10.1007/s11051-008-9435-7

Khan I, Saeed K, Khan I (2017) Nanoparticles: properties, applications and toxicities. Arab J Chem 12(7):908–931. https://doi.org/10.1016/j.arabjc.2017.05.011

Laure C, Khodja H, Herlin-Boime N, Brisset F, Flank AM, Fayard B, Chaillou S, Carriere M (2011) Investigation of titanium dioxide nanoparticles toxicity and uptake by plants. J Phys Conf Ser 304:012057. https://doi.org/10.1088/1742-6596/304/1/012057.012057

Lefevre CT, Abreu F, Schmidt ML (2010) Moderately thermophilic magnetotactic bacteria from hot springs in Nevada. Appl Environ Microbiol 76(11):3740–3743

Li W, Zamani R, Gil PR, Pelaz B, Ibanez M, Cadavid D, Shavel A, Alvarez-Puebla RA, Parak WJ, Arbiol J, Cabot A (2013) Phosphorus release from rock phosphate as influenced by organic acid loaded nanoclay polymer composites in an Alfisol. J Am Chem Soc 135:7098

Liu D, Li C, Zhou F, Zhang T, Zhang H, Li X, Duan G, Cai W, Li Y (2015) Rapid synthesis of monodisperse Au nanospheres through a laser irradiation-induced shape conversion, self-assembly and their electromagnetic coupling SERS enhancement. Sci Rep 5:7686

Mandal N, Datta SC, Manjaiah KM (2015) Synthesis, characterization and controlled release study of Zn from zincated nanoclay polymer composites (ZNCPCs) in relation to equilibrium water absorbency under zinc deficient typicHaplustept. Ann Plant Soil Res 17(2):187–195

Mandal N, Datta SC, Manjaiah KM, Dwivedi BS, Kumar R, Aggarwal P (2018) Zincated Nanoclay polymer composites (ZNCPCs): synthesis, characterization, biodegradation and controlled release behaviour in soil. Polym-Plast Technol Eng 57:1760–1770. https://doi.org/10.1080/03602559.2017.1422268

Mourdikoudis S, Pallares RM, Thanh NTK (2018) Characterization techniques for nanoparticles: comparison and complementarity upon studying nanoparticle properties. Nanoscale 10:12871

Plofino DJM, Brutas CCP, Alfafara CG, Maguyon-Detras MC, Migo VP (2019) Parametric and optimization studies on the entrapment of potassium fertilizer into chitosan-poly (methacrylic acid) carrier via ionic gelation. MATEC Web Conf Sect Process Energy Environ 268:06009

Rossi L, Fedenia LN, Sharifan H, Ma X, Lombardini L (2019) Effects of foliar application of zinc sulfate and zinc nanoparticles in coffee (Coffe arabica *L.*) plants. Plant Physiol Biochem 135:160–166

Roy T, Biswas DR, Datta SC, Sarkar A (2016) Phosphorus release from rock phosphate as influenced by organic acid loaded nanoclay polymer composites in an Alfisol. Proc Natl Acad Sci India Sect B Biol Sci 88(1):128–132

Sarkar A, Biswas DR, Datta SC, Roy T, Moharana PC, Biswas SS, Ghosh A (2018) Polymer coated novel controlled release rock phosphate formulations for improving phosphorus use efficiency by wheat in an Inceptisol. Soil Tillage Res 180:48–62

Sarkar DJ, Kumar J, Shakil NA, Walia S (2012) Release kinetics of controlled release formulations of thiamethoxam employing nano-ranged amphiphilic PEG and diacid based block polymers in soil. J Environ Sci Health A 47(11):1701–1712

Sarkar S, Datta SC, Biswas DR (2014) Synthesis and characterization of nanoclay–polymer composites from soil clay with respect to their water-holding capacities and nutrient-release behavior. J Appl Polym Sci 131:39951

Sempeho SI, Kim HT, Mubofu E, Pogrebnoi A, Shao G, Hilonga A (2015) Encapsulated urea-kaolinite nanocomposite for controlled release fertilizer formulations. J Chem 2015:237397. https://doi.org/10.1155/2015/237397

Servin AD, Castillo-Michel H, Hernandez-Viezcas JA, Diaz BC, Peralta-Videa JR, Gardea-Torresdey JL (2012) Synchrotron micro-XRF and micro-XANES confirmation of the uptake and translocation of TiO_2 nanoparticles in cucumber (*Cucumis sativus*) plants. Environ Sci Technol 46:7637–7643

Shrivastava M, Srivastav A, Gandhi S, Rao S, Roychoudhury TA, Kumar A, Singhal RK, Jha SK, Singh SD (2019) Monitoring of engineered nanoparticles in soil-plant system: a review. Environ Nanotechnol Monit Manag 11:100218

Sun T, Zhang YS, Pang B, Hyun DC, Yang M, Xia Y (2014) Engineered nanoparticles for drug delivery in cancer therapy. Angew Chem Int Ed 53(46):12320–12364

Yin L, Cheng Y, Espinasse B, Colman BP, Auffan M, Wiesner M, Rose J, Liu J, Bernhardt ES (2011) More than the ions: the effects of silver nanoparticles on *Lolium multiflorum*. Environ Sci Technol 45(6):2360–2367

Zhao L, Peralta-Videa JR, Varela-Ramirez A, Castillo-Michel H, Li C, Zhang J, Aguilera RJ, Keller AA, Gardea-Torresdey JL (2012) Effect of surface coating and organic matter on the uptake of CeO_2 NPs by corn plants grown in soil: insight into the uptake mechanism. J Hazard Mater 225:131–138

Soil Health Assessment

12

Kavya Krishnan, Robert Schindelbeck, K. S. M. Kurtz, and Harold van Es

Abstract

Soil health, or soil quality, is defined as the continued capacity of soil to function as a vital living ecosystem that sustains plants, animals, and humans. Traditional views of soil quality often considered soils from the perspective of crop productivity and soil chemical properties, which generally led to an undervaluation of soil biological and physical processes. Soil health is a complex concept that requires a broader understanding of soil properties. It can be implemented through the comprehensive assessment of soil health (CASH) approach, which measures 15 physical, biological, and chemical soil indicators that can be linked to management recommendations. This chapter discusses all 15 tests, including the soil sampling and sample preparation methodologies. The methodology of the eight biological and physical tests is also covered in detail. These measures are interpreted through scoring functions that provide an overall soil health score. A comparison of indicator values between India, Colombia, and three US regions illustrates the need for scoring functions to be regionally adapted to production environments.

Keywords

Soil health · Soil health assessment · Soil quality indicators · Soil physical properties · Soil biological properties · Comprehensive assessment of soil health

K. Krishnan (✉) · R. Schindelbeck · K. S. M. Kurtz · H. van Es
Soil and Crop Sciences Section, Cornell University, Ithaca, NY, USA
e-mail: kk798@cornell.edu

A. Rakshit et al. (eds.), *Soil Analysis: Recent Trends and Applications*,
https://doi.org/10.1007/978-981-15-2039-6_12

12.1 Introduction

Soil health, or soil quality, is defined as the continued capacity of soil to function as a vital living ecosystem that sustains plants, animals, and humans (USDA-NRCS 2018). Soils are dynamic and complex living systems, which needs to be recognized when developing sustainable land management systems. Healthy soils affect not just crop production but broader social and ecological services like supporting human health, habitation of animals, and enhancing water and air quality.

Traditional views of soil quality often consider soils from the sole perspective of crop productivity. This has led to a historical over-emphasis on soil chemical properties because nutrients were often limiting factors for crop yield. But it is not indicative of the long-term resilience of the soil, which requires a more comprehensive consideration of all physical, chemical, and biological processes as well as the interactions among them. Most scientists, farmers, and governments are now aware of the importance of the multitude of soil processes that influence soil services. However, soil health is a complex concept and most of these processes cannot be measured directly. Instead they are generally evaluated using soil indicators as proxies for specific soil processes. Choosing the appropriate soil indicator requires a thorough understanding of these dynamic soil properties, which often also vary by region, landscape position, or land use.

Aside from the ecosystem benefits of improved soil health, research has also shown increases in yield (Lal 2006), as soil degradation is a major reason for the gap between typical versus attainable yields. This is usually a result of overuse or mismanagement of resources. Human-induced degradation like soil erosion, intensive cultivation, over-grazing, land clearing, salinization, and desertification is estimated to affect almost 40% of the world's agricultural land (Doran and Zeiss 2000). Increasing overall yields where possible is especially important given the pressing need to increase crop production levels with less land. Improving soil health is similarly important in urban landscapes where soil and vegetation typically provide critical and highly valued ecosystem services like water intake from adjacent impervious surfaces, green and recreational spaces, tree shading, etc.

12.1.1 The Comprehensive Assessment of Soil Health

One method of appraising soil health is the comprehensive assessment of soil health (CASH) approach which measures fifteen soil properties, representing physical, biological, and chemical processes, that give a broad appraisal of soil health. These are soil pH, organic matter, extractable phosphorous (P), extractable potassium (K), four micro-nutrients, active carbon, wet aggregate stability, soil respiration, autoclave-citrate extractable (ACE) protein test, available water capacity, and surface and sub-surface hardness. The assessment framework also rates the measured value for each indicator based relative to its optimal range using a fuzzy normative scoring framework. CASH was developed at Cornell University, and has been rigorously tested in the USA (Fine et al. 2017). In addition, this approach has been evaluated for soils in Jharkhand, India (Frost et al. 2019), western Kenya (Moebius-Clune et al. 2011), and coffee farming regions in Columbia (Rekik et al. 2018). There are a few other soil health assessments that have been developed around the world, as discussed in a recent paper by Bünemann et al. (2018). Many, however, are particular to their country of origin and cannot be readily applied to different environments. They are also usually limited to one or two soil properties that are then used to make inferences about overall soil health. In contrast, CASH offers a more multifaceted and comprehensive approach to determining soil health (Bone et al. 2014).

12.1.2 Soil Parameters and Their Selection

The CASH approach is the result of careful study of 42 potential indicators that have been narrowed down to a minimum dataset that represents critical soil processes. They were also selected based on the cost of analysis, consistency and reproducibility, and ease and cost of sampling.

Soil physical properties assessed in CASH are available water capacity, wet aggregate stability, and surface and sub-surface hardness. Wet aggregate stability is a measure of the extent to which soil aggregates resist falling apart (i.e., slaking) when wetted and either agitated or exposed to the energy of rain drops. Aggregate stability is an important indicator since it is influenced by multiple soil properties, such as microbial activity, organic matter and clay contents, and mineralogy, and is highly sensitive to management (Van Eerd et al. 2018; Moebius et al. 2007). Available water capacity is a physical indicator of the soil's capacity to store plant available water. Being the fraction of total porosity between field capacity, 10 μm, and the permanent wilting point, 0.2 μm, available water capacity is measured using air pressure chambers and ceramic pressure plates (Reynolds and Topp 2008). Recent research also suggests that this measure may be predicted using data mining techniques once a regional training database has been established. Surface and sub-surface hardness are indicators of soil compaction, representing the potential for physical root proliferation. Soil penetrability is sensitive to many management practices like tillage, as well as soil moisture.

The biological properties considered in the CASH approach are the soil organic matter, autoclave-citrate extractable protein, and active carbon, as well as the soil respiration process. Organic matter is a measure of total carbon, consisting of living material and dead organic materials in all stages of decomposition. Active carbon (also known as permanganate-oxidizable carbon or POXC) is a measure of the labile carbon pools in the soil, i.e., the readily available energy source for the soil microbes. A simple colorimetric method is used to determine this indicator (Weil et al. 2003). Studies have shown it to be sensitive to changes in management and one of the best predictors of yield (Culman et al. 2013; van Es and Karlen 2019). Autoclave-citrate extractable protein is an indicator of the amount of protein present in soil organic matter. A sodium citrate buffer is used to dissolve soil protein under high heat and pressure and the protein content is quantified using a bicinchoninic acid protein assay (Hurisso et al. 2018). Soil respiration is a measure of the metabolic activity of the soil microbial community. The indicator is used as a proxy for general microbial activity in the soil which is extremely important as a diverse and active microbial community is responsible for a multitude of soil functions.

The chemical properties considered are soil pH, extractable P, extractable K, and other minor elements like manganese (Mn), iron (Fe), magnesium (Mg), and zinc (Zn). These are considered standard nutrient analyses and their effects and methodologies have been well studied.

12.1.2.1 Condensed Soil Health Test

The time and cost of analysis to measure all 15 indicators of the CASH test may be prohibitive in some cases. Research on a potential condensed version of the CASH approach that includes the most useful indicators for predicting overall soil health of a sample has shown active carbon to be the best predictor of overall CASH scores with soil respiration and organic matter content also being strong indicators (Fine et al. 2017; Rekik et al. 2018; Frost et al. 2019). Similarly, a study of ten European long-term experiments found active carbon being the fastest and most cost-effective among labile carbon tests (Bongiorno et al. 2019). In a recent CASH analysis of long-term tillage experiments, indicators of the labile carbon fractions, active carbon, ACE protein, and respiration, were found to be the most strongly related to corn and soybean yields (van Es and Karlen 2019). The Cornell Soil Health Laboratory offers a "Basic" version of the test measuring only surface and sub-surface hardness, the standard suite of chemical properties, wet aggregate stability, and active carbon.

12.2 Soil Sampling and Sample Preparation

Detailed guidance and videos are also available at http://soilhealth.cals.cornell.edu/testing-services/collecting-samples/.

12.2.1 Sampling

Equipment for Soil Sampling

1. A large bucket
2. Sample bags
3. Straight shovel or trowel
4. Soil penetrometer

Procedure

Sampling should be done when the field is neither too wet nor too dry. After light showers and when there is no standing crop in the field would be the best time of the year to sample. A complete soil sample consists of a composite of around ten samples from at least five representative locations around the field. Sampling locations should be spread uniformly across the field, ideally in a W pattern that covers the field.

Start by clearing the debris at each sampling point and dig a small hole about 20 cm deep with a spade or trowel (depending in the size of the plot). Take a vertical, rectangular slice of soil from the edge of the hole that is 15 cm deep and 5 cm uniform thickness. Continue sampling at five locations across the field with two adjacent soil samples per location. The material is accumulated in a bucket and mixed thoroughly before a composite sample is derived. A standard set of CASH tests require about 1 kg of composite soil sample. These samples should be stored in a cool, dry location as soon as possible after sampling. Notably, direct exposure to the sun, especially in sealed plastic bags, should be avoided. During transportation, care should be taken to maintain the physical integrity of the samples and avoid high sample stacking in boxes, especially if it occurs on rough roads.

12.2.2 In-Field Hardness Test (PR15 and PR45)

At each of the five sampling locations (W pattern), soil surface and sub-surface hardness data should be collected. Measure the maximum hardness at the surface (0–15 cm) and the sub-surface (15–45 cm) depths using a penetrometer. Given the high variability of penetrometer readings, a minimum of two measurements should be taken at each sampling location. Also, as penetrometer readings are highly impacted by soil moisture conditions, the measurements should be made when the soil is field-moist, and ideally at field capacity.

12.2.3 Soil Preparation

Equipment for Soil Preparation

1. Drying trays
2. 2-mm and 8-mm sieves
3. Wooden rolling pin

Procedure

Spread the bulk sample thinly on drying trays and air-dry for a few days, until the sample can pass through a sieve without smearing. Thoroughly mix the sample and pass the entire sample through an 8-mm sieve. Roll the sample with a wooden rolling pin to break up any large clods if needed, but do not crush the sample. Any material that still does not pass through an 8-mm sieve can be discarded. Separate two cups (150 g) of soil for the respiration and protein test. The remaining soil is homogenized and then passed through a 2-mm sieve for the remaining tests. Oven dry around two tablespoons of soil at 105 °C. Use the total wet weight and oven dry weight to calculate the percentage of moisture in the sample. The moisture loss is used to correct the lab analyses to oven dry soil weight.

12.3 Laboratory Methodology

While the CASH assessment covers 15 different tests, this chapter only details the methodology of the eight biological and physical tests. The remaining chemical tests are briefly discussed below. Detailed procedures are not included in this chapter, because these tests have already become widely implemented across the world based on regionally appropriate methodologies. Soil texture, while not considered one of the 15 CASH soil health indicators, is used to interpret soil health measurements and therefore a detailed procedure of rapidly assessing soil texture is included in this section.

a. Organic Matter Content: The organic matter (%) is determined by measuring the mass loss of soil sample on ignition at 500 °C in a furnace.
b. pH: Soil pH is measured in a 1:1 water solution.
c. Nutrient Testing: Macro- and micro-nutrients are determined using a modified Morgan extractant, or regionally appropriate method.

12.3.1 Soil Respiration (Resp)

Equipment/Supplies

1. Soil samples, air-dried sample, sieved through 8-mm
2. Weigh balance
3. Wide mouth, 500 mL glass jars
4. Assembly traps. These are fashioned out of 10 mL glass beakers affixed onto a table top pizza saver (plastic mini-stool) using double sided foam tape. These traps should be washed and reassembled after every experimental run.
5. Round filter paper. The size of filter paper is dependent on the radius of the canning jars. In case of uneven cover, use two filter papers to cover the bottom of the glass jar.
6. Aluminum weigh dishes—57 mm. Punch nine holes in each aluminum weighing dish using either a single or an assembly of nine dissecting needles.
7. Large forceps
8. Electrical Conductivity Meter. Ensure that the probe of the conductivity meter can be inserted into the 10 mL beakers.
9. 0.5M Potassium hydroxide (9 mL/jar)
10. ddH_2O (7.5 mL/jar)
11. 10 mL Eppendorf pipette

Procedure

This procedure can be adjusted for any number of samples in single batch. It is recommended that each sample is replicated at least once per run, along with a blank sample after every ten jars containing soil sample. A known sample should also be run for every replication as a form of quality control.

Fig. 12.1 Image of a respiration setup depicting the respiration trap inserted into the soil sample in a glass jar. The conductivity of the KOH is being read with an EC meter

1. Using the large forceps, place the rounded filter paper(s) in the bottom of each labeled jar so they span the whole surface.
2. Weigh 20 g (±0.03) of 8-mm, air-dried soil into aluminum weigh boat. Using large forceps, gently place weigh boat in jar on top of filter paper. Repeat this for all samples.
3. Position an assembled trap in each weigh boat of soil, pressing down enough to bend the aluminum slightly so it conforms to the bottom of the jar. A depiction can be seen in Fig. 12.1.
4. First, pipette 9 mL KOH into each small beaker.
5. Pipette 7.5 mL ddH2O into each jar so water runs down the side of the jar onto the filter paper, and not into the soil. Then close the jar tightly to ensure that the setup remains air-tight. It is recommended that the fourth and fifth step be done in batches of ten to twelve jars so as to not leave the KOH exposed to the atmosphere for long.
6. Incubate the jars for 4 days. Note the setup date and time. The samples can be read 3 h before or after the setup time but aim to start within a few minutes of the actual setup start time.
7. After 96 h of incubation, read the conductivity of the KOH for each jar.

   NOTE: The test was calibrated to a temperature of 23.5 °C, so be aware of changes to EC probe temperature sensor reading. It should not be below 22.5 °C or above 25 °C because conductivity responds to temperature. Higher temperature generates greater conductivity.
8. Soil respiration can then be calculated using the following equations:

   a. The capacity of the respiration trap to absorb CO_2 can be calculated based on the volume and concentration of KOH. One mole of KOH used will react with the evolved CO_2 to produce 0.25 moles of K_2CO_3 when fully saturated. Therefore the amount of CO_2 that can be stored by our trap is calculated using the following equation:

$$9\text{ mL KOH} * \frac{1\text{ L}}{1000\text{ mL}} * \frac{0.25\text{ moles CO2}}{\text{L KOH}} * \frac{44.01\text{ g CO2}}{\text{mole CO2}} * \frac{1000\text{ mg}}{\text{g}} = 99.025\text{ mg CO2}$$

b. The proportion of the trap capacity for CO_2 absorption that is actually used (P) is calculated by the following equation:

$$\left(EC_{raw} - EC_{sample}\right)/\left(EC_{raw} - EC_{sat}\right)$$

where EC_{raw} is the electrical conductivity of pure 0.5M KOH (usually 112 microsiemens), EC_{sat} is the electrical conductivity of 0.25M K_2CO_3 (usually 42.6 microsiemens), and EC_{sample} is the measured electrical conductivity of the sample.

c. The amount of CO_2 (mg) absorbed by the trap = P * Trap capacity (99.025 mg).

d. This final amount is corrected by subtracting the average amount of CO_2 absorbed by the blanks from the amount of CO_2 absorbed by each sample.

e. The respiration rate is calculated by dividing the amount of CO_2 absorbed by the trap by the amount of soil sample used (here 20 g).

f. This value is then averaged over each replication to produce the respiration amount of each sample.

g. The relationship between electrical conductivity and respiration is further discussed in Fig. 12.2.

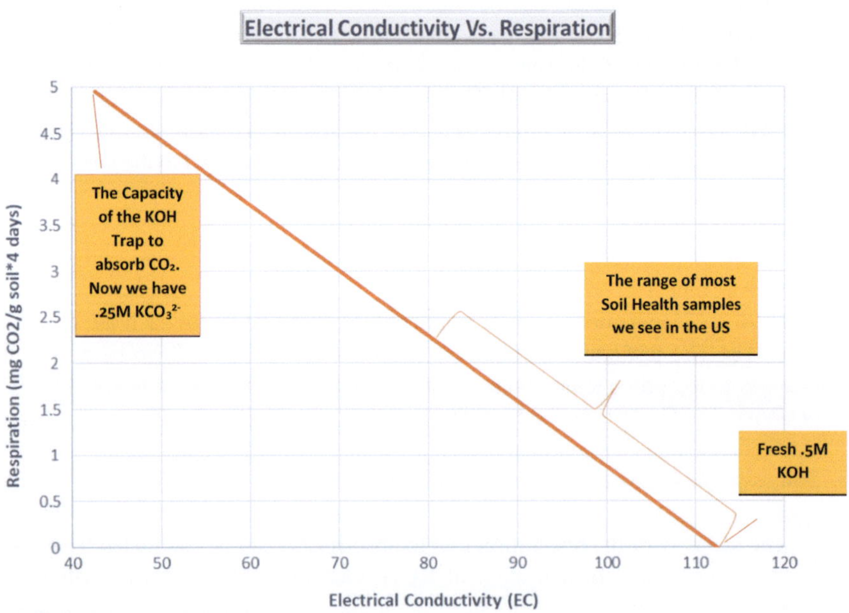

Fig. 12.2 Graph examining the relationship between electrical conductivity and respiration. Fresh KOH has an electrical conductivity of 112. KOH saturated with CO_2 forms K_2CO_3 and has an EC of 42.6

12.3.2 Active Carbon (AC)

Equipment/Supplies

1. 0.2M $KMnO_4$ solution. This solution is prepared by first dissolving 11.09 g $CaCl_2$ in ~750 mL distilled water in a beaker using a stir bar. Then, 31.61 g of $KMnO_4$ and ~200 mL of distilled water are added to the solution and dissolved completely (this takes about 1 h). While the solution is stirring, ensure that the beaker is covered with an opaque box or paper bag to prevent light exposure. The pH of the solution should, at this point, be around 7.2. If not, adjust the pH of the solution of with a dilute acid or base solution (such as 0.1 N HCl or KOH) before transferring the solution to a volumetric flask and bringing the final volume of the solution to 1000 mL. This solution is light sensitive and should be stored in an amber bottle. Solution should remain stable for 3–6 months.
2. Soil samples, air-dried sample, sieved through 2-mm
3. 50 mL centrifuge tubes in tube racks
4. Weigh balance
5. Colorimeter (w/550 nm setting)
6. 100–1000 μL and 1000–5000 μL pipettes and disposable tips
7. Platform shaker
8. Stop watch

Procedure

1. Preparing a standard curve: This step should be done every few days but can be used for multiple sets of samples.

   a. Ensure that the colorimeter is set to 550 nm and zero with distilled water.
   b. Dispense 48.75 mL distilled water into tube 1, 47.50 mL distilled water into tube 2, and 45 mL distilled water into tube 3.
   c. Add 0.2M $KMnO_4$ to the tubes in the following volumes: tube1: 1.25 mL; tube 2: 2.50 mL; tube3: 5.00 mL. Final concentrations of 50 mL $KMnO_4$ solutions are now 0.005M, 0.01M, 0.02M.
   d. Cap and shake for 10 s.
   e. Dispense 20 mL distilled water into nine centrifuge tubes—three for each standard solution.
   f. Add 0.2 mL of each standard to each respective triplicate set. Cap and shake for 10 s.
   g. Read and record the absorbance of each triplicate standard, rinsing the cuvette with one volume of standard and cleaning the outside to remove any liquid or smudges before each reading.

2. Measuring active carbon in soil samples.

   a. Soil samples should have previously been air-dried and sieved through 2 mm. It is recommended that each soil sample is run in duplicate.
   b. For each sample in the set, fill one centrifuge tube with 18 mL distilled water and one with 20 mL distilled water.
   c. Weigh 2.5 g (±0.005) for each soil sample into a weigh boat.
   d. Dispense 0.2 M KMnO4 solution into a beaker in small amounts as needed and cover with an opaque container to block light.
   e. In sequence, add a 2.5g of soil to each centrifuge tube with 18 mL distilled water. Then, in same sequence, begin redox reaction by adding 2 mL of 0.2M KMnO4 to each tube. Cap tightly.
   f. Place rack of tubes on the shaker at 120 rpm, start stopwatch and allow to shake for 2 min.
   g. After 2 min (do not stop stopwatch), remove samples from the shaker and lightly shake solution in tubes to ensure that soil is not stuck to the cap or top of the tube. Uncap tubes. On bench-top, allow settling and reaction to continue for a further 8 min.
   h. After 10 min of total reaction time, remove 0.2 mL supernatant from each reaction tube and transfer to the centrifuge tube with 20 mL distilled water.
   i. After all samples have been transferred, cap diluted sample tubes and shake by hand for 10 s.
   j. Read and record the absorbance of each sample at 550 nm.

      Note: Repeat duplicates with a difference in absorbance >5%. Repeat samples when duplicate sample absorbance readings fall outside the values of the standard curve, adjusting weight of sample used in reaction if necessary.

3. Calculating Active Carbon values:

   a. Plot a standard curve with concentration of the standards as the dependent variable (y) and absorbance as the independent variable (x).
   b. Determine the slope (b) and y-intercept (a) of the linear regression equation of the standard curve, where [Concentration $= a + b *$ (absorbance)].
   c. Determine concentration of sample using the equation.

      Active C (mg/kg) = [0.02 mol/L − (a + b ∗ absorbance)] ∗ (9000 mg C/mol) ∗ (0.02 L solution/0.0025 kg soil).

      Note: Where 0.02 mol/L is the initial solution concentration; 9000 is the mg of C (0.75 mol) oxidized by 1 mol of MnO4 changing from Mn^{7+} to Mn^{2+}; 0.02 L is the volume of KMnO4 solution reacted, and 0.0025 is the kg of soil used.

12.3.3 Autoclave: Citrate Extractable Protein (ACE Protein)

Equipment/Supplies

1. Sodium citrate solution—20 mM, pH 7.0. For a 20 L preparation of sodium citrate—dissolve 1.603 g anhydrous citric acid and 115.19 g sodium citrate, tribasic dihydrate in 500 mL ddH2O. This solution is then rinsed into a 20 L carboy and ddH2O is added to the carboy until the 20 L mark.
2. Air-dried soil sample, sieved through 8-mm
3. Autoclave
4. Shaker
5. Mini-centrifuge
6. Glass 50 mL extraction tubes with caps
7. Autoclavable tube rack
8. 2 mL micro-centrifuge tubes
9. 2 mL Transfer pipettes
10. 1000 μL pipettes, 8-channel pipettes with volumes 5–50 μL and 30–300 μL along with respective pipette tips
11. Strip tubes
12. Pipetting reservoir
13. 96-well clear flat bottom chimney well polystyrene plate
14. Microplate sealing tape
15. BCA reagents A and B and bovine serum albumin (BSA) standards set — Including a set of eight standards (0, 125, 250, 500, 750, 1000, 1500, and 2000 μg/mL)
16. Heat block
17. Microplate reader
18. Weigh balance

Procedure

1. Weigh 3 g (±0.03) air-dried soil sample into each glass tube. Depending on the consistency of results, the extraction of soil samples can be duplicated or limited to a single replication. The extractant will be replicated at the clarification step of the procedure. A sample of known concentrations should be run as a form of quality control with every 38 samples.
2. Add 24 mL of sodium citrate (pH 7) 20 mM in each tube, then secure lids on tubes.
3. Shake the samples at 180 rpm for 5 min.
4. Lightly shake each tube to ensure that soil is not stuck to the cap or top of the tube before loosening the lids to prevent the buildup of pressure in the tube.
5. Place the rack of tubes in the pre-heated autoclave. Run the autoclave at 120 °C for 35 min. When the pressure dial is zero and the temperature is below 100 °C, carefully open the autoclave and remove racks.

6. Once tubes have cooled to room temperature, the lids of the tubes are re-tightened. The tubes are then shaken to re-suspend the soil in the glass extraction tube.
7. Using a transfer pipette, remove about 1.5 mL of extract from the center of the column of liquid and transfer the extract to a labeled 2 mL micro-centrifuge-tube. Note: Avoid the lipid surface of the liquid, as well as near the soil. Discard transfer pipette and use a fresh one for each tube of soil.
8. Place micro-centrifuge tubes in the mini-centrifuge and run for 3 min at 11,641 rpm.
9. Samples can also be stored for 3 days before being analyzed. To do so, transfer 0.8 mL of the clarified extracts into a strip tube and store in a refrigerator.
10. When ready, transfer 10 μL of each soil extract into wells of a fresh microplate. Duplicate each sample preferably not next to each other, on the plate.
11. Include at least two columns of the eight BSA standards on each plate.
12. Make a working reagent by mixing Reagent A and Reagent B in a 50:1 ratio (For example, 25 mL of Reagent A and 0.5 mL of Reagent B would yield a working reagent). Any leftover reagent can be refrigerated and used up to 3 days.
13. Use the multichannel pipette to transfer 200 μL of the working reagent to each well.
14. When plate is filled, seal with a tape seal. Use the heat block to incubate the 96-well plate at 61.5 °C for an hour.
15. Let the plate cool down for at least 10 min undisturbed. Roll the plastic down again, and flip the plate upside down to remove bubbles.
16. Read the absorbance of the samples at 562 nm.
17. Calculate the protein concentration using the following equations:

    a. First create a standard curve to obtain coefficients for the parabolic (second order) regression line of best fit. Note the *a*, *b,* and *c* coefficients of the line $y = ax^2 + bx + c$, where *y* is the absorbance and *x* is the concentration of the standards.
    b. Protein concentration of samples can then be calculated from the absorbance values using the quadratic formula.

$$\text{Protein concentration in sample}\left(\frac{\mu\text{g}}{\text{mL}}\right) = \frac{-\text{b} + \sqrt{\text{b}^2 - 4\text{a}(\text{c} - \text{absorbance of sample})}}{2\text{a}}$$

    c. To determine protein content of soil sample (mg g soil^{-1}), adjust by the 24 mL extractant and 3 g soil samples used for the test.
    d. Average absorbance values for multiple replications of the same extract on the plate, prior to calculating protein concentration, and average concentration values across replicate extracts of the same soil sample after calculation. If the relative average deviation of replicates from their mean exceeds 5%, the sample is usually rerun.

12.3.4 Wet Aggregate Stability (WAS)

Equipment/Supplies

1. Soil samples, air-dried sample, sieved through 2-mm
2. Eijkelkamp (Eijkelkamp Soil and Water) wet sieving apparatus, complete set. The complete set includes eight sieves with sieve opening 0.25 mm and 16 sieve cans of surface of 10.2 cm^2. Note: this procedure only uses a single sieve size.
3. Spray bottle of ddH_2O
4. Sodium hexametaphosphate solution prepared by dissolving 2 g solute in 1 L distilled water
5. Weigh balance
6. Pre-weighed filter paper (Filter paper is usually weighed and grouped in rounded tenths of a gram)
7. Drying oven

Procedure

1. Weigh 4 g of air-dried, 2-mm aggregate soil onto each sieve (i.e., soil pushed through a 2-mm sieve under procedure in Sect. 12.2.3). Note the exact weight of each sample. Place aggregates onto 0.25-mm sieve. It is recommended that each soil sample is replicated at least twice, i.e., four samples per 8-sieve batch.
2. Pre-moisten soil with fine spray of distilled water.
3. First disperse soil using 100 mL distilled water for 3 min (fixed apparatus setting).
4. Switch aluminum weighing cans. Then disperse soil using 2 g/L sodium hexametaphosphate solution for 10 min (continuous setting on Eijkelkamp unit).
5. Filter both solutions using pre-weighed filter papers.
6. Oven dry the filter papers at 105 °C and weigh each one.
7. Calculate % stable aggregates using the following equation:

$$\%\text{stable aggregates} = \frac{\text{Weight of soil in dispersing solution}}{\text{Weight of soil in dispersing solution} + \text{weight of soil in water}}.$$

12.3.5 Available Water Capacity (AWC)

Equipment/Supplies

1. Soil samples, air-dried sample, sieved through 2-mm
2. Pressure plates
3. Rubber rings
4. Pressure chambers
5. Compressor
6. Pressure regulation system
7. Weigh balance
8. Aluminum weighing cans
9. Drying oven

Fig. 12.3 Image of an available water capacity setup

Procedure

1. Saturate 1 bar plates (for 0.1 bar samples) and 15 bar plates overnight. Place an equal number of rubber rings on each plate.
2. Fill a rubber ring on each of two plates (1 bar and 15 bar) with about 15 g of soil (as much as will fit into each ring). Add enough water to the plate to allow soil to saturate itself through suction. Let soils sit until they look fully saturated (Fig. 12.3).
3. Carefully pat the soils on the 15 bar plate (but NOT on the 1 bar plate), so that they are well packed and saturated.
4. Stack pressure plates into the appropriate pressure chamber, using plastic cylinders as spacers between plates. Connect outflow tubes of plates to pressure chamber outlet tubes, set pressure chamber outlet tubes into beakers.
5. Tighten lids down using the appropriate bolts.
6. Bring the appropriate chamber to 0.1 bar and 15 bar pressure slowly over 2 min. Let samples equilibrate for a week.
7. After a week, the samples are placed in tared moisture cans and weighed.
8. These cans are then placed in an oven at 105 °C overnight and weighed again.
9. Calculate the available water capacity of each sample:

    (a) ΔM = ((weight of wet soil + can) – (weight of dry soil + can))/((weight of dry soil + can) – weight of can)
    (b) AWC sample = ΔM 0.1 bar – ΔM 15 bar

12.3.6 Texture

Equipment/Supplies

1. Soil samples, air-dried sample, sieved through 2-mm
2. Weigh balance
3. Aluminum drying cans
4. Centrifuge tubes, 50 mL
5. 3% Hexametaphosphate (HMP) solution
6. Shaker
7. 20 cm diameter 0.053 mm sieve
8. 20 cm diameter funnel
9. Catch basin
10. 1000 mL beakers
11. Drying oven at 105 °C
12. Squeeze bottles with water
13. Tube rack

Procedure

1. Add 42 mL of 3% HMP solution to a labeled centrifuge tube.
2. Weigh and record 14 g (±0.1 g) of air-dried, 2-mm sieved soil to the labeled centrifuge tube. Cap and shake each tube vigorously to bring soil into suspension.
3. Place the rack of tubes onto a shaker for 2 h at 150 rpm. Samples can be stored for several weeks before or after shaking.
4. Once shaken, re-suspend soil (shake by hand until no soil is stuck to the sides or bottom of the Falcon tube) into solution before uncapping.
5. Use a squeeze bottle to rinse ALL material onto the 20 cm diameter 0.053 mm sieve assembly (place sieve in funnel over 1000 mL beaker inside catch basin).
6. Rinse contents of Falcon tube (including inside the cap) onto the sieve assembly. Use <1000 mL water to rinse tube and force all soil particles (while wearing gloves, using fingers and water) through the mesh. When the contents of the Falcon tube rinse clean, collect sand grains and organic matter (OM) to a corner of the sieve. Decant OM. Flush the sand into a pre-weighed aluminum can.
7. Dry at 105 °C until sample reaches complete dryness. Record aluminum can weight once samples have been dried.
8. Once all samples in the set have been rinsed, re-suspend all soil particles for every sample by emptying contents from labeled beaker to a temporary beaker and back into the labeled beaker.

9. Allow a 2 h settling period.
10. At the end of the settling period, decant the water and clay in beaker above the settled silt particles. Use caution when decanting the suspended clay from the silt layer at the bottom of the beaker. These silt particles at the bottom of the beaker are rinsed into another pre-weighed aluminum can.
11. Dry at 105 °C until samples reach constant weight. When samples have dried, record the weight of the aluminum can containing silt.
12. The sand, silt, and clay percentages are calculated by:

    (a) Sand % = (oven dry sand mass/original sample mass) $*$ 100%
    (b) Silt % = (oven dry silt mass/original sample mass) $*$ 100%
    (c) Clay % = 100 − (Sand % + Silt %)

12.4 Scoring Functions

The comprehensive soil health assessment approach converts the measured values of each of the indicators into a unit-less score before combining all the indicator scores into a single metric of soil health. This process of evaluation, originally described by the Soil Management Assessment Framework (Andrews et al. 2004) has since been extensively tested for the CASH approach (Fine et al. 2017). Based off samples submitted to the Cornell Soil Health Laboratory from across the US scoring functions were developed for each of the indicators used in the test.

12.4.1 Developing the Soil Health Indicator Scoring Functions

Most of the soil health indicators are appraised using either (1) established critical values (mostly with conventional indicators like nutrients and pH) or (2) a normative framework where individual soil test results are evaluated relative to a broader population (mostly with the newer biological and physical indicators). The scoring functions are generally based on fuzzy logic and involve a sigmoidal curve based on the cumulative normal distribution function. This generalized approach allows scoring (interpretations of the measured values) to be adapted to the local or regional soil/production environments. This section briefly explains the interpretations of the measured indicator values and details the steps taken to determine their scoring functions.

The development of scoring functions requires a representative sample of a regional population of soil health samples (we recommend at least 250 samples, possibly updated with further sample analyses). This requires samples that represent a range of management conditions (bad to good) that constitutes a reasonable reference base. The complete dataset is used for the development of scoring functions for the CASH indicators.

The scoring framework assumes that the soil health indicators follow a normally distributed probability density function (NDF), which greatly facilitates interpretations:

$$p = f(x, \mu, \sigma) = \frac{1}{\sigma\sqrt{2\pi}} \int_{-\infty}^{+\infty} e^{\frac{-(x-\mu)^2}{2\sigma^2}} dx$$

where p is the probability that a measured value x will fall at given position in the interval $(+\infty, -\infty)$ and μ is the indicator mean and σ is the standard deviation. Normally distributed results are generally the case with soil health data, but the data can be normalized through transformations if needed. This then allows for the simple calculation of the two parameters that define the NDF, the mean and standard deviation for each indicator. In Excel, the appropriate functions are:

$$\text{Mean} = \text{AVERAGE}(\text{cell1} : \text{celln})$$

$$\text{Standard deviation} = \text{STDEV}(\text{cell1} : \text{celln})$$

These two parameters also define the cumulative normal distribution function (CNDF) for the indicator.

The CND is then converted to a standardized, unit-less scoring function for each indicator that falls between 0 and 100. In Excel, the appropriate function is:

$$\text{NORMDIST}\ (x, \text{mean}, \text{standarddev}, \text{true}) * 100$$

The CNDF thereby converts any soil health indicator value (x) to a score between 0 and 100.

These scoring functions can be categorized in three ways—more is better, less is better, or an optimum curve (Fig. 12.4). For certain indicators, such as WAS, AWC, organic matter, ACE protein, soil respiration, and AC, the higher the measured values of the test, the higher the converted score is. This is calculated by the above CNDF $*$ 100. For indicators where lower measured values for tests, such as surface and sub-surface hardness, are associated with better soil health, the converted score is calculated as $100 * (1 - \text{CNDF})$. Indicators such as pH and P are both scored using an optimum curve. Here, any measured value in the optimum range is given a converted score of 100 and as the values increase or decrease away from the optimum range the converted score decreases proportionally. These converted scores can be further classified as very low, low, medium, high, and very high. In a more is better scenario, these categories could be very low (0–20), low (20–40), medium (40–60), high (60–80), and very high (80–100), respectively.

An overall soil health score for the sample can be derived by computing a mean of the individual indicator scores, either as an unweighted or weighted average (unweighted is preferred unless certain soil health indicators are considered more important).

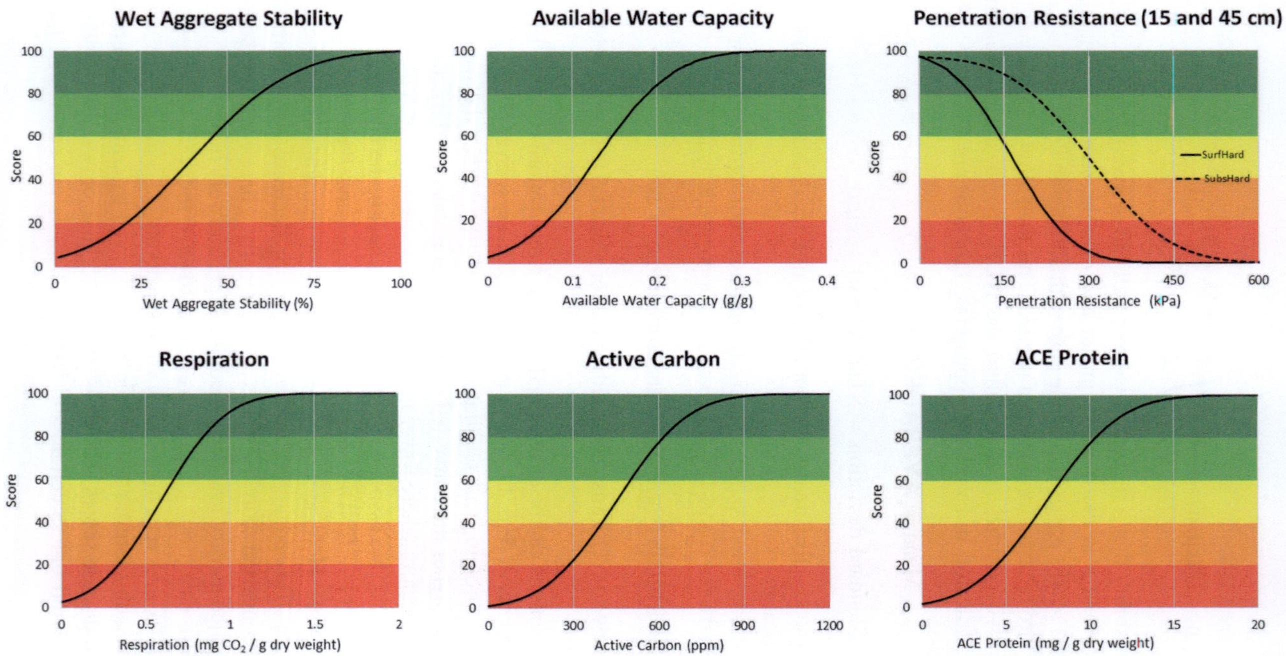

Fig. 12.4 Comprehensive assessment of soil health (CASH) scoring functions for wet aggregate stability, available water capacity, penetration resistance (15 and 45 cm), active carbon, and autoclave-citrate extractable protein. Functions are shown overlying the expanded 2016 CASH five color scheme (red, orange, yellow, light green, dark green) used to classify indicator scores as low, very low, medium, high, and very high. Source: Fine et al. (2017)

Table 12.1 Means for textural distribution of all soil samples ($n = 5767$) by textural group

Soil health indicator	Coarse	Medium	Fine
Sand (%)	65.6 A	30.6 B	16.5 C
Silt (%)	28.4 C	55.8 A	50.5 B
Clay (%)	6.0C	13.6 B	33.0 A
WAS (%)	52.2 A	42.2 B	41.8 B
AWC ($g\ g^{-1}$)	0.152 C	0.208 B	0.219 A
PR15 (kPa)	1158 A	1110 B	1110 AB
PR45 (kPa)	2199 A	241 B	208 B
OM (%)	3.26 C	3.74 B	4.42 A
AC ($mg\ kg^{-1}$)	86.1 C	531.2 B	608.7 A
Protein ($mg\ g^{-1}$)	10.2 A	7.0 B	5.6 C
Resp ($mg\ CO_2\ g^{-1}$)	0.64 A	0.62 A	0.61 A
pH	6.2 A AB	6.3 A	6.1B
P (ppm)	21.1 A	12.9 B	9.3 B
K (ppm)	122.8 B	126.7 B	207.6 A
Mg (ppm)	140.9 C	242.2 B	471.8 A
Fe (ppm)	8.6 A	5.9 B	5.6 AB
Mn (ppm)	10.5 B	14.4 A	10.2 B
Zn (ppm)	2.6 A	1.4 B	1.0 B

Source: Fine et al. (2017)

Abbreviations included in the table are *AC* active carbon, *AWC* available water capacity, *OM* organic matter, *PR15* penetration resistance 0–15 cm, *PR45* penetration resistance 15–45 cm, *Resp* soil respiration, and *WAS* wet aggregates stability

ANOVA results, represented as nonmatching capital letters, depict statically significant differences between textural groups at $\alpha = 0.05$

12.4.2 Variation in Results

CASH indicator values have been shown to vary among different textures as well geographical regions.

Comparisons between fine, medium, and coarse texture soils have been shown to have significant differences, whereas sand, loamy sand, and sandy loam classes are classified as "coarse" soils, sandy clay loam, loam, silt loam, and silt as "medium," and sandy clay, clay loam, silty clay loam, silty clay, and clay as "fine." Some indicators like WAS, PR15, PR45, Protein, P, Fe, and Zn were significantly higher for coarse samples, while AWC, OM, AC, K, and Mg were significantly higher for fine textured soils (Table 12.1; Fine et al. 2017). Given these strong differences in CASH values for different soil textures, scoring functions should be developed for coarse, medium, and fine soils separately.

Table 12.2 Means for textural distribution of all soil samples by geographical region

Soil health indicators	Jharkhand, India	Mid-Atlantic, USA	Midwest USA	Northeast USA	Colombia
Sand (%)	53	49.80	31.80	39.27	16.5
Silt (%)	28.7	41.55	48.70	43.27	58.5
Clay (%)	18.3	8.65	19.50	17.47	25.0
WAS (%)	17.0	43.75	31.83	51.33	94.3
AWC (g g^{-1})	0.20	0.17	0.20	0.18	0.3
PR15 (kPa)		1299.00	1241.67	1158.33	
PR45 (kPa)		2108.50	2091.67	2150.00	
OM (%)	1.8	3.16	3.28	4.18	17.4
AC (mg kg^{-1})	152	449.75	523.00	582.70	818.3
Protein (mg g^{-1})	2.2	8.20	5.33	9.17	9.2
Resp (mg CO_2 g^{-1})	0.14	0.69	0.49	0.69	1.0
pH	5.8	6.09	6.05	6.29	3.8
P (ppm)	4.3	18.70	20.33	13.10	1.5
K (ppm)	94	128.10	189.50	114.53	22.7
Mg (ppm)	224	130.10	350.93	251.83	4.6
Fe (ppm)	243	4.10	2.87	8.57	4.1
Mn (ppm)	146	12.30	10.07	12.70	0.7
Zn (ppm)	1.8	1.95	2.03	1.67	0.1

Sources: Frost et al. (2019), Fine et al. (2017), and Rekik et al. (2018)
Abbreviations included in the table are *AC* active carbon, *AWC* available water capacity, *OM* organic matter, *PR15* penetration resistance 0–15 cm, *PR45* penetration resistance 15–45 cm, *Resp* soil respiration, and *WAS* wet aggregates stability
This was averaged over three texture groups for the USA data

Scoring functions also need to be regionally adapted based on the soils and production environments. A comparison of indicator values between India, Colombia, and three US regions shows the importance of adapting scoring functions (Table 12.2). For example, average wet aggregate stability (WAS) was very low (17%) for degraded soils under predominantly rice production in Jharkhand, India, while Columbian coffee soils in a volcanic region averaged 94% WAS, with US regions having intermediate values. This suggests that interpretations of measured soil health indicators require appropriate scoring functions for each region.

References

Andrews SS, Karlen DL, Cambardella CA (2004) The soil management assessment framework. Soil Sci Soc Am J 68(6):1945

Bone J et al (2014) Prioritising soil quality assessment through the screening of sites: the use of publicly collected data. Land Degrad Dev 25(3):251–266

Bongiorno G et al (2019) Sensitivity of labile carbon fractions to tillage and organic matter management and their potential as comprehensive soil quality indicators across pedoclimatic conditions in Europe. Ecol Indic 99:38–50

Bünemann EK et al (2018) Soil quality–a critical review. Soil Biol Biochem 120:105–125

Culman SW, Snapp SS, Green JM, Gentry LE (2013) Short- and long-term labile soil carbon and nitrogen dynamics reflect management and predict corn agronomic performance. Agron J 105 (2):493

Doran JW, Zeiss MR (2000) Soil health and sustainability: managing the biotic component of soil quality. Appl Soil Ecol 15(1):3–11

Fine AK, van Es HM, Schindelbeck RR (2017) Statistics, scoring functions, and regional analysis of a comprehensive soil health database. Soil Sci Soc Am J 81(3):589–601

Frost PSD et al (2019) Soil health characterization in smallholder agricultural catchments in India. Appl Soil Ecol 138:171–180

Hurisso TT et al (2018) Soil protein as a rapid soil health indicator of potentially available organic nitrogen. Agric Environ Lett 3:180006

Lal R (2006) Enhancing crop yields in the developing countries through restoration of the soil organic carbon pool in agricultural lands. Land Degrad Dev 17(2):197–209

Moebius BN et al (2007) Evaluation of laboratory-measured soil properties as indicators of soil physical quality. Soil Sci 172(11):895–912. https://doi.org/10.1097/ss.0b013e318154b520

Moebius-Clune BN, van Es HM, Idowu OJ, Schindelbeck RR, Kimetu JM, Ngoze S, Lehmann J, Kinyangi JM (2011) Long-term soil quality degradation along a cultivation chronosequence in western Kenya. Agric Ecosyst Environ 141(1–2):86–99

NRCS, USDA (2018) Healthy soil for life

Rekik F et al (2018) Soil health assessment for coffee farms on andosols in Colombia. Geoderma Reg 14:e00176

Reynolds WD, Topp GC (2008) Soil water desorption and imbibition: tension and pressure techniques. Soil sampling and methods of analysis, 2nd edn. CRC Press, Boca Raton, pp 981–997

Van Eerd LL et al (2018) Quantitative and qualitative comparison of three wet aggregate stability methods using a long-term tillage system and crop rotation experiment. Can J Soil Sci 98 (4):738–742

Van Es HM, Karlen DL (2019) Reanalysis validates soil health Indicator sensitivity and correlation with long-term crop yields. Soil Sci Soc Am J 83(3):721–732. https://doi.org/10.2136/sssaj2018.09.0338

Weil R, Islam KR, Stine MA, Gruver JB, Samson-Liebig SE (2003) Estimating active carbon for soil quality assessment: a simple method for laboratory and field use. Am J Altern Agric 18 (1):3–17

Soil Health Indicators: Methods and Applications

13

M. Raghavendra, M. P. Sharma, A. Ramesh, A. Richa, S. D. Billore, and R. K. Verma

Abstract

Soil health indicators are a composite set of measurable physical, chemical, and biological attributes which relate to functional soil processes and are being used to evaluate soil health status. A range of soil health indicators have been developed to measure and asses changes in soil properties and functioning to understand soil health as a tool for sustainability. The physical, chemical, and biological indicators must be employed to verify soil status use and to undertake remedial management measures within a desired timescale. Soil properties which can change rapidly in response to natural or anthropogenic actions are considered as good soil health indicators. Among the physical indicators, bulk density, soil aggregate stability, and water holding capacity have been found ideal indicators. Chemical indicators such as pH, EC, soil organic carbon, and soil nutrient status are well established. However, most of them generally have a slow response, as compared to the microbiological and biochemical properties, such as soil enzymes, soil respiration, mycorrhiza, lipid profiling, and earthworms as they change rapidly due to perturbation caused by different agricultural management paradigm. Thus, systemic approaches based on different kinds of indicators (physical, chemical, and biological) in assessing soil health are discussed in this chapter.

Keywords

AMF · Earthworm · Enzymes · PLFA · Soil health indicators · Soil health management · Soil quality

M. Raghavendra (✉) · M. P. Sharma · A. Ramesh · A. Richa · S. D. Billore · R. K. Verma
Division of Crop Production, ICAR-Indian Institute of Soybean Research, Indore, MP, India

A. Rakshit et al. (eds.), *Soil Analysis: Recent Trends and Applications*,
https://doi.org/10.1007/978-981-15-2039-6_13

13.1 Introduction

Modern agricultural practices began to exploit soil by excessive use of chemical fertilizers devoid of organic sources, nature of the soil such as high pH, $CaCO_3$ and low organic carbon content; extensive tillage with heavy machinery and closely spaced cereal–cereal rotations. This has instead of created insert caused multifaceted deleterious effect on soil health by reducing time required by the soil health indicators (biological) to rejuvenate and perpetuate for maintaining ideal environment condition for crop growth without compromising on economic yield. Moreover, this situation has accelerated soil degradation process insidiously making roads into weakening of soil health indicators to become unproductive soil (Katyal et al. 2016). At present, demand for sustainable agricultural management practices mounting due to agricultural edges has already expanded near to the maximum all over the world. Feeding ever increasing population with maintaining optimum soil health indicators and sustainable environment is ever challenging task for present and future generations to come. In addition, public awareness and thrust on the need of environmental conservation, especially in the tropical region, claim for keeping forests as reserve of biodiversity, provider of environmental services, and needs for reclamation of degraded lands (Cardoso et al. 2013) is also a matter of great concerns. Therefore, sustainable agricultural practices to maintain optimum soil heath indicators with ideal soil fertility are needed for meeting the needs of the present without compromising the productive potential for the next generations. The rational soil use practices must allow economically and environmentally sustainable yields, and also quality of produce which will only be reached with the maintenance or recovery of the soil health indicators. Thus, a healthy soil has "the continued capacity of soil to function as a vital living system, within ecosystem and land-use boundaries, to sustain biological productivity, promote the quality of air and water environments, and maintain plant, animal and human health" (Doran and Safley 1997). To assess the sustainability of a production system, changes in soil health indicators (chemical, physical, and biological) and their effects on the soil's capacity to support plant growth and external environment functions must be monitored. Hence, in this chapter an impetus has been given to discuss soil health indicators in detail with methodologies to analyze them in the laboratory along with their potential applications in crop production and management aspects under field conditions.

13.2 General View of Soil Health Indicators

The soil consists of four major components such as air, water, mineral, and organic matter that are described in terms of soil health indicators, which can provide an assessment of how well the soil functions. Though the properties that constitute a healthy soil are not the same in all situations and locations, there are some important soil properties that indicate soil health. Soil health indicators are selected based on soil characteristics, soil use, and environmental circumstances along with their positive correlation with crop growth and yield under different management

Table 13.1 Soil health indicators selected based on certain criteria (modified from Arshad and Coen 1992; Idowu et al. 2008; Kelly et al. 1999; Paoletti et al. 2010; Griffiths et al. 2018)

Soil health indicators	Rationale for selection
Bulk density	Plant root penetration, porosity, adjust analysis to volumetric basis
Soil aggregate stability	Soil structure, erosion resistance, crop emergence an early indicator of soil management effect
Water holding capacity/infiltration	Runoff, leaching, and erosion potential
pH	Nutrient availability, pesticide absorption and mobility, process models
EC (electric conductivity)	Defines crop growth, soil structure, water infiltration; presently lacking in most process models
CEC (cation exchange capacity)	CEC represents the total amount of exchangeable cations that soil can absorb
Soil organic carbon/organic matter	Defines soil fertility and soil structure, pesticide and water retention, and use in process models
Soil nutrients status	Availability of crops, leaching potential, mineralization/immobilization rates, process modeling, capacity to support plant growth, environmental quality indicator
Suspected pollutants	Plant quality, and human and animal health
Soil respiration	Biological activity, process modeling; estimate of biomass activity, early warning of management effect on organic matter
Enzymes (dehydrogenase, β-glucosidase, acid and alkaline phosphatase, microbial biomass, and soil respiration)	Electron transferences in the respiratory chain in living cells, C oxidation, organic phosphorus cycling, source and/or drain of C and nutrients, microbial mineralization of organic carbon
Mycorrhiza	Nutrient mobilization, soil aggregation
Trichoderma	Residue decomposition
Lipid profiling	Diversity and biomass
Earthworm	Indicate relative change in soil structure, nutrient recycling, regulate soil water, aeration, and provide drainage

conditions (Cardoso et al. 2013). Some of the key soil health indicators for soil quality assessment are provided in Table 13.1 and the inter relationship between different soil health indicators are emphasized in Table 13.2. According to Bünemann et al. (2018), the most commonly used and frequently proposed soil health indicators by various authors across the globe are soil organic carbon and soil pH (Fig. 13.1), followed by available phosphorus, indicators of water storage, and bulk density. The soil texture, available potassium, and total nitrogen are also frequently used (>40%). For soil reclamation point of view, the important soil properties that indicate soil health could be physical, chemical, biological, or

Table 13.2 Interrelationship of soil indicators (Laishram et al. 2012)

Selected indicator	Other soil quality indicators
Aggregation	Organic matter, microbial (especially, fungal) activity, texture
Water holding capacity/ infiltration	Organic matter, aggregation, electrical conductivity, exchangeable sodium percentage (ESP)
Bulk density	Organic matter, aggregation, topsoil-depth, ESP, biological activity
Microbial biomass	Organic matter, aggregation, bulk density, pH, texture, ESP, and/or respiration
Available nutrients	Organic matter, pH, topsoil-depth, texture, microbial parameters (mineralization and immobilization rates)

biochemical within that average number of indicators selected based on their practical and economical feasibility as well as their relations with other indicators under all the conditions are described in this chapter.

13.3 Soil Health Indicators and Their Analytical Techniques

13.3.1 Soil Physical Health Indicators

Soil physical health indicators provide information related to water and air movement through soil, as well as conditions affecting germination, root growth, and erosion processes. Thus, soil physical health indicators form the foundation for other chemical and biological processes. Key soil physical indicators in relation to crop production include soil aggregate stability, water holding capacity, bulk density and are discussed below.

13.3.1.1 Water Holding Capacity and Bulk Density

Soil water holding capacity is the amount of water a given soil can hold for crop use. How much water a soil can hold is very important for crop production point of view. Soils which hold more water can support higher plant growth and development and reduce leaching losses of nutrients and pesticides. Hence, water holding capacity of soils is explained in terms of infiltration, soil available water and distribution. Soil water infiltration, the rate at which water enters the soil surface and moves through soil depth, is gaining increased interest (Dalal and Moloney 2000; Joel and Messing 2001). Since infiltration rate may change significantly with soil use, management, and time, it has been included as an indicator of soil health for assessments of land use change impacts (Arias et al. 2005; O'Farrell et al. 2010).

Bulk density is the weight of dry soil per unit of volume expressed in grams cm^{-3}. It is routinely assessed in agricultural systems to characterize the state of soil compactness in response to land use and management (Håkansson and Lipiec 2000). It has been considered as a useful indicator for the assessment of soil health with respect to soil functions such as aeration, infiltration (Reynolds et al. 2009), rooting depth/restrictions, available water capacity, soil porosity, plant nutrient availability, and soil microorganism activities influencing the key soil processes and productivity

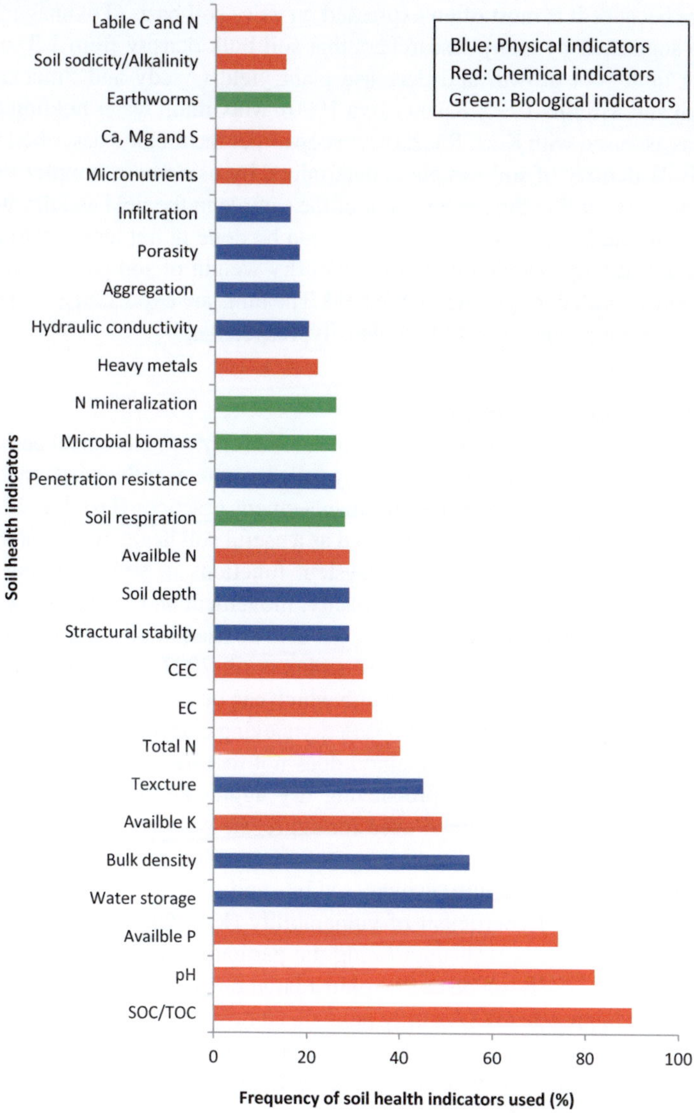

Fig. 13.1 Frequency of different indicators used all over the world (Modified from Bünemann et al. 2018)

(https://www.nrcs.usda.gov). Since bulk density in general is negatively correlated with soil organic matter (SOM) or SOC content (Weil and Magdoff 2004), loss of organic C from increased decomposition due to elevated temperatures (Davidson and Janssens 2006) may lead to increase in bulk density and hence making soil more prone to compaction through land management activities (Birkas et al. 2009). Bulk density directly measures compaction, and generally does not vary with other soil properties because it is most often expressed on a dry soil basis (Tokunaga 2006). It has been suggested by many researchers that soil bulk density from 1.3 to 1.7 mg m^{-3} may limit root growth and decrease plant yield (Asady and Smucker 1989; Bengough and Mullins 1990; Kuznetsova 1990). Maximum water holding capacity of soil was assessed with Keen Raczkowski cup as per the method described by Piper (1966). Bulk density of soil sample is determined by using core sampler technique (Black 1965), recording the fresh weight of the sample in the field and dry weight of the sample in the laboratory. Drying of soil can be done in hot air oven to constant weight. Bulk density calculation was done as dry weight of soil per unit volume of the core collect with core sampler in the field. The units are expressed as % and g cm^{-3} for water holding capacity and bulk density, respectively.

13.3.1.2 Aggregate Stability

Aggregate stability is an indicator of organic matter content, biological activity, and nutrient cycling in soil and is determined by soil structure as influenced by a range of chemical and biological properties and management practices (Dalal and Moloney 2000; Moebius et al. 2007). It is considered as a useful soil health indicator since it is involved in maintaining important ecosystem functions in soil including organic carbon (C) accumulation, infiltration capacity, movement and storage of water, and root and microbial community activity; it can also be used to measure soil resistance to erosion and management changes (Moebius et al. 2007; Rimal and Lal 2009). Aggregate stability is crucial for soil health which can be measured with the methods proposed by Kemper and Chepil (1965) (a dry sieving and wet sieving), Bissonnais (1996) and Six et al. (2000) (the method does not require the use of equipment to mechanically submerge sieves, pre-sieving dry aggregates but rather is done by hand). The most common method used for aggregate stability measurement is wet sieving (Haynes 1993). The disadvantage of the method proposed by Bissonnais (1996) is that aggregate stability is increased by sand particles that are not excluded from the calculation of coefficient of vulnerability (Kv). On the other hand, a big advantage of this method is distinguishing the particular mechanisms of aggregate breakdown. Therefore, it can be used within a large range of soils. In the assessment of water stable aggregate (WSA), only hexa-metaphosphate as a dispersing solution was used, because sodium hydroxide was too aggressive to the aluminum cans. An advantage of this method is that sand particles are excluded from the calculation of WSA index.

13.3.2 Soil Chemical Health Indicators and Their Analytical Techniques

Soil chemical health indicators are correlated with the capacity to provide nutrients for plants and/or retaining chemical elements or compounds harmful to the environment and plant growth. Soil pH, electrical conductivity, cation exchange capacity (CEC), soil organic carbon, and nutrient status are the main chemical indicators used in soil health assessment, especially when considering the soil capacity for supporting high yield crops (Kelly et al. 1999).

13.3.2.1 Soil pH, Electrical Conductivity, and Cation Exchange Capacity

Soil pH is one of the most indicative measurements of the chemical properties of soil. Whether a soil is acidic, neutral, or basic has much to do with solubility of various compounds, the relative bonding of ions to exchange sites, and the various microorganisms. Soil pH can be determined by an electrometric method (Jackson 1973) using a glass electrode pH meter in a 1:2 suspension of soil and water by using buffer solutions at pH 4.0 and 7.0, the pH read on pH meter. Soil electrical conductivity (EC), a measure of salt concentration, is considered an easily measured, reliable indicator of soil quality/health (Arnold et al. 2005). It can inform trends in salinity, crop performance, nutrient cycling (particularly nitrate), and biological activity and, along with pH, can act as a surrogate measure of soil structural decline especially in sodic soils (Dalal and Moloney 2000; Arnold et al. 2005). Electrical conductivity has been used as a chemical indicator to indicate soil biological quality in response to crop management practices (Vargas Gil et al. 2009). Clearly, there is a need for a comprehensive assessment of soil EC as an important soil health indicator in different ecosystems (Smith et al. 2002). Electrical conductivity of soil samples can be determined by the method suggested by Piper (1966) using a conductivity meter (Chemita 130) in 1:2 (soil:water ratio).

Cation exchange capacity (CEC) is also considered as an important determinant of soil chemical quality, particularly the retention of major nutrient cations Ca, Mg, and K and immobilization of potentially toxic cations Al and Mn; these properties can thus be useful indicators of soil health, informing of a soil's capacity to absorb nutrients, as well as pesticides and chemicals (Dalal and Moloney 2000; Ross et al. 2008). Ion exchange capacity mostly affects soil cation exchange capacity (CEC) binding to negative charge organic matter, clay, and soil colloid. CEC in soil can be measured by ammonium acetate method (Schollenberger and Dreibelbis 1930) at pH 7 and the barium chloride-triethanolamine method (Mehlich 1938) at pH 8.2.

13.3.2.2 Soil Organic Carbon

Soil organic carbon is a key attribute in assessing soil health, generally correlating positively with crop yield (Bennett et al. 2010). The soil organic carbon affects important functional processes in soil like the storage of nutrients, mainly N, water holding capacity, and stability of aggregates (Silva and SáMendonça 2007). In addition, the soil organic carbon also affects microbial activity. Hence, this is a key component of soil fertility, especially in tropical conditions, which interacts with chemical, physical, and biological soil properties and must be considered in assessments of soil health. Soil organic carbon content can be measured with help of Walkley and Black method. The method involves the oxidation of potassium dichromate solution in sulfuric acid medium and evaluating the excess of dichromate with titration against ferrous ammonium sulfate (Yeomans and Bremner 1988). Weil et al. (2003) reported a highly simplified method using slightly alkaline $KMnO_4$ to analyze oxidizable (active) forms of soil C. They showed that the active soil C measured was more sensitive to soil management practices than total organic C, and more closely related to biologically mediated soil properties, such as respiration, microbial biomass, and aggregation, than several other measures of soil organic C.

13.3.2.3 Available Nutrients (N, P, S, Zn, and Fe)

Available soil nutrients (N, P, K, S, Zn, and Fe) and their identification of basic soil properties to meet requirements of indicators for screening soil health, Doran and Safley (1997) proposed extractable nutrients as "they provide information on plant available nutrients and potential loss from soil providing indication of productivity and environment quality." Measurement of extractable nutrients may provide indication of a soil's capacity to support plant growth; conversely, it may identify critical or threshold values for environmental hazard assessment (Dalal and Moloney 2000). Nutrient cycling, especially N, is intimately linked with soil organic C cycling (Weil and Magdoff 2004) and possibly the cycling of other plant available nutrients. The mineralizable nitrogen in soil can be determined with help of alkaline permanganate method (Subbiah and Asija, 1956) using a Kjeldahl distillation unit. The available phosphorous can be extracted with Olsen's reagent (0.5M $NaHCO_3$, pH 8.5) in neutral to alkaline soils (Olsen et al. 1954), whereas under acid soils Brays P-1 (0.03N NH_4 F and 0.025N HCL) is widely followed (Bray and Kurtz 1945). The major drawback with blue color development (Dickman and Bray 1940) is that color starts fading soon and hence intensity has to be measured quickly. Therefore ascorbic acid method (Watanabe and Olsen 1965) provides stable blue color and therefore preferred over former methods to estimate available phosphorus in soil. Available sulfur can be extracted by using Morgan's universal extractant (pH 4.8) and is determined by turbidimetric method (Chesnin and Yien 1950) using UV/Visible spectrophotometer. For micronutrients extraction, neutral ammonium acetate and chelating agents like EDTA and DTPA have been used for extraction of Zn, Fe, Cu, and Mn from soil and the extracted amount is determined calorimetrically. Zn determination dithizone method (Shaw and Dean 1952) has been very popular until AAS become available. For those laboratories where AAS is not yet available the alternative (colorimetric) methods as described by Jackson (1973) are still employed. However, for rapid and accurate analysis of Zn, Fe, Cu, and Mn the DTPA method (Lindsay and Norvell 2010) is most widely used to estimate micronutrients.

13.3.3 Microbiological and Biochemical Health Indicators and Their Analytical Techniques

Soil microbial activity and diversity play an important role in the sustainability by keeping essential functions of soil health, involving carbon and nutrient cycling (Jeffries et al. 2003; Izquierdo et al. 2005). Microbial indicators are more sensitive than physical and chemical attributes to changes imposed to the environment like soil use and management (Masto et al. 2009). Some of the commonly used soil biochemical/biological parameters which depict the soil quality status of a given soil along with their analytical techniques are illustrated below:

13.3.3.1 Soil Microbial Biomass (Microbial Biomass Carbon (MBC) and Microbial Biomass Nitrogen (MBN))

The soil microbial biomass (MBC and MBN) is the active component of the soil organic pool and plays an important role in nutrient cycling, plant nutrition, and functioning of different ecosystems. It is responsible for organic matter decomposition thus affecting soil nutrient content and, consequently, primary productivity in most biogeochemical processes in terrestrial ecosystems (Gregorich et al. 2000; Haney et al. 2001). In the last 30 years, relatively rapid assessment of soil microbial biomass has been possible based on physiological, biochemical, and chemical techniques (Horwath and Paul 1994) such as chloroform fumigation incubation (CFI) (Jenkinson and Powlson 1976), chloroform fumigation extraction (CFE) (Brookes et al. 1985; Vance et al. 1987), substrate-induced respiration (SIR) (Anderson and Domsch 1978), and adenosine triphosphate (ATP) analysis (Jenkinson et al. 1979; Eiland 1983; Webster et al. 1984). Microbial biomass has even been proposed as a sensitive indicator of soil quality (Karlen et al. 1997) and soil health (Sparling 1997). Of these, the first two methods have been widely used to estimate microbial biomass in agricultural, pastoral, and forestry systems, rehabilitation of disturbed lands, and pesticide and heavy metals polluted soils. The methods are used to analyze microbial biomass carbon and nitrogen as explained in detail below.

Chloroform Fumigation Incubation (CFI)

In this method, a moist soil is fumigated with ethanol free chloroform for 24 h; chloroform is then removed by repeated evacuation; the soil is reinoculated with a small amount of unfumigated soil and then incubated at a constant temperature (usually 22 or 25 °C) for 10 days at field capacity or 50% of its water holding capacity (about −0.01 MPa). An additional soil sample is retained unfumigated and used as a control. The CO_2 evolved during incubation can be measured by gas chromatography, as a continuous flow or by sorption in alkali followed by titrimetric, conductometric, or colorimetric determination. As the net C mineralized as CO_2 is only a proportion of the total microbial biomass C, a kC factor is used to calculate total soil biomass C. As for as soil microbial biomass N determination, mineral N (NH_4-N and NO_3-N) from both fumigated and unfumigated (control) samples are extracted with 2Ml KCl after incubation. The mineral N in the extracts is then determined colorimetrically or by steam distillation. As for microbial biomass N, a kN factor is used to correct for incomplete mineralization of N from killed microorganisms for calculating total biomass N. Soil microbial biomass C and N are calculated from equations (1) and (2): Biomass C = (CO_2-C fumigated − CO_2-C control)/kC (1), Biomass N = (mineral N fumigated − mineral N control)/kN (2). The widely accepted kC value is 0.41 at 22 °C (Anderson and Domsch 1978) or 0.45 at 25 °C (Jenkinson and Powlson 1976). However, kN varies from 0.30 to 0.68 (Smith and Paul 1990). Jenkinson (1988) suggested a kN value of 0.57 at 25 °C, which is about 0.50 at 22 °C. Two basic assumptions of the CFI method are: (1) that CO_2-C evolved or mineral N produced during incubation in fumigated soil must exceed that from the corresponding unfumigated soil; and (2) that CO_2-C evolved or mineral N produced during incubation from the non-microbial source must be equal in both fumigated and unfumigated soil samples (Jenkinson 1988). In soils with relatively low microbial biomass but high respiration activity, subtraction of the CO_2 evolved from an unfumigated sample (control) often leads to low or even negative biomass estimates because unequal amounts of non-microbial biomass C is mineralized (Horwath et al. 1996). To overcome this problem, Jenkinson and Powlson (1976) suggested that CO_2-C released during the 10–20 day incubation rather than that from the initial 0–10 day incubation of unfumigated soil should be subtracted from the CO_2-C released from the fumigated soil. Horwath et al. (1996) suggested that the proportion of CO_2-C subtracted from the unfumigated (0–10 day incubation) soil should vary as a function of the ratio of CO_2-C fumigated/CO_2 control. When the ratio is large the proportion of CO_2-C subtracted from the unfumigated soil should be large and vice versa. They also suggested that equation (1) can be modified to: Biomass C = (0.71 × CO2-C fumigated – 0.23 × CO_2-C controls)/*k*C. However, the modified equation needs to be validated for soils under different land use and management and in different climates. The two basic assumptions mentioned above do not hold for soils with pH <5, air-dried soils, waterlogged soils, and soils that contain recently added organic materials or plant residues. In acidic soils, the re-establishment of a C and N mineralizing microbial population after fumigation and reinoculation is very slow. This causes a reduced mineralization of the killed microorganisms which makes the usual *k*C and *k*N factors invalid (Jenkinson 1988; Martens 1995). In air-dried soils, the amount of already dead microorganisms may constitute most of the microbial biomass in both fumigated and unfumigated soil samples, in addition to the less effective lysing of microbial cells by chloroform (Sparling and West 1989). In waterlogged soils, CO_2 and CH_4 are produced under conditions that restrict diffusion of gases (Jenkinson 1988). In soils with recently added organic materials or plant residues, the second assumption is not met since the mass of the re-established microbial population in the fumigated and reinoculated soil sample corresponds to only 10–20% of the original microbial biomass and consists mainly of bacteria. This can be avoided by either careful removal of the amendments such as roots, or a sufficient preincubation of at least 3 weeks (Martens 1995).

Chloroform Fumigation Extraction (CFE)

The above-mentioned limitations of the CFI method are mainly overcome by extraction of C and N with 0.5 mol K_2SO_4/L from the chloroform fumigated and the unfumigated soil samples. The proportions of C (*k*EC) and N (*k*EN) extracted from the fumigated (killed microbial biomass) soil vary from 0.2 to 0.68 (Jenkinson 1988; Martens 1995). However, most frequently used *k*EC values are in the range 0.36–0.45, while the *k*EN values are in the range 0.49–0.62. Likely limitations of the CFE method are differential extraction of released C from soils that differ in clay content and clay mineralogy, and variable *k* values (Martens 1995). The CFE method has been successfully used to estimate soil microbial biomass P (Hedley and Stewart 1982) and S (Saggar et al. 1981). Inorganic P is extracted with 0.5 mol Na_2HCO_3/L (pH 8.5) from both a fumigated and an unfumigated soil; the proportion of P is extracted from the killed microbial biomass, and the *k*P value is taken as 0.4. The allowance is also made for P sorption during fumigation and extraction by including an internal P standard. For strong P retention soils such as Ferrosols, Bray extractant (30 mmol NH_4F/L + 25 mmol HCl/L) appears to be more appropriate than 0.5 mol Na_2HCO_3/L extractant (Oberson et al. 1997). The procedure for microbial biomass S determination is similar to that for microbial biomass P but 0.15% $CaCl_2$ is used as an extractant and determined using turbidimetric method, the most commonly used *k*S value is 0.41 (Smith and Paul 1990).

Substrate-Induced Respiration (SIR)

An excess of substrate, usually glucose, is added to a soil, which is then incubated at constant temperature and moisture, and the respiration rate, CO_2 evolved per hour, is measured during a 0.5–2.5 h period, before the microorganisms start proliferating and actually increase microbial biomass (Anderson and Domsch 1978). Limitations of this method are: (1) that the pattern of soil microbial response to glucose differs between soils; (2) that only glucose responsive soil microbial biomass is measured; (3) that soils recently amended with organic materials or plant residues contain a large proportion of young cells, and, therefore, the conversion factor used, from mL CO_2/h to microbial biomass C of 40 (30 at 22 °C, Beck et al. 1997) for an average population in soil, is not valid (Martens 1995); (4) it measures only microbial activity which does not necessarily equate with microbial biomass; and (5) that microbial biomass N, P, and S cannot be measured (Smith and Paul 1990).

Adenosine Triphosphate Analysis (ATP)

Adenosine triphosphate is a universal constituent of living microbial cells. Although ATP can occur in dead microbial cells and extracellularly in soil, it is rapidly degraded by microorganisms. Therefore, ATP concentration in soil can be used to estimate the amount of living microbial biomass. It is usually extracted with acid reagents from moist, preincubated soil, and estimated by the luciferin–luciferase system. The C: ATP ratio is about 200 although it varies from 120 to 240 (Jenkinson et al. 1979; Eiland 1983; Martens 1995). The limitations of the ATP method are: (1) that ATP is decomposed by enzymatic and chemical hydrolysis during the extraction process; (2) after its release from microbial cells, ATP is strongly sorbed by soil constituents (Martens 1995); (3) biomass C: ATP ratio changes substantially over time in response to soil amendments such as organic materials and plant residues (Tsai et al. 1997); and (4) it cannot measure microbial biomass N, P, and S in soil (Smith and Paul 1990).

Phospholipid Fatty Acids

Phospholipid fatty acids with a chain length of <20 C atoms are considered to be of mainly bacterial origin (Harwood and Russel 1984). However, 18-C chain phospholipid fatty acid, 18: 2ω6 fatty acid constitute on average 43% of the total phospholipid fatty acid in soil fungi (Federle et al. 2010). Since ergosterol is specific to the fungal membrane (Seitz et al. 1979), the fungal biomass can be estimated from the correlation between the amounts of 18:2ω6 fatty acid and the ergosterol content. Frostegard and Baath (1996) observed a close correlation between the amounts of 18:2ω6 fatty acid and the ergosterol in soil ($r = 0.92$), thus, indicating that this phospholipid fatty acid can be used to estimate fungal biomass. The ratio of 18: 2ω6 fatty acid:bacterial phospholipid fatty acids is then used as a fungal:bacterial biomass ratio (Frostegard and Baath 1996). Phospholipid fatty acids can be extracted from soil with a one-phase mixture of chloroform, methanol, and citric acid buffer, fractionated into neutral, glyco- and phospholipids on columns containing silicic acid, methylated into fatty acid methyl esters, and then measured on a gas chromatograph/mass spectrometer. The advantage of the phospholipid fatty acid method, compared with other methods to estimate the microbial biomass of individual communities, is that both fungal and bacterial biomass can be estimated by the same technique in a single soil extract (Frostegard and Baath 1996). Currently PLFA analysis in soil and roots are being analysed using high throughput method, where PLFA is being eluted through 5:5:1 (chloroform, methanol, water) through column chromatography and eluted PLFA were transesterified and FAME profiles were identified using the MIDI PLFAD1 calibration mix and peak naming table through MIDI (MIDI, Inc., Newark, DE) system attached with GC (Buyer and Sasser 2012; Sharma and Buyer 2015). Although high throughput method is rapid, cost effective, and has added technical advantages than conventional method. However, its uses are limited due to high instrumentation costs and technical skills.

Ninhydrin Reaction Method

Amato and Ladd (1988) proposed to use ninhydrin reactive C and N compounds released during fumigation incubation as a measure of biomass. They specifically determined that fumigated soils retained protease but lost dehydrogenase activity required to decompose glucose and immobilize NH_4-N during the incubation period. They proposed to quantify ninhydrin reactive N compounds released in CFI (10 days incubation at 25 °C, extraction with 2N KCl) and determine biomass N by using a multiplication factor of 21. Thus the method differs from original CFI in which ninhydrin reactive C and N compounds rather than NH_4-N (or total mineral N) and CO_2 are taken into consideration while calculating biomass. Ocio and Brookes (1990) considered the ninhydrin method suitable for freshly amended soils (CFI gives unreliable results for such soils) and found good correlation with CFE and SIR. Sparling (1997) concluded that the ninhydrin method can give a reliable estimate of biomass in organic as well as mineral soils. Van Gestel et al. (1993) also determined biomass C indirectly by multiplying ninhydrin reactive extractable N of fumigated soils with 21 (Amato and Ladd 1988); they used 2N KCl for extraction. As compared with original CFI, the ninhydrin reaction method is less preferred due to its long processing time (at least 10 days is required for obtaining biomass values), nevertheless it has advantages due to its reliability in results particularly for freshly amended soils or soils rich in easily oxidizable C.

Microcalorimetry

Sparling (1981) proposed microcalorimetry as a method to assess microbial metabolism in soil on the pretext that the heat produced depends only on the initial and final energy states of the system and is independent of the types of organisms or reaction pathway. In addition, the total catabolic activity in the soil is closely related to the heat production; anabolic processes normally contribute a little to the heat. Sparling (1981) studied heat output from 12 soils and compared the results with CFI and SIR, ATP, dehydrogenase and amylase, and basal respiration. The rate of heat output from soil is closely related with the rate of respiration. Heat is found to be less correlated with most of parameters used. Hence, microcalorimetry method has not achieved popularity to a significant extent.

Microwave Irradiation

Microwave irradiation is an effective biocide treatment of soil which kills weeds, nematodes, and microorganisms; the effect on microorganisms being probably entirely thermal (Vela and Wu 1979), fungi being more susceptible (Wainwright et al. 1980). Spier et al. (1986) were probably the first to use microwave radiation for soil treatment to measure microbial biomass, an approach akin to $CHCl_3$ fumigation. In spite of its simplicity, this method has not gained widespread acceptability.

13.3.4 Comparison of Different Methods to Estimate Soil Microbial Biomass

Currently, all methods used to analyze soil microbial biomass have some limitations since these were developed for soils with microbial biomass in a relatively steady state. The soil microbial biomass has been measured through various methods in which values are variable due to having different *k* factors, soils at different moisture contents, different incubation temperatures, soils containing variable amounts of organic materials or plant residues, and different instrumentation and analytical techniques. Therefore, it is difficult to compare and get reproducible soil microbial biomass values obtained by different methods in different laboratories (Dalal 1998; Azam et al. 2003).

13.3.5 Soil Enzymes

Soil enzymes play a key role in the energy transfer through decomposition of soil organic matter and nutrient cycling, and hence play an important role in agriculture. Soil enzymes, being necessary catalysts for organic matter recycling, strongly influence on soil fertility and agronomic productivity (Rao et al. 2014). Soil enzymes are highly sensitive and quickly respond to any changes in soil management practices and environmental conditions. Their activities are closely related to physio-chemical and biological properties of the soil. Hence, soil enzymes are used as sensors for soil microbial status, for soil physio-chemical conditions, and for the influence of soil treatments or climatic factors on soil fertility. Understanding the possible roles of different soil enzymes in maintaining soil health can help in the soil health and fertility management, particularly in agricultural ecosystems (Rao et al. 2017). Some of the frequently analyzed soil enzymes for soil health point of view are discussed.

Phosphomonoesterase, i.e., acid and alkaline phosphatase activity in rhizosphere soil sample is determined using the procedure of Tabatabai (1994) with the following modification as suggested by Schinner et al. (1996). Arylsulfatase activity is measure by adopting the method of Sarathchandra and Perrott (1981). ß-Glucosidase is determined using *p*-nitrophenyl-ß-D-glucopyranoside (PNG, 0.05M) as substrate. This assay is based on the release and detection of p-nitrophenol (PNP) (Tabatabai 1982). Dehydrogenase activity is measure with reduction of 2,3,5-triphenyl-tetrazolium chloride (TTC) to triphenyl formazan (TPF) using colorimetric procedure of Tabatabai (1994). Fluorescein diacetate (FDA) hydrolysis is determined by the method of Schnürer and Rosswall (1982) and Aseri and Tarafdar (2006). Urease activity (urea amidohydrolase) is determined by the non-buffer method of Zantua and Bremner (1975).

13.3.6 Arbuscular Mycorrhizal Fungi

Arbuscular mycorrhizal fungi (AMF) establish a symbiotic relationship with more than 80% of terrestrial plants (Brundrett 2002). In order to establish a new mycorrhizal association, AMF forms infectious propagules such as spores, extraradical phase consisting of hyphae that develops into the soil, and intraradical phase consisting of arbuscules and vesicles (Linderman 1997) where its colonizing ability varies from species to species (Klironomos and Hart 2002). Spores proved efficient for infecting roots for *Gigaspora* and *Scutellospora* species whereas for *Glomus* and *Acaulospora* all inoculum forms were found to be equally efficient (Klironomos and Hart 2002). Several factors come into play while shaping the AMF community composition such as agricultural management practice (Jansa et al. 2006; Oehl et al. 2010; Curaqueo et al. 2011); soil type (Oehl et al. 2010); and concentration of nutrients (Gosling et al. 2013) and host species (Lovelock et al. 2003; Gosling et al. 2013), etc. AMF draws nutrients from the soil with the help of its extraradical hyphae for the use of the plant and receives photosynthates from plant in the root cortex as well as in the rhizospheric region (Smith and Read 2008). AMF together with fibrous roots facilitates the formation of sticky string bag where it mechanically binds soil aggregates together forming macroaggregates (Miller and Jastrow 2000). Practices such as tillage cause the mechanical disruption of hyphae (Boddington and Dodd 2000). AMF has also been credited with the production of heat-stable glycoprotein called glomalin (Wright and Upadhyaya 1996). Glomalin acts a soil particle cementing agent and its concentration strongly relates with soil aggregate stability (Wright and Upadhyaya 1998). Hence AMF are integral component of plant rhizosphere where array of microbial activities are taking place. The stabilized crop and soil conservation practices enhance AMF biomass (Sharma et al. 2012). Therefore

AMF can be used as potential indicator to assess the sustainability of long-term farming systems. The AMF biomass can be determined through microscopic and biochemical methods in terms of spore's density (Gerdemann and Nicolson 1963), root colonization (Phillips and Hayman 1970), and 16:1ω5cis PLFA and NLFA as AMF signature fatty acids (Sharma and Buyer 2015; Olsson 1999). Signature fatty acid analysis provides a more promising approach over the conventional methods. Glomalin has also been used as an indicator of AMF (Krivtsov et al. 2004). In the following sections we have provided a comprehensive assessment of techniques used for the quantification of AMF biomass. Quantification of AMF biomass has mainly been done through microscopic methods (Gerdemann and Nicolson 1963; Phillips and Hayman 1970).

13.3.6.1 Microscopic Methods of AMF Quantification

The quantification of AMF biomass is performed conventionally through extracting spores by wet sieving and decantation method (Gerdemann and Nicolson 1963). The suspension obtained can be observed directly or filtered through a filter paper disc and spores are counted under a microscope. For the assessment of root colonized by AMF, the techniques used include the root staining (Phillips and Hayman 1970) followed by quantification using the gridline intersect method (Giovannetti and Mosse 1980) that provides an estimate of root length colonized by AMF. Other important parameters include the measurement of hyphal dry weight and microscopic examination of stained hyphae for the study of extraradical hyphal length and hyphal connections (Miller et al. 1995; Mosse 2009).

13.3.6.2 Signature Fatty Acid Analysis

The intensity of response unveiled by the membrane lipids to instabilities/disturbances is highest (Denich et al. 2003). For the quantification of AMF signature fatty acid PLFA 16:1ω5cis has been extensively used (Olsson et al. 1995). Phospholipid 16:1ω5cis is a reflection of AMF extraradical hypha length and neutral lipid 16:1ω5cis portrays storage lipids that include spore copiousness (Olsson et al. 1997). Ester-linked fatty acids (ELFAs) include all the three major classes of lipids such as phospholipid, neutral lipid, and glycolipids (Sharma and Buyer 2015). ELFA 16:1ω5cis and 18:1ω5cis have also been used to study AMF dynamics (Grigera et al. 2007). Lipids are extracted through the Bligh–Dyer extraction method (Bligh and Dyer 1959) which is followed by division of lipids into phospholipids, neutral lipids, and glycolipids, which are later exposed to mild alkaline methanolysis and analyzed on a gas chromatograph (Frostegard et al. 1993). The use of solid phase extraction (SPE) technique by means of column chromatography further improves the extraction efficiency (Zelles et al. 1992; Zelles 1999). To advance further, a high throughput method was introduced that permitted the analysis of a batch of 96 samples within 48 h (Buyer and Sasser 2012). This high throughput technique implicates the Bligh–Dyer extraction of overnight dried samples and subsequent drying and dissolution of samples in chloroform followed by extraction using a 96 well solid phase extraction column. Elution of phospholipids is performed using 5:5:1 methanol: chloroform: H_2O in a 96 well format glass vial microplate after which drying, transesterification, and GC analysis are performed subsequently (Buyer and Sasser 2012). For the elution of neutral lipids, chloroform fraction obtained from the SPE column is used (Sharma and Buyer 2015). This method is applicable for both soil and roots (Buyer and Sasser 2012; Sharma and Buyer 2015). The biochemical method analyzing signature fatty acids provides an edge over the error-prone methods such as microscopic visualization of AMF structures. Nevertheless, the incidence of PLFA 16:1ω5cis in bacteria (Nichols et al. 1986) necessitates the need for confirmation of results using microscopic and molecular methods as well.

13.3.6.3 Glomalin

Glomalin is a thermostable glycoprotein formed on the hyphal walls of arbuscular mycorrhizal fungi (Wright and Upadhyaya 1996; Driver et al. 2005). Large quantity of glomalin remains attached to the hyphae and spores and as small as 20% becomes a part of the released fraction (Driver et al. 2005). Upon its release into the soil, it becomes a component of the stable organic matter (Wright and Upadhyaya 1996). Apparently, glomalin exists in two pools. Easily extractable glomalin is believed to be newly formed fraction belonging to young hyphae (Wright and Upadhyaya 1996, 1998; Wright 2000) whereas total glomalin fraction is considered to be a relatively recalcitrant fraction and is often referred to as older glomalin (Lovelock et al. 2004). As it is difficult to extract glomalin from the soil in pure form, Rillig (2004) recommended a new terminology for it, where it was called "glomalin-related soil protein" or "GRSP."

13.3.6.4 Prominence of Glomalin

It plays a key role in soil carbon sequestration as a constituent of the soil organic carbon pool (Rillig et al. 2001) and indirectly by enhancing soil aggregation by acting as a soil particle binding agent (Rillig et al. 2002; Wilson et al. 2009). It has been used as a proficient indicator to elucidate the effect of land use management (Rillig et al. 2003); soil quality and agricultural management approaches (Fokom et al. 2012); assessment of variations in AMF biomass (Krivtsov et al. 2004).

13.3.6.5 Extraction from Soil

Easily extractable glomalin fraction is extracted with 20 mM sodium citrate and 30–60 min autoclaving followed by centrifugation at 5000 xg, and total glomalin fraction is extracted with 50 mM sodium citrate and 60–90 min autoclaving followed by centrifugation at 5000 xg (Wright and Upadhyaya 1996, 1998). Bradford protein assay (Bradford 1976) is extensively used for the quantification of glomalin. The immunoreactive fraction of glomalin is quantified using ELISA (Wright and Upadhyaya 1996).The current extraction protocol rests on the fact that the harsh conditions of temperature and pressure employed for glomalin extraction destroy the vast majority of protein except for glomalin and to get higher recovery depending on soil types, samples may require many cycles of extraction (Agnihotri et al. 2015). The persistence of polyphenols (Whiffen et al. 2007), added glycoproteins and proteins from plant sources (Rosier et al. 2006) in glomalin extracts and their successive binding to Bradford reagent Coomassie brilliant blue G-250 (CBB) during quantification questions the presently used procedures of its extraction and quantification (Koide and Peoples 2013). Intraradically produced glomalin has been efficaciously used as a signal of AMF root colonization (Rosier et al. 2008).

13.3.7 Earth Worm

Earthworms belong to macrofauna (4–200 mm in size) but some species can reach the dimension attributed to megafauna (>200 mm) (Bachelier, 1986) and are considered soil engineers, as they are able to modify soil structure and features by their etho-physiological action (Gavinelli et al. 2018). Earthworm sampling should preferably be carried out during cool and wet seasons; sampling of dry soils (dry seasons) or of frozen soils should always be avoided. In temperate areas, sampling studies in autumn, spring, and some of the winter months give the best results (Paoletti 1999). Earthworm sampling can be done by hand sorting. It is the traditional method, in which active collection of earthworms from standard soil volumes advocated (Valckx et al. 2011). In detail, this technique consists of extracting a soil bulk (30 × 30 × 20 cm) with a spade fork (Paoletti 1999; Fusaro et al. 2018). Afterwards, a visual examination of soil bulk takes place for 15 min upon a white cloth and each earthworm is picked up. In order to collect deep burrower species, an effective recommendation is the use of an irritant suspension (Bouché 1972; Lee 1985) poured into the soil. The mustard powder water suspension (30 g L^{-1}) acts as an expellant for earthworms and it is a natural substance without toxic or dangerous consequences for the operator and the environment (Pelosi et al. 2009; Valckx et al. 2011). In the humid tropical forests some species are arboriculous and live in suspended soils, such as the soil that accumulates in the leaves rosette of bromeliads, in the tree canopy. These earthworms can be collected by photo-eclectors, a special trap that catches all moving invertebrates on the surface of trunks (Adis and Righi 1989).

13.4 Applications of Soil Health Indicators

Soil health encompasses the physical, chemical, and biological features, but the use of biological indicators is the least well advanced (Griffiths et al. 2018). Hence, for sustainable crop production, the application of different soil health indicators and their analytical techniques used have paramount significance. Lists of application of these indicators along with their analytical methods used in different laboratories are enlisted in Table 13.3.

Table 13.3 Different soil health indicators, analytical techniques and their applications

Soil health indicators	Protocol	References	Application	References
Soil physical health indicators				
Bulk density (Mg m^{-3})	Core sampler	Black (1965)	Soil management, provides information regarding soil compaction that might help in planning of modern farming techniques, Geochemical studies	Sharma and Bhattacharya (2017)
Water holding capacity	Keen Raczkowski cups	Piper (1966)	Availability of water to crops and helps to decide how much quantity of irrigation water to apply for crop production	Bhavya et al. (2018), Water Conservation Factsheet (2015)
Aggregate stability	Wet sieving techniques	Haynes (1993)		
Soil chemical health indicator				
pH	pH meter (supernatant suspension of 1:2 soil to water ratio)	Jackson (1973)	Indication of the acidity or alkalinity of soil and application of soil amendments such as lime/gypsum, crop suitability for location, rough indicator of plant nutrients availability	Kadam (2016), Hanlon (2015)
EC	EC meter (supernatant suspension of 1:2 soil to water ratio)	Jackson (1973)	Helps to know the nature of soil and soluble salts status	Kadam (2016)
CEC	Ammonium acetate method, barium chloride-triethanolamine method	Schollenberger and Dreibelbis (1930), Mehlich (1938)	Indicates the availability of cations in the soil for crop growth and development	Agronomy Fact Sheet Series (2007)
SOC	Walkley–Black or dry combustion/rapid titration method	Walkley and Black (1934)	Main indicator for soil fertility, maintains sustainable productivity and agro ecosystem	Hijbeek (2017), Moharana et al. (2017)
Available N	Micro-Kjeldahl procedure	Subbiah and Asija (1956), Chapman and Pratt (1961)	Imparts greenness to plant, improves growth and yield, and occupies a	Leghari et al. (2016)

			conspicuous place in plant metabolism system	
Available P	Olsen's reagent, Brays solution, ascorbic acid method	Olsen et al. (1954), Bray and Kurtz (1945), Watanabe and Olsen (1965)	Plays important role in plant physiological and biochemical processes and improves plant root architecture	Johnston and Steen (2000), Syers et al. (2008)
Available K	Flame photometrically/ ammonium acetate method	Piper (1966), Hanway and Heidal (1952)	Increases the tolerance level in the plants to biotic and abiotic stresses	Hasanuzzaman et al. (2018)
Available S	Calcium chloride method	Chesnin and Yien (1950)	Plays significant role in biosynthesis oil and enhances oil content in seed crops	Skwierawska et al. (2016)
Micronutrients (Zn, Fe, Mn, and Cu)	DTPA extraction (atomic absorption spectrophotometer)	Lindsay and Norvell (2010)	Acts as a catalyst in various oxidation–reduction reactions, plays integral part in chlorophyll synthesis and activators of several enzymes	Lohry (2007)
Soil biological and biochemical health indicators				
Soil microbial biomass carbon (μg g soil^{-1})	Chloroform fumigation extraction (CFE), dichromate digestion	Nunan et al. (1998), Vance et al. (1987)	Ecosystem services such as carbon cycling, nutrient turnover, production of trace gases, or pollutant degradation	Ananyeva et al. (2008), Makova et al. (2011)
Soil microbial N	Ninhydrin reactive N	Joergensen and Brookes (1990)	----------------do--------------------------	----------------do--------
Soil respiration (mg CO_2 kg^{-1} soil day^{-1})	Carbon dioxide release	Anderson and Domsch (1980)	----------------do--------------------------	
Substrate-induced respiration	Substrate addition (glucose)	Anderson and Domsch (1978)	----------------do--------------------------	----------------do--------
Enzymes				
Acid and alkaline phosphates	p-Nitrophenol release	Tabatabai (1994), Tabatabai and Bremner (1969)	Organic phosphorus cycling	Cardoso et al. (2013)

(continued)

Table 13.3 (continued)

Soil health indicators	Protocol	References	Application	References
ß-Glucosidase	p-Nitrophenol release	Tabatabai (1982)	C oxidation	Cardoso et al. (2013)
Dehydrogenase (µgTPF g $soil^{-1}$ day^{-1})	TTC reduction method	Tabatabai (1994), Klein et al. (1971)	Electron transferences in the respiratory chain in living	Cardoso et al. (2013)
FDAse (µg fluorescein g $soil^{-1}$ h^{-1})	Fluorescein diacetate method	Green et al. (2006), Schnürer and Rosswall (1982)	Total microbial activity	Cardoso et al. (2013)
Urease	Non-buffer method	Zantua and Bremner (1975)	Organic N mineralization to ammonia/ ammonium	Cardoso et al. (2013)
Soil organisms				
Mycorrhiza	Microscopic methods (spore density, root colonization)	Gerdemann and Nicolson (1963), Philips and Hayman	Crop production, impact of farming practices, reclamation of stressed ecosystems, drought tolerance	Sharma et al. (2012), Ruiz-Lozano (2003), Berruti et al. (2016), Kabir (2005)
Lipid profiling	Biochemical (16:1ω5cis) PLFA and NLFA PLFA	Sharma and Buyer (2015), Buyer and Sasser (2012), Olsson (1999)	Crop production systems, AMF quality inocula, soil microbial community diversity	Ferrari et al. (2018), Butler et al. (2012)
Earthworm	Hand sorting	Valckx et al. (2011)	Ecological engineers and soil fertility indicators	Van Groenigen et al. (2014)

13.5 Strategies for Management of Health Indicators

The different strategies employed to manage soil health indicators are varied with location, climate, soil type, and land use. But several general principles that focus on sustainable soil health management practices may suit in most of the situations to bring significant improvement in soil health indicators which are increased organic matter, decreased erosion, better water infiltration, more water holding capacity, less subsoil compaction, and less leaching of agrochemicals to groundwater (Rosa and Sobral 2008). The detailed management strategies are listed in Table 13.4.

13.6 Effects of Crop and Soil Management Practices on Soil Health Indicators: Previous Reports

The key crop and soil management practices such as crop rotation, nutrient management, and tillage practices influence the soil physical, chemical, and biological health indicators (Sharma et al. 2010). Crop rotation is a very ancient cultural practice (Howard 1996) that has a strong influence on soil structure, organic matter, and microbial communities (Janvier et al. 2007). Traditionally, it has been used to disrupt disease cycles (Curl 1963) and fix atmospheric nitrogen by legumes for subsequent non-leguminous crops (Pierce and Rice 1998). Sharma et al. (2012) showed the importance of including maize in rotation with soybean under conventional reduced tillage that helped in enhancing soybean yield, AM inoculum load, and organic carbon. Studies on tillage indicate that many critical soil quality indicators and functions can be improved by decreasing tillage intensity (Govaerts et al. 2007a). Compared to conventional tillage, reduced tillage practices offer not only long-term benefits to soil stability, reducing erosion, but also enhance soil microbial diversity (Welbaum et al. 2004; Govaerts et al. 2008). No till practices combined with crop residue retention increase soil organic matter content in the surface layer, improve soil aggregation, and preserve the soil resources better than conventional till practices (Govaerts et al. 2007b). Increased soil organic matter content associated with no till practices not only improves soil structure and water retention but also serves as a nutrient reservoir for plant growth and a substrate for soil microorganisms. Sharma et al. (2012) evaluated the impact of tillage practices and crop sequences on AM fungal propagules and soil enzyme activities in a 10-year long-term field trial in vertisols of soybean–wheat–maize (S–W–M) cropping system where S–M–W or S–W–M–W rotations under reduced-reduced tillage system showed higher soil dehydrogenase activity and fluorescein diacetate hydrolytic activity compared to other combinations. The inclusion of maize in the rotation irrespective of tillage systems showed comparatively higher mycorrhizal and higher phosphatase activities and organic carbon and maintained higher soybean yield. Organic amendments cover a wide range of inputs, including animal manure, solid waste, and various composts, and often improve soil health indicators and productivity. Girvan et al. (2004) and Melero et al. (2006) showed that these amendments, as well as crop residues, resulted in significant increases in total organic carbon

Table 13.4 Strategies of soil health management as per NRCS-USDA (2016)

Management strategies	What does it do?	How does it do?
(I) Conservation crop rotation		
Growing a diverse number of crops in a planned sequence in order to increase soil organic matter and biodiversity in the soil	– Increases nutrient cycling – Manages plant pests (weeds, insects, and diseases) – Reduces sheet, rill, and wind erosion and holds soil moisture – Adds diversity so soil microbes can thrive	– Improves nutrient use efficiency – Decreases use of pesticides – Improves water quality – Conserves water improves plant production
(II) Cover crop		
An un-harvested crop grown as part of planned rotation to provide conservation benefits to the soil	– Increases soil organic matter – Prevents soil erosion and conserves soil moisture – Increases nutrient cycling – Provides nitrogen for plant use, suppresses weeds, and reduces compaction	– Improves water quality and crop production – Conserves water and improves nutrient use efficiency – Decreases use of pesticides – Improves water efficiency
(III) No till		
A way of growing crops without disturbing the soil through tillage	– Increases organic matter and improves water holding capacity of soils – Reduces soil erosion and energy use – Decreases soil compaction	– Conserves water and improves water quality and efficiency – Improves air quality and crop production – Saves renewable resources – Increases productivity
(IV) Mulch tillage		
Using tillage methods where the soil surface is disturbed but maintains a high level of crop residue on the surface	– Reduces soil erosion from wind and rain – Increases soil organic matter, moisture and reduces energy use	– Improves water quality – Conserves water – Saves renewable resources – Improves air quality and crop production
(V) Mulching		
Applying plant residues or other suitable materials to the soil surface to compensate for loss of residue due to excessive tillage	– Reduces erosion from wind and rain and moderates soil temperatures – Increases soil organic matter and conserve soil moisture	– Conserves water, improves air and water quality – Improves crop productivity – Increases crop production

(continued)

Table 13.4 (continued)

Management strategies	What does it do?	How does it do?
	– Reduces dust and control weeds	– Reduces pesticide usage
(VI) Nutrient management		
Managing soil nutrients to meet crop needs while minimizing the impact on the environment and the soil	– Increases plant nutrient uptake – Improves physical, chemical, and biological properties of soil – Budgets, supplies, and conserves nutrients for plant production	– Improves water quality – Improves plant production – Improves air quality

(TOC), Kjeldahl-N, available-P, soil respiration, microbial biomass, and enzyme activities (e.g., protease, urease, and alkaline phosphatase). Microbial diversity and crop yields also increased as compared to conventional management. Khan et al. (2017) reported that integrated nutrient management practices (NPK+FYM) significantly increased soil organic matter and available water holding capacity but decreased the soil bulk density, creating a good soil condition for enhanced crop growth. Microbial population (bacteria, fungi, and actinomycetes) were very responsive to organic manure application. The long-term application of organic manures in rice-brown sarson cropping system increased the index value because it increased the nutrient index (NPKS and micronutrients), microbial index, and crop index of soils. Chemical indicators (pH, EC, and CEC) also improved with integrated nutrient management practices. The use of only chemical fertilizers in the rice–brown sarson cropping system resulted in poor soil microbial index and crop index. Soil pH decreased significantly over the initial values due to the application of organic manures in combination with chemical fertilizers. The lowering of soil pH toward the neutral range favors the availability of different major and micronutrients, viz. N, P, K, Fe, Cu, Mn, Zn, etc. which helps in optimum growth of plants. The highest organic carbon content (0.88%) found in 4 t ha^{-1} manure+ NPK and Zinc at 0.5 kg ha^{-1} applied plot. Hence, there was a great role of INM in augmenting the soil fertility build-up with respect to both major and micronutrients as well as in maintaining soil health indicators (Sur et al. 2010). Crop residue retention along with application of 50% recommended dose of potassium plus seed inoculation of potassium solubilizing bacterial has brought significant improvement in soil physical, chemical, and biological indicators under zero till maize–wheat cropping system and that intern helped in increasing productivity of maize and wheat crops (Raghavendra et al. 2018).

13.7 Conclusion

Soil health indicators are key elements required for maintaining the soil quality. The soil health indicators are dynamic in nature; some of soil health indicators (biological and chemical) are more prone to change in a shorter period whereas some (physical) may take longer period to change due to its management practices. Developing sustainable soil health indicators management practices by using a systematic approach that integrates soil physical, chemical, and biological principles into management practices will help in optimizing the sustainable crop production. There is a need for developing critical levels for some of the soil health indicators to which information is limited. Our research experiments should be planned in such a way that must include three aspects such as soil health indicators restoration, improvement, and maintenance. Systematic research is needed to study soil health indicators for diversity of edaphic, climatic, and management conditions. Conservation agricultural practices such as zero tillage, residue recycling, soil cover management, appropriate crop rotations, and integrated nutrient management practices along with addition of organic amendments have shown the proven benefit to improve soil health indicators.

References

Adis J, Righi G (1989) Mass migration and life cycle adaptation—a survival strategy of terrestrial earthworms in central Amazonian inundation forests. Amazoniana 11(1):23–30

Agnihotri R, Pandey A, Ramesh A, Billore SD, Sharma MP (2015) Contribution of native AM fungi to soil carbon sequestration assessed in the form of glomalin and C-stocks in different soil and crop management practices of soybean-based cropping system. In: Symbiotic Lifestyle (Interdisciplinary approach to mycorrhizal symbiosis): Proceedings of 8th International Congress of Symbiosis Society held in Lisbon, Portugal (Eds. Munzi et al.) from July 12–18, 2015, p 310

Agronomy Fact Sheet Series (2007) Cation exchange capacity (CEC), Department of Crop and Soil Sciences, College of Agriculture and Life Sciences, Cornell University Cooperative extension Fact Sheet 22

Amato M, Ladd JN (1988) Assay for microbial biomass based on ninhydrin reactive nitrogen in extracts of fumigated soils. Soil Biol Biochem 20:107–114

Ananyeva ND, Susyan EA, Chernova OV, Wirth S (2008) Microbial respiration activities of soils from different climatic regions of European Russia. Eur J Soil Biol 44:147–157

Anderson JPE, Domsch KH (1978) A physiological method for the quantitative measurement of microbial biomass in soils. Soil Biol Biochem 10:215–221

Anderson TH, Domsch KH (1980) Quantities of plant nutrients in microbial biomass of selected soils. Soil Sci 130:211–216

Arias ME, González-Pérez JA, González-Vila FJ, Ball AS (2005) Soil health - a new challenge for microbiologists and chemists. Int Microbiol 8:13–21

Arnold SL, Doran JW, Schepers J et al (2005) Portable probes to measure electrical conductivity and soil quality in the field. Commun Soil Sci Plant Anal 36:2271–2287. https://doi.org/10.1080/00103620500196689

Arshad MA, Coen GM (1992) Characterization of soil quality: physical and chemical criteria. Am J Altern Agric 7:25–31. https://doi.org/10.1017/s0889189300004410

Asady GH, Smucker AJM (1989) Compaction and root modifications of soil aeration. Soil Sci Soc Am J 53:251–254. https://doi.org/10.2136/sssaj1989.03615995005300010045x
Aseri GK, Tarafdar JC (2006) Fluorescein diacetate: a potential biological indicator for arid soils. Arid Land Res Manag 20(2):87 99. https://doi.org/10.1080/15324980500544473
Azam F, Farooq S, Lodhi A (2003) Microbial biomass in agricultural soils-determination, synthesis, dynamics and role in plant nutrition. Pak J Biol Sci 6(7):629–639
Bachelier G (1986) La vie animaledans le sol. O.R.S.T.O.M, Paris
Beck T, Joergensen G, Kandeler E, Makeschin E, Nuss H, Oberholzer R, Scheu S (1997) An inter-laboratory comparison of ten different ways of measuring soil microbial biomass C. Soil Biol Biochem 29:1023–1032
Bengough AG, Mullins CE (1990) Mechanical impedance to root growth: a review of experimental techniques and root growth responses. J Soil Sci 41:341–358. https://doi.org/10.1111/j.1365-2389.1990.tb00070.x
Bennett LT, Mele PM, Annett S, Kasel S (2010) Examining links between soil management, soil health, and public benefits in agricultural landscapes: an Australian perspective. Agric Ecosyst Environ 139:1–12. https://doi.org/10.1016/j.agee.2010.06.017
Berruti A, Lumini E, Balestrini R, Bianciotto V (2016) Arbuscular mycorrhizal fungi as natural biofertilizers: let's benefit from past successes. Front Microbiol 6:1559. https://doi.org/10.3389/fmicb.2015.01559
Bhavya VP, Anil Kumar S, Ashok A, Shivanna M, Shiva KM (2018) Changes in soil physical properties as a result of different land use systems with depth. Int J Curr Microbiol App Sci 7(1):2319–7706
Birkas M, Dexter A, Szemok A (2009) Tillage-induced soil compaction, as a climate threat increasing stressor. Cereal Res Commun 37:379–382. https://doi.org/10.1556/crc.37.2009.suppl.1
Bissonnais Y (1996) Aggregate stability and assessment of soil crustability and erodibility: I. Theory and methodology. Eur J Soil Sci 47:425–437. https://doi.org/10.1111/j.1365-2389.1996.tb01843.x
Black CA (1965) Methods of soil analysis: part 1, physical and mineralogical properties. American Society of Agronomy, Madison
Bligh EG, Dyer WJ (1959) A rapid method of total lipid extraction and purification. Can J Biochem Physiol 37:911–917
Boddington C, Dodd JC (2000) The effect of agricultural practices on the development of indigenous arbuscular mycorrhizal fungi. I. Field studies in an Indonesian ultisol. Plant Soil 218:137–144
Bouché MB (1972) Lombriciens de France ecologieet systématique. Institut National de la recherche Agronomique, Paris
Bradford MM (1976) A rapid and sensitive method for the quantitation of microgram quantities of protein utilizing the principle of protein-dye binding. Anal Biochem 72:248–254. https://doi.org/10.1016/0003-2697(76)90527-3
Bray RH, Kurtz LT (1945) Determination of total, organic and available forms of phosphorus in soils. Soil Sci 59:39–45
Brookes PC, Landman A, Pruden G, Jenkinson DS (1985) Chloroform fumigation and the release of soil nitrogen: a rapid direct extraction method to measure microbial biomass nitrogen in soil. Soil Biol Biochem 17:837–842
Brundrett MC (2002) Coevolution of roots and mycorrhizas of land plants. New Phytol 154:275–304. https://doi.org/10.1046/j.1469-8137.2002.00397.x
Bünemann EK, Bongiorno G, Bai Z et al (2018) Soil quality – a critical review. Soil Biol Biochem 120:105–125. https://doi.org/10.1016/j.soilbio.2018.01.030
Butler E, Whelan M, Ritz K, Sakrabani R, Van Egmond R (2012) The effect of triclosan on microbial community structure in three soils. Chemosphere 89:1–9
Buyer JS, Sasser M (2012) High throughput phospholipid fatty acid analysis of soils. Appl Soil Ecol 61:127–130

Cardoso EJBN, Vasconcellos RLF, Bini D et al (2013) Soil health: looking for suitable indicators. What should be considered to assess the effects of use and management on soil health? Sci Agric 70:274–289. https://doi.org/10.1590/S0103-90162013000400009

Chapman HD, Pratt PF (1961) Methods of analysis for soils, plants and waters. University of California, Los Angeles, pp 60–61

Chesnin L, Yien CH (1950) Turbidimetric determination of available sulphates. Soil Sci Soc Am J 15:149–151. https://doi.org/10.2136/sssaj1951.036159950015000C0032x

Curaqueo G, Barea JM, Acevedo E et al (2011) Effects of different tillage system on arbuscular mycorrhizal fungal propagules and physical properties in a Mediterranean agroecosystem in central Chile. Soil Tillage Res 113:11–18. https://doi.org/10.1016/j.still.2011.02.004

Curl E (1963) Control of plant diseases by crop rotation. Bot Rev 29:413–479

Dalal RC (1998) Soil microbial biomass-wheat do the numbers really mean? Aust J Exp Agric 38:649–665

Dalal RC, Moloney D (2000) Sustainability indicators of soil health and biodiversity. In: Hale P, Petrie A, Moloney D, Sattler P (eds) Management for sustainable ecosystems. Centre for Conservation Biology, The University of Queensland, Brisbane, pp 101–108

Davidson EA, Janssens IA (2006) Temperature sensitivity of soil carbon decomposition and feedbacks to climate change. Nature 440:165–173. https://doi.org/10.1038/nature04514

Denich TJ, Beaudette LA, Lee H, Trevors JT (2003) Effect of selected environmental and physico-chemical factors on bacterial cytoplasmic membranes. J Microbiol Methods 52:149–182. https://doi.org/10.1016/S0167-7012(02)00155-0

Dickman SR, Bray RH (1940) Colorimetric determination of phosphate. Ind Eng Chem Anal Ed 12:665–668

Doran JW, Safley M (1997) Defining and assessing soil health and sustainable productivity. In: Pankhurst C, Doube B, Gupta V (eds) Biological indicators of soil health. CAB International, Wallingford, pp 1–28

Driver JD, Holben WE, Rillig MC (2005) Characterization of glomalin as a hyphal wall component of arbuscular mycorrhizal fungi. Soil Biol Biochem 37:101–106. https://doi.org/10.1016/j.soilbio.2004.06.011

Eiland F (1983) A simple method for quantitative determination of ATP in soil. Soil Biol Biochem 15:665–670

Federle TW, Livingston RJ, Wolfe LE, White DC (2010) A quantitative comparison of microbial community structure of estuarine sediments from microcosms and the field. Can J Microbiol 32:319–325. https://doi.org/10.1139/m86-063

Ferrari AE, Ravnskov S, Wall LG (2018) Crop rotation in no-till soils modifies the soil fatty acids signature. Soil Use Manag 34(3):427–436

Fokom R, Adamou S, Teugwa MC et al (2012) Glomalin related soil protein, carbon, nitrogen and soil aggregate stability as affected by land use variation in the humid forest zone of south Cameroon. Soil Tillage Res 120:69–75. https://doi.org/10.1016/j.still.2011.11.004

Frostegard A, Baath E (1996) The use of phospholipid fatty acid analysis to estimate bacterial and fungal biomass in soil. Biol Fertil Soils 22:59–65

Frostegard A, Baath E, Tunlid A (1993) Shifts in the structure of soil microbial communities in limed forests as revealed by phospholipid fatty acid analysis. Soil Biol Biochem 25:723–730. https://doi.org/10.1016/0038-0717(93)90113-P

Fusaro S, Gavinelli F, Lazzarini F, Paoletti MG (2018) Soil biological quality index based on earthworms (QBS-e). A new way to use earthworms as bioindicators in agroecosystems. Ecol Indic 93:1276–1292. https://doi.org/10.1016/j.ecolind.2018.06.007

Gavinelli F, Barcaro T, Csuzdi C, Blakemore RJ, Fernandez Marchan D, De Sosa I, Dorigo L, Lazzarini F, Nicolussi G, Dreon AL, Toniello V, Pamio A, Squartini A, Concheri G, Moretto E, Paoletti MG (2018) Importance of large, deep-burrowing and anecic earthworms in forested and cultivated areas (vineyards) of northeastern Italy. Appl Soil Ecol 123:751–774

Gerdemann J, Nicolson T (1963) Spores of mycorrhizal endogone species extracted from soil by wet sieving and decanting. Trans Br Mycol Soc 46:235–244. https://doi.org/10.1016/S0007-1536(63)80079-0

Giovannetti M, Mosse B (1980) An evaluation of techniques for measuring vesicular arbuscular mycorrhizal infection in roots. New Phytol 84:489–500. https://doi.org/10.1111/j.1469-8137.1980.tb04556.x

Girvan MS, Bullimore J, Ball AS, Pretty JN, Osborne AM (2004) Responses of active bacterial and fungal communities in soils under winter wheat to different fertilizer and pesticides regimes. Appl Environ Microbiol 70:2692–2701

Gosling P, Mead A, Proctor M et al (2013) Contrasting arbuscular mycorrhizal communities colonizing different host plants show a similar response to a soil phosphorus concentration gradient. New Phytol 198:546–556. https://doi.org/10.1111/nph.12169

Govaerts B, Sayre KD, Lichter K, Dendooven L, Deckers J (2007a) Influence of permanent raised bed planting and residue management on physical and chemical soil quality in rain fed maize/wheat systems. Plant Soil 291:39–54

Govaerts B, Fuentes M, Sayre KD, Mezzalama M, Nicol JM, Deckers J, Etchevers J, Figueroa Sandoval B (2007b) Infiltration, soil moisture, root rot and nematode populations after 12 years of different tillage, residue and crop rotation managements. Soil Tillage Res 94:209–219

Govaerts B, Mezzalama M, Sayre KD, Crossa J, Lichter K, Troch V, Vanherck K, De Corte P, Deckers J (2008) Long-term consequences of tillage, residue management, and crop rotation on selected soil micro-flora groups in the subtropical highlands. Appl Soil Ecol 38:197–210

Green VS, Stott DE, Diack M (2006) Assay for fluorescein diacetate hydrolytic activity: optimization for soil samples. Soil Biol Biochem 38:693–701

Gregorich EG, Liang BC, Drury CF, Mackenzie AF, McGill WB (2000) Elucidation of the source and turnover of water soluble and microbial biomass carbon in agricultural soils. Soil Biol Biochem 32:581–587

Griffiths B, Faber J, Bloem J (2018) Applying soil health indicators to encourage sustainable soil use: the transition from scientific study to practical application. Sustainability 10(9):3021

Grigera MS, Drijber RA, Wienhold BJ (2007) Increased abundance of arbuscular mycorrhizal fungi in soil coincides with the reproductive stages of maize. Soil Biol Biochem 39:1401–1409. https://doi.org/10.1016/j.soilbio.2006.11.025

Håkansson I, Lipiec J (2000) A review of the usefulness of relative bulk density values in studies of soil structure and compaction. Soil Tillage Res 53:71–85. https://doi.org/10.1016/S0167-1987(99)00095-1

Haney RL, Franzluebbers AJ, Hons FM, Hossner LR, Zuberer DA (2001) Molar concentration of K_2SO_4 and soil pH effect estimation of extractable C with chloroform fumigation extraction. Soil Biol Biochem 33:1501–1507

Hanlon EA (2015) Soil pH and Electrical Conductivity: A County Extension Soil Laboratory Manual This document is CIR1081, one of a series of the Soil and Water Science Department, UF/IFAS Extension. Original publication date April 1993. Reviewed August 2015. Visit the EDIS website at http://edis.ifas.ufl.edu

Hanway JJ, Heidal H (1952) Soil analysis method as used in Iowa State College Soil Testing Laboratory. Iowa Agric 57:1–31

Harwood JL, Russel NJ (1984) Lipids in plants and microbes. Springer, Cham. https://doi.org/10.1007/978-94-011-5989-0

Hasanuzzaman M, Bhuyan MHMB, Nahar K, Hossain M, Mahmud J, Hossen M, Fujita M (2018) Potassium: a vital regulator of plant responses and tolerance to abiotic stresses. Agronomy 8:31

Haynes RJ (1993) Effect of sample pretreatment on aggregate stability measured by wet sieving or turbidimetry on soils of different cropping history. J Soil Sci 44:261–270. https://doi.org/10.1111/j.1365-2389.1993.tb00450.x

Hedley MJ, Stewart JWB (1982) Method to measure microbial biomass phosphorus in soils. Soil Biol Biochem 14:377–385

Hijbeek R (2017) On the role of soil organic matter for crop production in European arable farming PhD Thesis Submitted at Wageningen University Netherlands by the authority of the Rector Magnificus

Horwath WR, Paul EA (1994) Microbial biomass. In: Weaver RW, Angle JS, Bottomley PS (eds) Methods of soil analysis. Part 2. Microbiological and biochemical properties. SSSA, Madison, pp 753–773

Horwath WR, Paul EA, Harris D, Norton J, Jagger L, Horton KA (1996) Defining a realistic control for the chloroform fumigation-incubation method using microscopic counting and 14C-substrates. Can J Soil Sci 76:459–467

Howard RJ (1996) Cultural control of plant diseases: a historical perspective. Can J Plant Pathol 18:145–150

Idowu OJ, Van Es HM, Abawi GS et al (2008) Farmer-oriented assessment of soil quality using field, laboratory, and VNIR spectroscopy methods. Plant Soil 307:243–253. https://doi.org/10.1007/s11104-007-9521-0

Izquierdo I, Caravaca F, Alguacil MM et al (2005) Use of microbiological indicators for evaluating success in soil restoration after revegetation of a mining area under subtropical conditions. Appl Soil Ecol 30:3–10. https://doi.org/10.1016/j.apsoil.2005.02.004

Jackson ML (1973) Soil chemical analysis. Prentice Hall of India Pvt. Ltd., New Delhi

Jansa J, Wiemken A, Frossard E (2006) The effects of agricultural practices on arbuscular mycorrhizal fungi. In: Frossard E, Blum W, Warkentin B (eds) Function of soils for human societies and the environment. Geological Society, London, pp 89–115

Janvier C, Villeneuve F, Alabouvette C, Edel-Hermann V, Mateille T, Steinberg C (2007) Soil health through soil disease suppression: which strategy from descriptors to indicators? Soil Biol Biochem 39:1–23

Jeffries P, Silvo G, Silva P et al (2003) The contribution of arbuscular mycorrhizal fungi in sustainable maintenance of plant health and soil fertility. Biol Fertil Soils 37:1–16. https://doi.org/10.1007/s00374-002-0546-5

Jenkinson DS (1988) Determination of microbial biomass carbon and nitrogen in soil. In: Wilson JR (ed) Advances in nitrogen cycling in agricultural ecosystems. CAB International, Wallingford, pp 368–386

Jenkinson DS, Powlson DS (1976) The effects of biocidal treatment on metabolism in soil. V. A method for measuring soil biomass. Soil Biol Biochem 8:209–213

Jenkinson DS, Davidson SA, Powlson DS (1979) Adenosine triphosphate and microbial biomass in soil. Soil Biol Biochem 11:521–527

Joel A, Messing I (2001) Infiltration rate and hydraulic conductivity measured with rain simulator and disc permeameter on sloping arid land. Arid Land Res Manag 15:371–384

Joergensen RG, Brookes PC (1990) Ninhydrin-reactive nitrogen measurements of microbial biomass in 0.5 M K_2SO_4 soil extracts. Soil Biol Biochem 22(8):1023–1027

Johnston AE, Steen I (2000) Understanding phosphorus and its use in agriculture. European Fertilizer Manufacturers Association, Brussels

Kabir Z (2005) Tillage or no-tillage: impact on mycorrhizae. Can J Plant Sci 85:23–29

Kadam PM (2016) Study of pH and electrical conductivity of soil in Deulgaon Raja Taluka, Maharashtra. Int J Res Appl Sci Eng Technol 4(4):399–402

Karlen DL, Mausbach MJ, Doran JW, Cline RG, Harris RF, Schuman GE (1997) Soil quality: a concept, definition and framework for evaluation. Soil Sci Soc Am J 61:4–10

Katyal J, Datta S, Golui D (2016) Global review on state of soil health. Bull Indian Soc Soil Sci 30:1–33

Kelly JJ, Häggblom M, Tate RL (1999) Changes in soil microbial communities over time resulting from one time application of zinc: a laboratory microcosm study. Soil Biol Biochem 31:1455–1465. https://doi.org/10.1016/S0038-0717(99)00059-0

Kemper WD, Chepil WS (1965) Size distribution of aggregates. In: Black CA et al (eds) Methods of soil analysis, part I. Agronomy, vol 9, pp 499–510

Khan AM, Kirmani NA, Wani FS (2017) Effect of INM on soil carbon pools, soil quality and sustainability in rice-brown Sarson cropping system of Kashmir valley. Int J Curr Microbiol App Sci 6:785–809. https://doi.org/10.20546/ijcmas.2017.607.098

Klein DA, Loh TC, Goulding RL (1971) A rapid procedure to evaluate the dehydrogenase activity of soils low in organic matter. Soil Biol Biochem 3:385–387

Klironomos JN, Hart MM (2002) Colonization of roots by arbuscular mycorrhizal fungi using different sources of inoculum. Mycorrhiza 12:181–184. https://doi.org/10.1007/s00572-002-0169-6

Koide RT, Peoples MS (2013) Behavior of Bradford-reactive substances is consistent with predictions for glomalin. Appl Soil Ecol 63:8–14. https://doi.org/10.1016/j.apsoil.2012.09.015

Krivtsov V, Griffiths BS, Salmond R et al (2004) Some aspects of interrelations between fungi and other biota in forest soil. Mycol Res 108:933–946. https://doi.org/10.1017/S0953756204000516

Kuznetsova IV (1990) Optimum bulk density. Soviet Soil Sci 22:74–87

Laishram J, Saxena KG, Maikhuri RK, Rao KS (2012) Soil quality and soil health: a review. Int J Ecol Environ Sci 38:19–37

Lee KE (1985) S.l. In: Earthworms – their ecology and relationships with soils and land use. Academic, Sydney, p 411

Leghari SJ, Niaz AW, Ghulam ML, Abdul HL, Ghulam MB, Khalid ST, Tofique AB, Safdar AW, Ayaz AL (2016) Role of nitrogen for plant growth and development: a review. Adv Environ Biol 10:209–218

Linderman RG (1997) Vesicular-arbuscular mycorrhizal (V AM) fungi. In: Carroll GC (ed) Plant relationships part B. The mycota (a comprehensive treatise on fungi as experimental systems for basic and applied research). Springer, Berlin

Lindsay WL, Norvell WA (2010) Development of a DTPA soil test for zinc, iron, manganese, and copper. Soil Sci Soc Am J 42:421–428. https://doi.org/10.2136/sssaj1978.03615995004200030009x

Lohry R (2007) Micronutrients: functions, sources and application methods. 2007 Indiana CCA Conference Proceedings

Lovelock CE, Andersen K, Morton JB (2003) Arbuscular mycorrhizal communities in tropical forests are affected by host tree species and environment. Oecologia 135:268–279. https://doi.org/10.1007/s00442-002-1166-3

Lovelock CE, Wright SF, Clark DA, Ruess RW (2004) Soil stocks of glomalin produced by arbuscular mycorrhizal fungi across a tropical rain forest landscape. J Ecol 92:278–287. https://doi.org/10.1111/j.0022-0477.2004.00855.x

Makova J, Javoreková S, Medo J, Majerčíková K (2011) Characteristics of microbial biomass carbon and respiration activities in arable soil and pasture grassland soil. J Cent Eur Agric 12 (4):752–765

Martens R (1995) Current methods for measuring microbial biomass C in soil: potentials and limitations. Biol Fertil Soils 19:87–99

Masto RE, Pramod K, Singh CD, Patra AK (2009) Changes in soil quality indicators under long-term sewage irrigation in a sub-tropical environment. Environ Geol 56:1237–1243

Mehlich A (1938) Use of triethanolamine acetate-barium hydroxide buffer for the determination of some base Exchange properties and lime requirement of soil. Soil Sci Soc Am Proc 29:374–378

Melero S, Porras JCR, Herencia JF, Madejon E (2006) Chemical and biochemical properties in a silty loam soil under conventional and organic management. Soil Tillage Res 90:162–170

Miller RM, Jastrow JD (2000) Mycorrhizal fungi influence soil structure. In: Kapulnik Y, Douds DD (eds) Arbuscular mycorrhizas: physiology and function. Kluwer Academic Publishers, Dordrecht, pp 3–18

Miller R, Reinhardt D, Jastrow J (1995) External hyphal production of vesicular-arbuscular mycorrhizal fungi in pasture and tallgrass prairie communities. Oecologia 103:17–23. https://doi.org/10.1007/BF00328420

Moebius BN, Van Es HM, Schindelbeck RR et al (2007) Evaluation of laboratory-measured soil properties as indicators of soil physical quality. Soil Sci 172:895–912. https://doi.org/10.1097/ss.0b013e318154b520

Moharana PC, Sharma BM, Biswas DR (2017) Changes in the soil properties and availability of micronutrients after six-year application of organic and chemical fertilizers using STCR-based targeted yield equations under pearl millet-wheat cropping system. J Plant Nutr 40(2):165–176

Mosse B (2009) Observations on the extra-matrical mycelium of a vesicular-arbuscular endophyte. Trans Br Mycol Soc 42:439–448. https://doi.org/10.1016/s0007-1536(59)80044-9

Nichols P, Stulp BK, Jones JG, White DC (1986) Comparison of fatty acid content and DNA homology of the filamentous gliding bacteria Vitreoscilla, Flexibacter, Filibacter. Arch Microbiol 146:1–6. https://doi.org/10.1007/BF00690149

NRCS-USDA (2016) Natural resources conservation service-United States Department of Agriculture. Healthy, productive soils checklist for growers. www.nrcs.usda.gov

Nunan N, Morgan MA, Herlihy M (1998) Ultraviolet absorbance (280 nm) of compounds released from soil during chloroform fumigation as an estimate of the microbial biomass. Soil Biol Biochem 30:1599–1603

O'Farrell PJ, Donaldson JS, Hoffman MT (2010) Vegetation transformation, functional compensation, and soil health in a semi-arid environment. Arid Land Res Manage 24:12–30. https://doi.org/10.1080/15324980903439263

Oberson A, Friesen DK, Morel C, Tiessen H (1997) Determination of phosphorus released by chloroform fumigation from microbial biomass in high P sorbing tropical soils. Soil Biol Biochem 29:1579–1583

Ocio JA, Brookes PC (1990) An evaluation of methods for measuring the microbial biomass in soils following recent additions of wheat straw and the characterization of the biomass that develops. Soil Biol Biochem 22:685–694

Oehl F, Laczko E, Bogenrieder A et al (2010) Soil type and land use intensity determine the composition of arbuscular mycorrhizal fungal communities. Soil Biol Biochem 42:724–738. https://doi.org/10.1016/j.soilbio.2010.01.006

Olsen BC, Cole CV, Watenabe FS, Dean LA (1954) Estimation of available phosphorus by extraction with sodium carbonate. USDA Circ Number 939:19

Olsson PA (1999) Signature fatty acids provide tools for determination of the distribution and interactions of mycorrhizal fungi in soil. FEMS Microbiol Ecol 29:303–310

Olsson PA, Bååth E, Jakobsen I, Söderström B (1995) The use of phospholipid and neutral lipid fatty acids to estimate biomass of arbuscular mycorrhizal fungi in soil. Mycol Res 99:623–629. https://doi.org/10.1016/j.ecoleng.2010.11.01110.1016/S0953-7562(09)80723-5

Olsson PA, Bååth E, Jakobsen I (1997) Phosphorus effects on the mycelium and storage structures of an arbuscular mycorrhizal fungus as studied in the soil and roots by analysis of fatty acid signatures. Appl Environ Microbiol 63:3531–3538

Olsson PA, Thingstrup I, Jakobsen I, Bååth E (1999) Estimation of the biomass of arbuscular mycorrhizal fungi in a linseed field. Soil Biol Biochem 31:1879–1887. https://doi.org/10.1016/S0038-0717(99)00119-4

Paoletti MG (1999) The role of earthworms for assessment of sustainability and as bioindicators. Agric Ecosyst Environ 74:137–155

Paoletti MG, D'Incà A, Tonin E et al (2010) Soil invertebrates as bio-indicators in a natural area converted from agricultural use: the case study of Vallevecchia-Lugugnana in north-eastern Italy. J Sustain Agric 34:38–56. https://doi.org/10.1080/10440040903396698

Pelosi C, Bertrand M, Roger-Estrade J (2009) Earthworm collection from agricultural fields: comparisons of selected expellants in presence/absence of hand-sorting. Eur J Soil Biol 45:176–183

Phillips J, Hayman D (1970) Improved procedures for clearing roots and staining parasitic and vesicular-arbuscular mycorrhizal fungi for rapid assessment of infection. Trans Br Mycol Soc 55:158–161. https://doi.org/10.1016/S0007-1536(70)80110-3

Pierce FJ, Rice CW (1998) Crop rotation and its impact on efficiency of water and nitrogen use. In: Hargrove WL (ed) Cropping strategies for efficient use of water and nitrogen, special publication no 51. American Society of Agronomy, Madison, pp 21–36

Piper CS (1966) Soil and plant analysis. Hans Publisher, Bombay

Raghavendra M, Singh YV, Gaind S, Meena MC, Das TK (2018) Effect of potassium and crop residue levels on potassium solubilizers and crop yield under maize wheat rotation. Int J Curr Microbiol App Sci 7(6):424–435. https://doi.org/10.20546/ijcmas.2018.706.048

Rao CHS, Lal R, Kundu S, Prasad BMBB, Venkateswarlu B, Singh AK (2014) Soil carbon sequestration in rainfed production systems in the semiarid tropics of India. Sci Total Environ 487:587–603. https://doi.org/10.1016/j.scitotenv.2013.10.006

Rao CHS, Grover M, Kundu S, Desa S (2017) Soil enzymes. Encyclopedia of soil science, 3rd edn. Taylor & Francis, Boca Raton

Reynolds WD, Drury CF, Tan CS et al (2009) Use of indicators and pore volume-function characteristics to quantify soil physical quality. Geoderma 152:252–263. https://doi.org/10.1016/j.geoderma.2009.06.009

Rillig MC (2004) Arbuscular mycorrhizae, glomalin, and soil aggregation. Can J Soil Sci 84:355–363

Rillig MC, Wright SF, Nichols KA et al (2001) Large contribution of arbuscular mycorrhizal fungi to soil carbon pools in tropical forest soils. Plant Soil 233:167–177. https://doi.org/10.1023/A:1010364221169

Rillig MC, Wright SF, Eviner VT (2002) The role of arbuscular mycorrhizal fungi and glomalin in soil aggregation: comparing effects of five plant species. Plant Soil 238:325–333

Rillig MC, Ramsey PW, Morris S, Paul EA (2003) Glomalin, an arbuscular-mycorrhizal fungal soil protein, responds to land-use change. Plant Soil 253:293–299. https://doi.org/10.1023/A:1024807820579

Rimal BK, Lal R (2009) Soil and carbon losses from five different land management areas under simulated rainfall. Soil Tillage Res 106:62–70. https://doi.org/10.1016/j.still.2009.09.014

Rosa DDL, Sobral R (2008) Soil quality and methods for its assessment. In: Braimoh AK, Vlek PLG (eds) Land use and soil resources. Springer, Dordrecht

Rosier CL, Hoye AT, Rillig MC (2006) Glomalin-related soil protein: Assessment of current detection and quantification tools. Soil Biol Biochem 38:2205–2211. https://doi.org/10.1016/j.soilbio.2006.01.021

Rosier CL, Piotrowski JS, Hoye AT, Rillig MC (2008) Intraradical protein and glomalin as a tool for quantifying arbuscular mycorrhizal root colonization. Pedobiologia 52:41–50. https://doi.org/10.1016/j.pedobi.2008.02.002

Ross DS, Matschonat G, Skyllberg U (2008) Cation exchange in forest soils: the need for a new perspective. Eur J Soil Sci 59:1141–1159. https://doi.org/10.1111/j.1365-2389.2008.01069.x

Ruiz-Lozano JM (2003) Arbuscular mycorrhizal symbiosis and alleviation of osmotic stress. New perspectives for molecular studies. Mycorrhiza 13:309–317

Saggar S, Bettany JR, Stewart JWB (1981) Measurement of microbial sulphur in soils. Soil Biol Biochem 13:493–498

Sarathchandra SU, Perrott KW (1981) Determination of phosphatase and sulphatase activity in soil. Soil Biol Biochem 13:543–545

Schinner F, Öhlinger R, Kandeler E, Margesin R (1996) Methods in soil biology. Springer, Berlin

Schnürer J, Rosswall T (1982) Fluorescein diacetate hydrolysis as a measure of total microbial activity in soil and litter. Appl Environ Microbiol 43:1256–1261

Schollenberger C, Dreibelbis ER (1930) Analytical methods in base-exchange investigations on soils. Soil Sci 30:161–174

Seitz LM, Sauer DB, Burroughs R, Mohr HE, Hubbard JD (1979) Ergosterol as a measure of fungal growth. Phytopathology 69:1202–1203

Sharma B, Bhattacharya S (2017) Soil bulk density as related to soil texture, moisture content, Ph, electrical conductivity, organic carbon, organic matter content and available macro nutrients of Pandoga sub watershed, Una District of H.P (India). Int J Eng Res Dev 13(12):72–76

Sharma MP, Buyer JS (2015) Comparison of biochemical and microscopic methods for quantification of mycorrhizal fungi in soil and roots. Appl Soil Ecol 95:86–89

Sharma SK, Ramesh A, Sharma MP, Joshi OP, Govaerts B, Kerri LS, Douglas LK (2010) Microbial community structure and diversity as indicators for evaluating soil quality. In: Lichtfouse E (ed) Biodiversity, biofuels, agroforestry and conservation agriculture, vol 5. Springer, Cham

Sharma MP, Gupta S, Sharma SK, Vyas AK (2012) Effect of tillage and crop sequences on arbuscular mycorrhizal symbiosis and soil enzyme activities in soybean (Glycine max L. Merrill) rhizosphere. Indian J Agric Sci 82:25–30

Shaw E, Dean LA (1952) The use of dithizone as an extractant to estimate the zinc nutrient status of soils. Soil Sci 73:341–344

Silva IR, SáMendonça E (2007) Matériaorgânica do solo = Soil organic matter. In: Novais RF, Alvarez VH, Barros NF, Fontes RLF, Cantarutti RB, Neves JC (eds) Fertilidade do solo Soil fertility. Sociedade Brasileira de Ciência do Solo, Viçosa, pp 275–374

Six J, Elliott ET, Paustian K (2000) Soil macroaggregate turnover and microaggregate formation: a mechanism for C sequestration under no-tillage agriculture. Soil Biol Biochem 32:2099–2103. https://doi.org/10.1016/S0038-0717(00)00179-6

Skwierawska M, Benedycka Z, Jankowski K, Skwierawski A (2016) Sulphur as a fertiliser component determining crop yield and quality. J Elem 21(2):609–623. https://doi.org/10.5601/jelem.2015.20.3.992

Smith JL, Paul EA (1990) The significance of soil biomass estimations. In: Bollog JM, Stotzky G (eds) Soil biochemistry, vol 6. Marcel Dekker, New York, pp 357–396

Smith SE, Read DJ (2008) Mycorrhizal symbiosis, 3rd edn. Academic, New York

Smith JL, Halvorson JJ, Bolton H Jr (2002) Soil properties and microbial activity across a 500 m elevation gradient in a semi-arid environment. Soil Biol Biochem 34:1749–1757. https://doi.org/10.1016/S0038-0717(02)00162-1

Sparling GP (1981) Microcalorimetry another methods to assess biomass and activity in soil. Soil Biol Biochem 13:93–98

Sparling GP (1997) Soil microbial biomass, activity and nutrient cycling as indicators of soil health. In: Doube BM, Gupta VVSR (eds) Biological indicators of soil health Pankhurst CE. CAB International, Wallingford, pp 97–119

Sparling GP, West AW (1989) Importance of soil water content when estimating soil microbial C, N and P by the fumigation–extraction methods. Soil Biol Biochem 21:245–253

Spier TW, Cowling JC, Sparling GP, West AW, Corderoy DM (1986) Effect of microwave radiation on the microbial biomass, phosphatase activity and levels of extractable N and P in low fertility soil under pasture. Soil Biol Biochem 18:377–382

Subbiah BV, Asija GL (1956) A rapid procedure for the determination of available nitrogen in soils. Curr Sci 25:259–260

Sur P, Mandal M, Das DK (2010) Effect of integrated nutrient management on soil fertility and organic carbon in cabbage (Brassica oleracea var. capitata) growing soils. Indian J Agri Sci 80:695

Syers JK, Johnston AE, Curtin D (2008) Efficiency of soil and fertilizer phosphorus use. FAO Fertilizer and Plant Nutrition Bulletin 18. FAO, Rome, p 108

Tabatabai MA (1982) Soil enzymes. In: Page AL, Miller RH, Keeney DR (eds) Methods of soil analysis, part 2, vol 9, 2nd edn. American Society of Agronomy, Madison, pp 903–947

Tabatabai MA (1994) Soil enzymes. In: Weaver RW, Angle S, Bottomley P et al (eds) Methods of soil analysis part 2: microbiological and biochemical properties. Soil Science Society of America, Madison, pp 775–833

Tabatabai MA, Bremner JM (1969) Use of p-nitrophenyl phosphate for assay of soil phosphatase activity. Soil Biol Biochem 1:301–307

Tokunaga A (2006) Effects of bulk density and soil strength on the growth of blue wildrye (Elymus Glaucus Buckl.). The Faculty of Humboldt State University, Arcata

Tsai CS, Killham K, Cresser MS (1997) Dynamic response of microbial biomass, respiration rate and ATP to glucose additions. Soil Biol Biochem 29:1249–1256

Valckx J, Govers G, Hermy M, Muys B (2011) Optimizing earthworm sampling in ecosystems. In: Karaca A (ed) Biology of earthworms. Soil biology, vol 24. Springer, Berlin

Van Grestel MR, Mercxk R, Vlassak K (1993) Soil drying and rewetting and the turnover of 14C-labelled plant residue: first order decay rates of biomass and non biomass 14C. Soil Biol Biochem 25:125–134

Van Groenigen JW, Lubbers IM, Vos HM, Brown GG, De Deyn GB, VanGroenigen KJ (2014) Earthworms increase plant production: a meta-analysis. Sci Rep 15(4):6365

Vance ED, Brookes PC, Jenkinson DS (1987) An extraction method for measuring soil microbial biomass C. Soil Biol Biochem 19:703–707

Vargas Gil S, Meriles J, Conforto C et al (2009) Field assessment of soil biological and chemical quality in response to crop management practices. World J Microbiol Biotechnol 25:439–448. https://doi.org/10.1007/s11274-008-9908-y

Vela GR, Wu JF (1979) Mechanism of lethal action of 2450-MHz radiation on microorganisms. Appl Environ Microbiol 37:550–553

Wainwright M, Killham K, Diprose MF (1980) Effect of 2450-MHz radiation on nitrification, respiration and S-oxidation in soil. Soil Biol Biochem 12:489–491

Walkley A, Black TA (1934) An examination of the Degtjareff method for determination of soil organic matter and a proposed modification of chromic acid titration. Soil Sci 37:29–38

Watanabe F, Olsen SR (1965) Test of an ascorbic acid method for determining P in water and NaHCO3 extracts from soil. Soil Sci Soc Proc 29:677–678. https://doi.org/10.2136/sssaj1965.03615995002900060025x

Water Conservation Factsheet (2015) Soil water storage capacity and available soil moisture. Ministry of Agriculture Factsheet British Columbia 619.000-1

Webster JJ, Hampton GJ, Leach FR (1984) ATP in soil: a new extractant and extraction procedure. Soil Biol Biochem 16:335–342

Weil R, Magdoff F (2004) Significance of soil organic matter to soil quality and health. In: Weil R, Magdoff F (eds) Soil organic matter in sustainable agriculture. CRC Press, Boca Raton, pp 1–43

Weil RR, Islam KR, Stine MA, Gruver JB, Samson-Liebig SE (2003) Estimating active carbon for soil quality assessment: a simplified method for laboratory and field use. Am J Altern Agric 18 (1):3–17

Welbaum GE, Sturz AV, Dong Z, Nowak J (2004) Managing soil microorganisms to improve productivity of agro-ecosystems. Crit Rev Plant Sci 23:175–193

Whiffen LK, Midgley DJ, McGee PA (2007) Polyphenolic compounds interfere with quantification of protein in soil extracts using the Bradford method. Soil Biol Biochem 39:691–694. https://doi.org/10.1016/j.soilbio.2006.08.012

Wilson GWT, Rice CW, Rillig MC et al (2009) Soil aggregation and carbon sequestration are tightly correlated with the abundance of arbuscular mycorrhizal fungi: results from long-term field experiments. Ecol Lett 12:452–461. https://doi.org/10.1111/j.1461-0248.2009.01303.x

Wright SF (2000) A fluorescent antibody assay for hyphae and glomalin from arbuscular mycorrhizal fungi. Plant Soil 226:171–177

Wright SF, Upadhyaya A (1996) Extraction of an abundant and unusual protein from soil and comparison with hyphal protein of arbuscular mycorrhizal fungi. Soil Sci 161:575–586

Wright SF, Upadhyaya A (1998) A survey of soils for aggregate stability and glomalin, a glycoprotein produced by hyphae of arbuscular mycorrhizal fungi. Plant Soil 198:97–107

Yeomans JC, Bremner JM (1988) A rapid and precise method for routine determination of organic carbon in soil. Commun Soil Sci Plant Anal 19:1467–1476

Zantua MI, Bremner JM (1975) Comparison of methods of assaying urease activity in soils. Soil Biol Biochem 7(4-5):291–295

Zelles L (1999) Fatty acid patterns of phospholipids and lipopolysaccharides in the characterisation of microbial communities in soil: a review. Biol Fertil Soils 29:111–129. https://doi.org/10.1007/s003740050533

Zelles L, Bai QY, Beck T, Beese F (1992) Signature fatty acids in phospholipids and lipopolysaccharides as indicators of microbial biomass and community structure in agricultural soils. Soil Biol Biochem 24:317–323. https://doi.org/10.1016/0038-0717(92)90191-Y

Indexing Methods of Soil Quality in Agro-Ecosystems: An Overview of Indian Soils and Beyond

14

Debarati Bhaduri, Tapan Jyoti Purakayastha, Pragati Pramanik, Ajoy Saha, Bibhash Chandra Verma, Mohammad Shahid, and Debasish Saha

Abstract

Soil quality research has been started in late 1990s to early 2000 in different parts of the world. The initial understanding was tough and confined to a group of researchers only as revealed from the publications, but slowly the interest was engulfed among soil scientists. Different authors define soil quality in number of ways. Soil quality is not limited to particular parameters or attributes hence it is difficult to calculate the soil quality in a simple step. To assess soil quality there is a need of soil quality index (SQI) which is the function of indicators termed as soil quality indicators. Mostly the SQIs are aimed to particular problems like soil erosion, soil pollution and soil nutrient depletion, crop production and productivity, etc. To identify the best suitable indicators there are few simple but important steps to keep in mind, like level of significance for various soil indicators as influenced by various management practices, cropping system, etc., selecting the representative minimum data set (MDS) with the help of suitable statistical techniques, correlation analysis among soil variables to reduce spurious grouping among highly weighted variables, scoring of the minimum data set indicators, and finally the computation of soil quality index. Here in this review we have discussed various progresses made so far, mostly in various Indian conditions. A vivid description of SQI developed under various agro-ecosystems (including

D. Bhaduri (✉) · M. Shahid
ICAR-National Rice Research Institute, Cuttack, Odisha, India

T. J. Purakayastha · P. Pramanik
ICAR-Indian Agricultural Research Institute, New Delhi, India

A. Saha
Regional Centre, ICAR-Central Inland Fisheries Research Institute, Bengaluru, Karnataka, India

B. C. Verma
CRURRS, ICAR-NRRI, Hazaribagh, Jharkhand, India

D. Saha
Biosystems Engineering and Soil Science, University of Tennessee, Knoxville, TN, USA

A. Rakshit et al. (eds.), *Soil Analysis: Recent Trends and Applications*,
https://doi.org/10.1007/978-981-15-2039-6_14

cereal, pulse, horticultural and value added crops like cotton, tea, medicinal plants) has been emphasized along with spatial approaches for SQI and the effect of ecosystems on SQI. The understanding of SQI is bit complex, but considering the immense possibility of this useful concept and progression required in future, we trust this compilation will serve a beneficial document for the years to come.

Keywords
Soil quality · Sustainable agriculture · Soil quality index · Indexing approaches · Soil health · Ecosystems · Climate change

14.1 Introduction and Concept of Soil Quality

Increasing population coupled with demand for food results in over-exploitation of the natural resources under intensive agricultural production systems which cause adverse impact on natural resources and lead to stagnation of the system productivity. These situations impart a serious concern on the sustainability, viability, and profitability of agriculture based system. To meet the requirement of large population, the biggest challenge in sustainable agriculture is to maintain a balance among natural resources, ecosystem, and demand (food, fiber, fuel, etc.). Sustainability of ecosystem in general and agriculture in particular largely depends on different natural resources and among them soil is one of the most important natural resource. Soil is a dynamic, living, porous, three dimensional natural body which is vital to the function of terrestrial ecosystems and represents a balance among physical, chemical, and biological factors. Importance of soil is well known to all of us; however, its importance in terms of agriculture or crop production is mainly judged by its inherent capacity to support the crop or plant and measured in terms of soil quality and health. In this context, soil quality/health acquires an important dimension related to agro-ecosystem sustainability.

There is very thin margin between soil quality and soil health that is defined and differentiated by different workers in various ways. Broad definition of soil quality as proposed by The Soil Science Society of America (SSSA) is "The ability of a specific type of soil to function within natural or managed ecosystem boundaries, to sustain plant and animal productivity, maintain or improve air quality and water to support human health and livable" (Karlen et al. 1997). In other way, soil quality has been simply defined as the "fitness for use" (Pierce and Larson 1993; Acton and Gregorich 1995) and as the "capacity of a soil to function" (Doran and Parkin 1994; Karlen et al. 1997). Karlen et al. (1992) defined soil quality as "the ability of the soil to serve as a natural medium for the growth of plants that sustain human and animal life" or soil quality refers to its ability to sustain productivity and maintain environmental quality (Lal 1993). Other researchers also summarized the concept of soil quality as a resource for food production, to support human life and to preserve or improve the soil for future generations. Soil quality of any region depends on climatic conditions, soil characteristics, vegetation, anthropogenic influence, and interactions among them. Assessment of soil quality is major concern because soil is a complex system which is governed by several (physical, chemical, and biological) factors that interacts with each other in a very complex manner and then makes the soil suitable to function. As a complex functional state, soil quality cannot be measured directly; however, it may be inferred from management-induced changes in soil properties, which is better known as soil quality indicators. Soil quality indicators are measurable soil attributes that influence the capacity of soil to perform crop production or environmental functions and are also sensitive to change in land use management and conservation practices. Soil quality indicators have been defined from ecological, economic, and social development perspectives; they usually take into consideration soil properties and associated crops that can be used in response to the dynamic changes in agro-ecosystems. These indicators are neither well defined nor accepted or approved parameters to characterize or to define soil quality (Bouma 2002). As soil quality is a function of physical, chemical, and biological properties hence its indicator must take care of these properties. However, it would be unrealistic and impossible to use all soil attributes as indicators, so a minimum set of soil attributes encompassing chemical, physical, and biological soil properties are selected for soil quality assessment (Larson and Pierce 1991). USDA classified soil quality indicators mainly into four categories, i.e., visual, physical, chemical, and biological indicators. Soil quality assessment can be done in a proper and systematic way only when individual parameters will be collected and combined in a meaningful manner. Hence, integrated soil quality indicators based on a combination of soil properties could better reflect the status of soil quality than individual parameters. However, it is not possible to collect all individual parameters for a particular soil, so a minimum data set is required to measure soil quality and the minimum data set should be sensitive enough to be changed due to change in management practices and have significant link with soil and plant properties (Larson and Pierce 1994). For selection of minimum data set, suitable technique needs to be used so that it can identify the indicators that best represent variability in a large existing data and affords less opportunity for disciplinary bias. Mechanistically, the data set must have sufficient number of observations and variables (Andrews et al. 2001). After getting the key indicators it is important to interpret them and give proper weightage or score through linear and non-linear scoring functions.

Different approaches to quantify soil quality are descriptive approach which emphasized on the characterization of different facets or attribute of soil quality, while the other one is indicative approach which identifies the ability or capacity of an attribute in a desired manner.

14.2 Importance of Soil Quality

14.2.1 Linking Soil Health to Soil Quality

Soil is strategically located in this universe linking lithosphere with hydrosphere and atmosphere. The maintenance and sustenance of quality of all these three spheres are extremely vital for existence of living creatures in this mighty earth. Soil (lithosphere) quality is one of the three components of environmental quality (atmosphere), besides water (hydrosphere) and air (atmosphere) quality (Andrews et al. 2002a). Water and air quality are defined mainly by their degree of pollution that impacts directly on human and animal consumption and health, or on natural ecosystems (Carter et al. 1997; Davidson 2000). In contrast, soil quality is not limited to the degree of soil pollution but is commonly defined much more broadly as "the capacity of a soil to function within ecosystem and land-use boundaries to sustain biological productivity, maintain environmental quality, and promote plant and animal health" (Doran and Parkin 1994, 1996).

There are various concepts/approaches/methodologies used for assessment of soil quality. The suitability of soil for agricultural production is embedded in the concept of soil fertility, originating from the German literature on "Bodenfruchtbarkeit" that is predominantly aligned to crop yields (Patzel et al. 2000). Accordingly, the FAO describes soil fertility as "the ability of the soil to supply essential plant nutrients and soil water in adequate amounts and proportions for plant growth and reproduction in the absence of toxic substances which may inhibit plant growth" (www.fao.org). Nevertheless, the concept of soil fertility is generally operationalized chemically and partly physically in terms of the provision to crops of nutrients and water only. To address physical and/or biological characteristics of soil, other concepts are more commonly used. One of the earliest is land quality, which integrates characteristics of soil, water, climate, topography, and vegetation (Carter et al. 1997; Dumanski and Pieri 2000) in the context of land evaluation, which aims to assess the use potential of land, based on its attributes (Rossiter 1996). Similarly, soil capability, i.e., the intrinsic capacity of a soil to contribute to ecosystem services (Bouma et al. 2017), provides a neutral assessment of what soils can do and how their potential can be reached.

Soil quality is indirectly related with human and animal health and their welfare is associated with it; stimulated by this perception recent call for development of a "soil health index" was given (Haberern 1992). However, defining and assessing soil quality or health is complicated by the fact that soils perform multiple functions in maintaining productivity and environmental well-being. Identifying and integrating the physical, chemical, and biological soil attributes which define soil functions is the challenge (Papendick and Parr 1992; Rodale Institute 1991). Although mostly these two terms "soil quality" and "soil health" are used synonymously (Karlen et al. 2001; Doran and Zeiss 2000), with scientists, in general, preferring "soil quality" and producers preferring "soil health" (Harris and Bezdicek 1994), nevertheless, their definitions must be differentiated. In some cases, the term soil health is preferred because it portrays soil as a living, dynamic body that functions holistically rather than as an inanimate mixture of sand, silt, and clay. While others used the term soil quality and descriptors of its innate quantifiable physical, chemical, and biological characteristics. Soil quality is related to possible functions and uses of soil, but also to the location and scale of study. In contrast, soil health represents a holistic approach for understanding the soil system, independent of soil use and soil users (Garriguesa et al. 2012).

Soil quality is affected by a number of physical, chemical, and biological factors, such as soil pH, climate and temperature, moisture, carbon content, mineral content, microbial diversity, and geographical conditions. These factors play an important role in the maintenance of soil health or quality and thus play a vital role in plant growth. But majorly the fundamental role is played by soil carbon. Soil organic carbon is the most important factor in the soil ecosystem which directly or indirectly governs almost all the aspects of the biosphere and is essential for the growth of all living organisms. Soil organic carbon improves almost all the properties of soil and has numerous beneficial effects on soil quality. Soil organic carbon is the principal component of soil organic matter (SOM) and is the key factor of soil which governs most of the soil properties. Soil organic carbon is very important for maintaining soil quality or soil health and is established as one of the most important factors to govern productivity and sustainability of entire ecosystem. It is the central element to govern soil fertility, productivity, and quality, hence, maintaining and improving its level is very important to ensure soil quality, future productivity, and sustainability (Katyal et al. 2001). Being a direct source of plant nutrients, SOM also indirectly influences nutrient availability in soil. Carbon sequestration is an important aspect for maintaining or enhancing the soil carbon level which is the process of transferring carbon dioxide from the atmosphere into the soil by different agricultural management practices in a form that is not immediately emitted.

The debate about soil quality vs. soil health arose quickly after the concept of soil quality was criticized in the 1990s. In contrast to soil quality, soil health would "capture the ecological attributes of the soil which have implications beyond its quality or capacity to produce a particular crop. These attributes are chiefly those associated with the soil biota; its biodiversity, its food web structure, its activity and the range of functions it performs" (Pankhurst et al. 1997). These authors further consider "that the term soil health encompasses the living and dynamic nature of soil, and that this differentiates it from soil quality." They therefore "adopt the view that although the concepts of soil quality and soil health overlap to a major degree and that in many instances the two terms are used synonymously, soil quality focuses more on the soil's capacity to meet defined human needs such as the growth of a particular crop, whilst soil health focuses more on the soil's continued capacity to sustain plant growth and maintain its functions." Meanwhile, the debate subsided and partly changed focus. For example, Moebius-Clune et al. (2016) consider that soil quality includes both inherent and dynamic soil properties and that soil health is equivalent to dynamic soil quality. The differential usage may also link to the observation of Romig et al. (1996), that, whereas soil quality is the preferred term of researchers, soil health is often preferred by farmers.

14.2.2 Relevance of Soil Health Card

Soil health card (SHC) based management is needed for efficient management of resources and sustainable crop productivity. Soil health index (SHI) could be used as a yardstick for judging the health of the soil. Researchers in many parts of the world have used the "Soil Management Assessment Framework (SMAF)" for development of soil health indices (Andrews et al. 2004; Bhaduri and Purakayastha 2014; Masto et al. 2007, 2008a, b; Purakayastha et al. 2019). At present soils are analyzed for only a few chemical parameters (e.g., pH, EC, SOC, available N, P, K) on a routine basis in soil testing laboratories. This type of approach only provides a snap shot of a few soil chemical parameters, and that ignores vital physical and biological soil parameters that should otherwise be of paramount importance to various soil functions (e.g., physical stability and support, nutrient cycling and biodiversity and habitat of organisms) (Purakayastha et al. 2019). The current need is to take a holistic view in assessing soil health by analyzing chemical, physical, and biological parameters. Soil health index is the vital parameter that could be used as a yardstick for assessing soil health as impacted by different management practices. By using the principal component analysis tool, they identified SOC, available Zn, available Mn, available K, aggregate stability, microbial biomass carbon, and potentially mineralizable nitrogen as the seven most sensitive soil health indicators in alluvial soils of northwest India. The soil health card thus developed had information on the farmer, the land (altitude, longitude, and latitude), soil type, various soil health indicators, along with soil health index and appropriate recommendation practices on fertilizers, manures, and reclamation measures like gypsum application, leaching of salts of salt affected soils. Instead of analyzing large number of soil parameters, the above seven most important parameters could be used for judging the soil health on temporal and spatial basis in similar types of soil and cropping system.

Online assessment of soil health is also an important area of research and development. In this direction, Purakayastha et al. (2016) developed a conceptual framework based decision support system (DSS) for online assessment of soil health for alluvial soils and hill soils of India (http://ssacdss.iari.res.in/home.php). The DSS has 10–11 chemical, 2–4 physical, and 3–5 biological parameters for the above two soils. By inserting the values of analyzed data in respective listed soil health indicator parameters, the in-built software computes the soil health index value. This type of DSS could be useful for the researchers, soil testing laboratories, farmers, agricultural department, policy makers, etc. The major challenge is to integrate this DSS with fertilizer and manure recommendations and amelioration measures, if any required.

Importantly, use of spectroscopic techniques, e.g., near-infrared spectroscopy and remote sensing, facilitates measurement of various soil chemical, physical, and biological parameters in a fast and inexpensive way (Cecillon et al. 2009; Gandariasbeitia et al. 2017; Kinoshita et al. 2012; Paz-Kagan et al. 2014) and can be used directly in the field or in the laboratory (McKenzie et al. 2003), and commercial providers increasingly offer spectroscopy-based analyses (e.g., www.soilcares.com, www.eurofins.com). Spectroscopic techniques face the criticisms that hamper their routine use in soil quality assessment. First, when applied to the soil surface in the field, information is gained only about the first millimeters of the soil (Bünemanna et al. 2018). Second, sample characteristics such as moisture content, particle size distribution, and roughness of the soil surface can influence the outcome of the analysis (Baveye and Laba 2015; Stenberg et al. 2010). Third, a calibration step is used to relate the spectral information to soil characteristics (Gandariasbeitia et al. 2017) and the prediction is as good as the calibration data set. But if it is established that soil health indicator data generated by spectroscopic techniques equally match with wet chemical analysis, it would be very fast and non-destructive technique to generate data to be linked with soil health card for efficient management of soil and assessment of the health status of the soil.

14.3 Approaches of Soil Quality Assessment Methods

14.3.1 Goal-Oriented Soil Quality Assessment

There are number of approaches to ascertain the goal of soil quality assessment. These are purpose-oriented and end users' based. Often the goals are system-based, i.e., sole agro-system, horticulture, forestry, or an integrated farming system. Otherwise the goals may be function-based consisting of few specific soil indicators. Broadly, goals of SQ may be:

1. Productivity.
2. Environmental protection.
3. Sustainability and food security.

Soil quality indices (SQIs) designed for ecologically based approach was attempted under a poultry litter management case study. In the same study site-specific indices were developed for two sites with different soil types but similar climatic regimes. SQI design framework (encompassing >40 assays) was evaluated for both land applications of fresh vs. composted poultry litter (Andrews and Carroll 2001). To compare the soil fitness for agricultural in the southern Italy (Mediterranean environment) the developed SQI revealed some interesting facts like: the fertilizer level appeared to drive SQI results and the higher NPK-rates promoted higher SQI values; moreover, no significant effects on SQI were visible owing to cropping sequence and stubble management (Armenise et al. 2013). Another goal-oriented SQI evaluation was done at three sites in central and southern Brazil to judge the suitability of sugarcane production to meet the increasing biofuel demand in the country. Both expert opinion based and PCA based approaches were followed to prioritize 38 soil quality indicators and revealed the importance of soil physical and biological indicators for long-term sustenance of soil quality in sugarcane producing areas (Cherubin et al. 2016). Besides agriculture, few reports were also focused on SQI in terms of environmental sustainability. Mukhopadhyay et al. (2016) developed SQI, using PCA approach, on reclaimed mine soils of Jharia coalfield, Dhanbad, Jharkhand, and compared the changes with reference site along with remediation potentials of four tree species (*Acacia auriculiformis*, Cassia siamea, *Dalbergia sissoo*, and *Leucaena leucocephala*).

14.3.2 Timeline of SQ Indexing Methods

Over the time there are number of approaches developed for SQ indexing; some may be region specific while some others are purpose-oriented. Say for example, while the agriculturists deal, productivity is their main concern so they prefer to choose soil indicators which have a direct impact on crop yield, whereas the environmentalists more carefully choose the indicators implanting their views towards protecting environment.

Use of a soil quality index helped combine the information obtained by measuring several indicators into a tool that could be used to improve decisions related to soil management. Different responses to the critical soil functions by the various management systems were clarified by using functional components of an overall soil quality index (Hussain et al. 1999). Several approaches from last two decades had proved their relevance; some of them are discussed here.

Parr et al. (1992) suggested that a soil quality index could take the form of the equation:

$$\mathrm{SQI} = f\ (\mathrm{SP, P, E, H, ER, BD, FQ, MI})$$

where, SQI is a function of soil properties (SP), potential productivity (P), environmental factors (E), human and animal health (H), erodibility (ER), biological diversity (BD), food quality (FQ), and management inputs (MI). Determination of specific measurable indicators of each variable and the interactions among these diverse variables is a daunting task. The inclusion of BD, FQ, and MI makes this as land quality index as suggested by FAO (1997).

Larson and Pierce (1994) proposed three different functions associated with good soil quality. These included the ability of a soil to: (1) function as a medium for plant growth, (2) regulate and partition water flow through the environment, and (3) serve as an environmental filter. To perform these functions, they stated that a high quality soil accepts, holds, and releases nutrients and water, promotes and sustains root growth, maintains suitable soil biotic habitat, responds to management, and resists degradation and soil quality is defined as the sum of individual soil qualities qi and expressed as:

$$Q = f\ (q_i \ldots\ldots\ldots q_n)$$

Karlen et al. (1994) calculated individual ratings for four soil functions: (1) accommodating water entry into the soil (we), (2) facilitating water transfer, adsorption, and delivery (wt), (3) resisting degradation (rd), and (4) supporting plant growth (spg). The individual scores were multiplied by weighting factors and combined by addition into an overall soil quality index. Bhaduri et al. (2014) have also implemented this indexing method for judging soil quality in rice–wheat system of Indo-Gangetic plains using two goal-oriented approaches of productivity and environmental sustainability.

Doran and Parkin (1994) described a performance-based index of soil quality that could be used to provide an evaluation of soil function with regard to the major issues of: (1) sustain production, (2) environmental quality, and (3) human and animal health. They proposed soil quality index consisting of six elements:

$$\mathrm{SQ} = f\ (\mathrm{SQE1}, \mathrm{SQE2}, \mathrm{SQE3}, \mathrm{SQE4}, \mathrm{SQE5}, \mathrm{SQE6})$$

where, SQE1 is food and fiber production, SQE2 is erosivity, SQE3 is groundwater quality, SQE4 is surface water quality, SQE5 is air quality, and SQE6 is food quality.

Harris et al. (1996) developed a soil quality index using a framework that included three soil functions: (1) to resist erosion (water relations), (2) to provide plant nutrients (nutrient relations), and (3) to provide favorable root environment (rooting relations). Later, Hussain et al. (1999) modified this framework and demonstrated that adjusting threshold limits for local conditions can make the function ratings more or less sensitive to the management practices being evaluated.

Wang and Gong (1998) introduced the concept of relative soil quality index (RSQI); with the help of geographical information system (GIS), 12 indicators were combined into an RSQI:

$$\mathrm{RSQI} = (\mathrm{SQI}/\mathrm{SQI_m}) \times 100$$

where, SQI is the soil quality index and $\mathrm{SQI_m}$ is the maximum value of SQI.

The PCA is also an appropriate method to examine data because the analysis produces uncorrelated indices from the linear combination of potentially correlated variables. Andrews et al. (2002a, b) demonstrated the use of PCA in assessing soil quality by the following equation:

$$\text{Soil Quality Index (SQI)} = \sum_{i=1}^{n} W_i \times S_i$$

where, S is the score for the subscripted variable and W is the weighing factor derived from PCA. Many of the recent studies followed this approach and found the most inevitable approach for quantifying soil quality (Chaudhury et al. 2005; Masto et al. 2008a, b; Shahid et al. 2013; Bhaduri and Purakayastha 2014).

Kang et al. (2005) developed "sustainability index" based on soil quality where all three components viz. microbial index (calculated by determining various soil microbial and biochemical activities), crop index (by measuring of crop yield parameters), and nutrient index were taken into account. In this connection, Nannipieri et al. (2002) discussed two approaches to assess the soil microbial activity: one is the use of single index and the other is based on the measurements of number of soil enzyme activities to estimate microbial functional diversity.

While this approach integrated enzyme activities limiting the rate of metabolic processes the other one by Stefanic et al. (1984) used a weighted average to calculate the biological index of fertility (BIF).

There are very few reports that solely focused on the soil physical parameters to assess soil quality either in the Vertisols of central India (Mohanty et al. 2007) or the Inceptisol of north India (Sinha et al. 2014a) and both the studies further concluded their findings with impact of SQ on crop yields. Another interesting study compared the soil physical quality under three different soil types and agro-ecosystems (sub-humid: Pantnagar; semi-arid: Ludhiana; and arid: Hisar) by analyzing 13 soil quality indicators across the 25 cropping systems following both linear and non-linear scoring functions with PCA approach and revealed the impact of cropping systems on developed SQISs (Sinha et al. 2014b).

A typical study comparing the index methods of soil quality revealed that SQI-3 can be regarded as the best and easiest model given its relatively higher success to predict crop yield and objectivity approach while dealing with three methods for estimating SQI viz. (1) simple additive SQI (SQI-1), (2) weighted additive SQI (SQI-2), and (3) statistically modeled SQI (SQI-3) based on principal component analysis (PCA) (Mukherjee and Lal 2014).

14.4 Comparative Assessment of SQI

For crop production and productivity, soil fertility is the key function of soil quality assessment and it was assessed by the concept of integrated fertility index (IFI) in southeastern China by Bo et al. (1995). Fertility indices were selected and were divided into two group like states of soil nutrients (N) and environments of nutrient supplication (E). Based on these parameters, soil fertility index was calculated. Soil quality is assessed with the help of physical, chemical and biological parameters depending on the availability of resources and objectives of the study. Microbial parameters are always considered as the potential indicators of soil quality as these changes are fast and seasonal variation is also observed.

14.4.1 Cereal-Based System

Cereal-based system has been most sought out to study SQI, both in Indian perspectives as well as beyond India. Hence many reports were found date back to early 2000, and till date research is being enriched.

Kang et al. (2005) calculated sustainability index to assess soil quality under the influence of different fertilizer management practices in Typic Ustochrept soils of Punjab. It is based on the area of the triangle in which nutrient index, microbial index, and crop index of soil represented three vertices of a triangle and found that the sustainability index values were 2.43 and 0.93 for rice–wheat and corn–wheat cropping systems, respectively.

Chaudhury et al. (2005) observed that 100% NPK + FYM and 100% NPK showed positive change in soil quality (SQI = 26.50 and 14.05, respectively) that is aggradation of soil quality, but the other three treatments, 100% N, 100% NP, and control, showed negative change of soil quality and indicates degradation of the system in long-term experiment of rice–wheat–jute cropping system in Eutrochrept soil of Barrackpore, West Bengal.

Mohanty et al. (2007) developed a soil quality index by regressing bulk density, penetration resistance, water stable aggregates, and organic matter on crop yield for a rice–wheat cropping system on a Vertisol in India, and showed optimum ranges of 0.84–0.92, 0.88–0.93, and 0.86–0.92, for the rice, wheat, and combined (rice + wheat) phases, respectively. Use of zero tillage (ZT) for wheat had the most positive effect on soil quality, but if conventional tillage (CT) is used, direct seeding of rice with CT and residue returned was predicted to be the best for long-term sustainability.

Masto et al. (2007) reported SQI ratings ranged from 0.552 (unfertilized control) to 0.838 for the combined NPK fertilizer plus manure treatment (100% NPK + FYM) followed by 0.777 (150% NPK) and 0.729 (100% NPK) in maize–wheat cropping system in Inceptisol of New Delhi. Sharma et al. (2005) concluded that to maintain the yield as well as soil quality in sorghum–castor bean cropping system in dryland semi-arid Alfisols, primary tillage along with organic residue and nitrogen application are needed and so $CTGLN_{90}$ treatment (conventional tillage + *gliricidia loppings* + N @ 90 kg ha^{-1}) got highest SQI (1.27) in contrast to the treatment $MTNRN_{30}$ (minimum tillage + no residue + N @ 30 kg ha^{-1}; SQI = 0.90).

For rice–wheat system the management goal based soil quality indices were derived by Bhaduri et al. (2014), Bhaduri and Purakayastha (2014) under integrated tillage–water–nutrient management, while Sinha et al. (2014a) focused on soil physical indicators at maize–wheat system to calculate SQI.

14.4.2 Pulse and Oilseed Based System

Acosta-Martínez et al. (2008) showed that the microbial parameters can be used to assed the soil condition in the peanut based cropping system. This study demonstrated that soil microbial biomass, fungal and bacterial FAMEs, and the enzymes activities can be a good indicator of soil quality. Based on this study, it was revealed that monoculture peanut was not sustainable for sandy soils because of yield reduction and higher production costs due to higher diseases and nematodes when compared to peanut in rotations. Hence the management decisions targeting the selection of cropping systems for peanut production on sandy soils should consider the positive effects of cropping systems and the microbial component to maintain and/or improve soil quality, functionality, and sustainability of agricultural production. The assessment of soil quality indicators under different cropping system, clay loam soil, and under arid ecosystem showed that the physical condition of soil is influenced by the cropping system. Pearl millet—wheat—fallow cropping system deteriorated the physical condition of soil as is expressed by very high BD under this system, also inclusion of vegetables in the cropping system were not desirable from soil structure point of view as they did not result in optimum soil aggregation. Smith and Elliot (1990) have emphasized the need to adopt appropriate soil and nutrient-management practices that avert the effects of soil degradation or maintain soil quality at a desirable level in rain fed regions. Supplementing the nutrient requirement of crops through organic manures, especially the farm-based organics, plays a key role in sustaining soil fertility and crop productivity, reducing use of fossil fuels, and restoring overall soil quality (Patra et al. 2000).

14.4.3 Horticultural System

Understanding the response of soil to agricultural use and practices over time is an important step for successful management of soil quality (Gregorich et al. 1994). Therefore the methods to quantify soil quality must evaluate changes in selected soil attributes over time together with the behavior of soil under defined conditions. Furthermore, there is no single measurement that can quantify soil quality (Stewart 1995). When certain well attributing soil properties are clubbed together, they can be used as a set of potential soil indicators.

In the context of this study, horticulture can be defined as the science and art of cultivating garden plants, including vegetables, fruits, flowers, and ornamentals. In regard to horticulture, management strategies have shown high potential for enhancing soil quality, by levering on natural soil processes and complex biological interactions and synergies (Altieri and Nicholls 2012; Diacono et al. 2016). Soil management can support functional biodiversity (e.g., soil biota, natural enemies, pollinators), which contribute to enhance the immunity of the agro-ecosystem (e.g., natural pest control mechanisms) and its regulatory processes (e.g., nutrient cycling). Horticultural crops like fruit tree ecosystems have the potential to reverse soil degradation trend by the adoption of sustainable orchard management practices that increase the sequestration of atmospheric CO_2 into soil, tree biomass, and litter, enhancing soil organic carbon (SOC) and biodiversity (Montanaro et al. 2017; IPCC 2006). Soil fertility is mainly related to the variability, abundance, and richness of micro-organisms (Zornoza et al. 2015). They are responsible for the cycling of organic matter and the generation of nutrients for plants through enzymatic processes (Nannipieri et al. 1990). Studies on the land use-induced changes on soil quality in different agro-ecosystems of India have focused only on the problems of the land use change from forest to seasonal agriculture (Panwar et al. 2011; Sharma et al. 2009; Singh et al. 2011). Very few reports are available on the changes in soil quality due to the conversion of native forest into horticultural fruit orchards and their long-term management effects in India (Wanshnong et al. 2013).

Hazarika et al. (2014) revealed that land use change (conversion of native forest land to guava and sapota orchards) and the long-term existing orchard soil management practices negatively impacted the soil physical, chemical, and biological properties under humid sub-tropical south western climate of India; and the extent of deterioration of soil quality increased with the increase of orchard age.

The strongest influencing factor of soil quality attributes was in the order of land use change > orchard age > orchard type. Soil organic carbon, pH, and earthworm populations were strongly influenced by orchard age, whereas soil enzyme activities were strongly influenced by orchard type. Araujo et al. (2018) develop soil quality index (SQI) for cacao cropping systems to meet the nutritional criteria of the crop, the environmental safety of the cropping sites, and the increasing demand for the production and quality of cocoa. Available water function (AWF), root growth function (RGF), mineral nutrition of plants function (MNF), and environmental safety function (ESF) for potentially toxic elements were included in the additive model of SQI for cacao cropping systems. Over 66% of the cacao fields cropping sites were classified as regular SQI with a range of scores between 0.42 and 0.61. Two important agricultural aspects of cacao cultivation have been taken into account when they defined the indicators and functions of the SQI. (1) Soils suitable for cacao cultivation require the use of simple management and conservation technologies; (2) Less suitable soils require the use of more complex technologies, involving mechanical techniques of management and conservation, such as sub-soiling and nutritional management of the crop with irrigation and fertigation systems. Despite its implicit character of diagnosis, the SQI for cacao was developed to support phytotechnical decisions before and during the establishment of the crop, interpreting the diagnosis of soil conditions in line with the economic reality and techniques of farmers.

Surchand-Singh et al. (2017) assessed the impact of long-term (20 years) management practices on soil quality under peach (*Prunus persica* L.) orchard by comparing soil properties of the orchard relative to the adjacent forest soil which demonstrated that the differences in management practices between drip circle and inter-row spaces seem to be one of the causes of spatial variability in soil quality within the orchard. The long-term impact of the existing recommended management practices significantly altered the quality characteristics of the orchard soil on hilly slope under high rainfall humid subtropics and that there is a need for modification to existing orchard management practice to improve the orchard soil quality for sustainability. Glover et al. (1998) used a quantitative index to assess the effects of conventional, integrated, and organic apple production systems on soil quality. They used weighted additive model of soil quality index to accommodate water entry, facilitate water transfer and absorption, resist degradation, and sustain crop productivity and quality. They also determined critical threshold values and relative importance for each soil property based on published data and soil conditions in adjacent permanent grass sites. Their study indicates that the soil quality was higher under both the integrated and organic apple production systems, as these systems resulted in increased surface water infiltration, higher microbial biomass carbon and nitrogen, greater surface aggregate stability, and more earthworms than the conventional system. Their goal was to develop site-specific soil quality indices for evaluating long-term effects of specific farm management systems on soil quality. Sofi et al. (2016) developed the soil quality index for evaluating the soil quality indicators under different cropping systems in northwest Himalaya, India and found that inclusion of legumes in the apple orchard floor recorded highest soil quality rating across the treatments. Cereal-based cropping systems were found in lower soil quality rating; however, the incorporation of peas in the system improved soil health.

The soil quality index presented in these studies seems to be useful to assess soil quality of horticultural production systems, but further testing to assess its range of application is necessary. The setting up of a soil quality surveillance system in horticultural production system is needed imperatively and remains an important management tool for the conservation and sustainability of the production system. Soil quality research can be expanded from small fruit crops into the many diverse specialty areas of horticulture. The concept of sustainability is becoming more widely accepted by society and as overall societal interest in sustainability increases, so will the opportunities for research in those areas.

14.4.4 Value-Added Crop Based System

14.4.4.1 Cotton

There are few reports found in cotton and cotton-based systems both in India and other countries. Study conducted in Mahabubnagar district, Telangana, India by Vasu et al. (2016) showed that among the different indexing methods of soil quality, weighted index by both principal component analysis (PCA) and expert opinion (EO) was highly correlated with cotton crop yields. Study conducted by Sharma et al. (2011) showed that the SQIs as influenced by different long-term conjunctive nutrient-management treatments varied from 1.46 to 2.10 across the management treatments under cotton-based cropping system in rainfed semi-arid tropical Vertisol of Maharashtra. They also observed that of all the treatments, application of 25 kg P_2O_5 ha^{-1} + 50 kg N ha^{-1} through *leucaena* significantly maintained higher soil quality with SQI of 2.10, which was at par with application of 25 kg N + 25 kg P_2O_5 + 25 kg N ha^{-1} through FYM (2.01). Yao et al. (2013) showed that cotton-based crop rotation system had a better soil quality over rice based rotation system as the former had higher SQI values than the later in coastal areas of China. Gui et al. (2010) showed that the cotton field had an obviously positive effect on soil quality of the topsoil with a great SQI and relative SQI values.

14.4.4.2 Tea

There are few reports found in the similar aspect from tea plantations of Assam. A minimum data set (MDS) was developed for soil quality index (SQI) assessment in the long-term tea cultivation systems (Baruah et al. 2017). Soil quality index calculated for different categories of tea soils under continuous tea cultivation was observed to be highest in the category where tea cultivation was practiced in less than 15 years. In deep, fine loamy well-drained soil, SQI was 14.11 for less than 15 years of continuous tea cultivation, 10.35 for 15–30 years, 12.15 for 30–45 years, 10.28 for 45–60 years of continuous tea cultivation, and 8.04 for more than 60 years. The most sensitive soil quality indicators identified in deep, fine loamy, well-drained soil were available nitrogen (Av N) for less than 15 years and 45–60 years of continuous tea cultivation, total nitrogen (TN) for 15–30 years and 30–45 years, and exchangeable Ca (Ex Ca) for more than 60 years of continuous tea cultivation. Soil quality index, a good indicator to represent overall soil function, was observed to be high in less than 15 years age group of tea cultivation irrespective of soil types studied. Continuous tea cultivation for >60 years led to reduce SQI compared to less than 60 years where exchangeable calcium had the major role to play towards SQI. Moeskops et al. (2012) explored the value of ergosterol, phospholipid fatty acid (PLFA) profiles, and neutral lipid fatty acid (NLFA) 16:1ω5c as soil quality indicators for the intensive systems of vegetable production in the humid tropical climate of West Java, by comparing organic and conventional management. They developed soil quality index by stepwise canonical discriminant analysis and based on the absolute amount of PLFA 16:0, the relative amount of PLFAs 10Me16:0 and 10Me18:0, and dehydrogenase activity and was successfully validated.

14.4.4.3 Protected Horticultural Cultivation

Over dosing of agricultural soils with fertilizers, pesticides, and other toxic chemicals makes the way for adverse impacts on soil health and environment pollution (Govindarajan and Thangaraju 2001; Muhammad et al. 2013). Nutrient imbalance, soil contamination and injudicious application of fertilizers, and pesticides are the main reason for degradation of soil health and quality. Modern intensified industrialization led to shrinkage of agricultural land and it has almost become a compulsion to increase the per unit yield levels from the available land under agriculture which, however, can be achieved by using modern technology such as cultivation in polyethylene houses. The research data showed that production potential of horticultural crops in protective cultivation may increase from 20 to 30% which shows the potential of protected cultivation system to improve the economy of the growers as well as fulfill the demand for horticultural crops throughout the year from the small piece of land (Baghel et al. 2003). Protected cultivation has emerged as an alternative to open field production. Biswas et al. (2017) investigated the impact of intensive cultivation on soil health by growing floricultural crops carnation (*Dianthus caryophyllus* Linn.) under polyhouses condition. In light of the soil health index values, 36.7, 46.7, and 16.6% of samples were categorized under the very high, high, and medium soil health, respectively, under polyhouse conditions. Such values for open field condition were noted to be 10, 70, and 20%, respectively. Soil health was found to be affected by the management practices adopted by the farmers and the degree of manure and fertilizer usage over a period of time. The soil indicators like pH, N, K, Ca, Mg, S, micronutrients, and chloride had less effect on soil health, while, EC, phosphorus, organic carbon, porosity, and microbial biomass significantly influenced the soil health both under polyhouse and open field conditions.

Munoz and Anton (2008) used life cycle assessment (LCA) tool to compare the environment burden associated with greenhouse as opposed to open field production process for spring season tomato crop. Structure, irrigation equipment, fertilizers, pesticides, cultural task, and irrigation were analyzed as subsystem and revealed the environmental burden is more in open field condition. Chandel et al. (2017) developed an index of soil health under protected cultivation of vegetable crops in northwest Himalayas by integrating physical, chemical, and biological properties of soils into a single number that serves as the soil's "vital sign" of overall soil health/quality. The results revealed that soil health index show that polyhouse soils in the mid hill zone of Himachal Pradesh were rated to be 40, 57, and 3% as very high, high, and medium health level, respectively. Contrary to this, 27, 53, and 20% soils in open conditions fall in very high, high, and medium soil health conditions, respectively. Soil health status under polyhouse condition is better as compared to corresponding open conditions. This may be attributed to proper adoption of crop rotation, which increases or maintains the quantity and quality of soil organic matter, and improve soil chemical and physical properties. Adequate application of fertilizers combined with farmyard manure may increase soil nutrients and soil organic carbon content. It has been observed that soil indicators viz. pH, N, K, Ca, Mg, S, micronutrients, and bicarbonates had less influence on soil health; while EC, phosphate, sulfur, organic carbon, porosity, chloride, and microbial biomass greatly influence the soil health. Although these studies must be considered as a preliminary, the results suggest that polyhouse production could have a smaller environmental impact than open field crops in most of the evaluation categories considered. There is a need for regular soil health monitoring in polyhouse production system and timely employment of corrective measures to maintain good soil health for sustainable productivity.

14.4.4.4 Medicinal Plant Based Cropping System

The use of medicinal plants is limited by the quality of active substances they contain. This quality depends on many ecological factors that affect the photophilous, but also the geophilous plant organs (Lombini et al. 1999). The distribution and the degree of presence of medicinal plants are directly correlated with the ecosystem conditions, especially the soil quality. However, limited information is available about the relationship between the bioactive compounds in medicinal plants and the soil characteristics where the medicinal plants grow. Hanudin et al. (2015) studied the relationship of soil quality index and vitexicarpine content in the leaves of *Vitex trifolia*, an important medical plant that grows in Indonesia as this has a capability for survival through a certain metabolism mechanism in the plant cell if put under a high environmental pressure. They found a negative correlation between the soil quality index and vitexicarpine content in the leaves of *V. trifolia*. Lower soil quality tends to result in higher content of vitexicarpine. However, individually nutrient indicated a positive correlation with vitexicarpine content. Higher nitrogen and magnesium content in the soil resulted in higher vitexicarpine content in the leaves. This may be related to the role of the both nutrients in biosynthesis of vitexicarpine. Obratov-Petković et al. (2006) studied the relations between the responses of medicinal plant populations developed at the particular localities to chemical processes of soil degradation in order to identify the potentials for further exploitation of medicinal plants. They have assessed the soil quality based on the calculation of indicator values of available nitrogen, heavy metals, and the sensitivity to acidification. According to their study, N (12.1–17.5), acidification (7–12), and indicator values for some heavy metals (0.3–46.5) show a low biological availability. Medicinal plant species at the investigated areas have low values of ecological indexes: N (2.41–2.82), moisture (2.45–2.70), and soil acidity (3.35–3.70).

So, before introducing a medicinal plant species in a new area, soil suitability analysis is a prerequisite to achieve an optimum exploitation of soil resources for a sustainable agricultural production with quality herb. To evaluate the soil suitability it is important to take into account the habitats of the plant species. Moreover, agronomic, logistic, and product quality aspects have to be considered. Soil quality index with reference to the growing of bioactive medicinal plants is changing from region to region, and thus development of region-specific soil quality index is need of the hour.

14.5 How Indexing of Soil Quality Varied for Each Ecosystem?

Soil quality is a complex concept that integrates soil physical, chemical, and biological properties to assess soil's functional ability, which is difficult to measure directly. However, dynamic soil properties or soil quality indicators can be measured and integrated into soil quality index to evaluate changes in the functional state of soil in response to anthropogenic interferences. Soil as an integrated part of the ecosystem performs an array of direct and indirect functions that are broadly categorized into four groups: productivity, environmental quality, biodiversity and habitat, and human welfare (Doran and Parkin 1994; Lal 1998). When the overall management goal is to balance productivity and environmental quality, resulting soil quality index may be viewed as the sustainability index of the larger agro-ecosystem.

Defining the management goal to target any particular or a set of soil functions is the first step in developing a soil quality index which leads to indexing of soil quality indicators (Christensen et al. 1996; Meyer and Swank 1996). The necessity of creating minimum data sets raises the important question on what soil quality indictors should be used to evaluate soil quality. The criteria for an ideal soil quality indicator have been defined as (Doran and Parkin 1996): (1) it should have a functional relationship with the ecosystem processes, (2) encompasses soil physical, chemical, and biological properties and processes, (3) user-friendly and accessible to broad range of users and field conditions, (4) sensitive to management and climate conditions, and (5) component of the existing database. However, often indicators influence multiple soil functions through interconnections. For example, while soil sodium (Na) content being a chemical indicator influences plant physiological toxicity and water uptake, it also affects soil physical quality through dispersion of soil particles. The search for universal soil quality indicator has only been successful with soil organic matter content which has considerable influence on soil physical, chemical, and biological processes and often related to the other indicators in the minimum data set (Larson and Pierce 1991; Gregorich et al. 1994; Doran and Parkin 1996; Sikora and Stott 1996). However, the sensitivity of soil quality index to its physical, chemical, and biological indicators is proposed to be a function of soil function of interest and site-specific limiting factors (Andrews et al. 2004;

Paz-Ferreiro and Fu 2013). For example, while higher clay fraction may help in retaining more water for plant growth in a semi-arid region, it can impair drainage in humid region and limit plant growth and field operations. Therefore, the scoring functions for the minimum data set indicators may change across management goals, climatic regions, and soil types. Furthermore, contrasting soil environmental conditions within a cropping sequence such as anaerobic vs. aerobic conditions in a rice–wheat cropping system have been proposed to have different set of indicators affecting soil quality in each crop (Bhaduri and Purakayastha 2014).

Diversity in soil quality indexing as influenced by management goals, soil type, and climate has been summarized in Table 14.1. While soil pH and salt content are often the most important indicators for saline-sodic soils (Gong et al. 2015; Nabiollahi et al. 2017), soil profile thickness and water retention properties drive the soil quality index outcomes in arid regions as they are the limiting factors of production in their respective geographic locations (Rezaei et al. 2006). The soil quality index has been extensively used to assess the production goal of the system and very few included other soil functions such as environmental protection (Andrews and Carroll 2001; Bhaduri and Purakayastha 2014). Most significant indicator of the minimum data set is often different when both production and environmental goals are simultaneously considered within the same system indicating only yield-dependent management practices may lead to environmental degradation. For example, aggregate stability has been identified as an important parameter to influence both production and environmental goals and hence informs the overall sustainability of rice–wheat systems in the Indo-Gangetic Plains of south Asia (Bhaduri and Purakayastha 2014). Therefore, a holistic approach of identifying the common important indicators influencing both production and environmental goals to develop unified soil quality index will represent multi-functionality and sustainability of the broader ecosystem.

14.6 Spatial Aspects of Soil Quality Studies, Applicability, and Justification

The concept of soil quality (SQ) is complex in nature and cannot be quantified directly in the field or laboratory (Stocking 2003), but can simply be determined from numerous indicators related to soil properties or features (Diack and Stott 2001). For precise assessment of SQ, appropriate soil indicators should be identified. The SQ indicators should be determined at local and regional scale for implementing proper management options at small and large scale for enhancing SQ. The mutual effect of soil physical, chemical, and biological processes in soil differs at different scale and intensities. Therefore, the soils exhibit greater degree of spatial variability (Goovaerts 1998). Information on spatial variability of SQ indicators is very much required in various disciplines of agricultural science including precision farming and site-specific nutrient management. Understanding geographical distribution and precise mapping of soil properties at large scale are very important for soil conservation and environmental modeling.

Table 14.1 Variations in soil quality indexing methods as influenced by different factors

References	Climate and/or soil type	Management goal	Top minimum data set indicators
Andrews and Carroll 2001	Temperate—Alfisols	Productivity	Zn, Ca, total N
		P runoff potential	Total N
	Temperate—Ultisols	Productivity	Available soil water, NO_3^-
		P runoff potential	pH, NO_3^-
Andrews et al. 2002	Mediterranean	Productivity	Total N, pH, Na
Andrews et al. 2002	Mediterranean	Productivity	Electrical conductivity, soil organic matter, aggregate stability
Askari and Holden 2014	Temperate	Productivity	Bulk density, penetration resistance, Mg
Bhaduri and Purakayastha 2014	Sub-tropical—alkaline soil	Productivity of rice	Fe, aggregate stability, hydraulic conductivity
		Environmental protection in rice	Aggregate stability, metabolic quotient, water holding capacity
		Productivity of wheat	Cu, aggregate stability, available P
		Environmental protection in wheat	Microbial biomass, Zn, potentially mineralizable N
Gong et al. 2015	Arid—saline and alkaline soil	Productivity	pH, soil water content, total salt content
Mandal et al. 2017	Semi-arid—Alfisols	Productivity	Soil moisture retention at field capacity, available N, P
Masto et al. 2008	Sub-tropical	Productivity	Available N, soil organic C, microbial biomass C
Nabiollahi et al. 2017	Semi-arid—saline sodic soil	Productivity	pH, bulk density, cation exchange capacity
Sayed Ata Rezaei et al. 2006	Semi-arid rangeland	Productivity	Soil profile thickness, total N, aggregate stability
Vasu et al. 2016	Semi-arid—sodic soil	Productivity	Clay, sodium adsorption ratio, base saturation

The study on spatial variability analysis of various soil properties has great influence on topography, plant cover, soil microclimate, several cropping system and their management (Chaneton and Avado 1996). For designing and execution of any sustainable land management schemes, complete information on spatial variability of SQ indicators is a crucial prerequisite (Zhang et al. 2012). Spatial variability of soil attributes is also essential for planning any irrigation and drainage systems, for selecting tillage operation and optimizing farm operations and nutrient management. Evaluation and understanding of the spatial and temporal variability of the biological, physical, and chemical properties of soils and crop yields across a field are required for precisely determining the best crop and soil management options and improvements required to enhance crop quantity and quality besides maintaining environmental sustainability (Awe et al. 2015, Gadja et al. 2016, Aranyos et al. 2016).

Scientific advancement in geographical information systems (GIS) and global positioning system (GPS) have made possible for the scientific community in handling massive amount of spatial data at different scales. Together with GIS, the use of GPS and remote sensing image offers data at different resolutions very reasonably and rapidly, use of repeatable approaches, perfection in error detection and accuracy measurement and generation of huge information (Aderonke and Gbadegesin 2013). The within-field variability is a significant basis of uncertainty in crop yield (Diacono et al. 2013). The geostatistical analysis of soil properties at unsampled location enables the detection of soil spatial variability (Nielsen and Wendroth 2003). Geostatistical analysis of spatial variability of soil attributes helps in increasing the precision of modeling soil variation at different scale (Serrano et al. 2010; Behera et al. 2011).

There are a number of traditional statistical techniques available for quantifying the spatial distribution of soil properties. Geostatistics is an efficient method of study for spatial distribution of soil properties and their inconsistency (Behera and Shukla 2015; Liu et al. 2014). Estimation through spatial statistical tools aids in forecasting values at unsampled sites by fascinating in the geographical association between projected and sampled points and reducing the variance of assessment error as well as execution costs (Behera and Shukla 2015; Saito et al. 2005). Geostatistics is a branch of applied statistics that quantifies the spatial dependence and spatial structure of a measured property and, in turn, uses the spatial structure to predict values of the property at unsampled locations. Geostatistics have proved useful for assessing the spatial variability of soil properties and have increasingly been utilized by the soil scientists and agricultural engineers in recent years (Webster and Oliver 2001; Iqbal et al. 2005). Furthermore, geostatistical methods have been adopted and used for site-specific management applications, soil sampling strategies, and assessment of farm management styles and decisions.

Spatial variability is expressed by a semivariogram $\widehat{\gamma}(h)$, which measures the average dissimilarity between data separated by a vector h. It was computed as half the average squared difference between the components of data pairs:

$$\widehat{\gamma}(h) = \frac{1}{2N(h)} \sum\nolimits_{i=n}^{N(h)} [z(\mathrm{xi}) - z(\mathrm{xi} + h)]^2$$

where $N(h)$ is the number of data pairs within a given class of distance and direction, z(xi) is the value of the variable at the location xi, and z(xi + h) is the value of the variable at a lag of h from the location xi.

Semivariograms and cross-semivariograms have been used to characterize and model spatial variance of data to assess how data points are related with separation distances while kriging uses modeled variance to estimate values between samples (Journel and Huijbregts 1987). Currently, various geostatistical methods such as inverse distance weighted (IDW), kriging, co-kriging, etc. are widely being used to prepare continuous spatial distribution using point observations of various variables (Lin and Chen 2004). The different spatial interpolation techniques estimate parameter values such as SOC, at unsampled locations using data from point observations and provide us with an ideal tool for meeting our requirement for spatial distribution data (Lin and Chen 2004). Kriging and co-kriging are common geostatistical procedures that have been used for optimal estimation and spatial interpolation of values at unsampled locations. Estimating semivariogram parameters of soil properties using geostatistical tools and further applying them to predict other soil properties using ordinary kriging is the general procedure to prepare soil maps. Co-kriging uses more than one variable in spatial interpolation process. It employs a second variable to estimate values of primary variable of interest that were assumed to be spatially dependent (McBratney and Webster 1983; Davis 1986). Previous studies have applied to assess spatial association in soils and to evaluate the geographical changeability of soil characteristics (Wei et al. 2008; Zare-mehrjardi et al. 2010). Both reported that kriging and co-kriging are more suitable techniques in comparison to inverse distance weighting (IDW) for acquiring precise information of the geographical distribution of soil properties. Weller et al. (2002) conducted different geostatistical techniques for spatial variability of soil properties and reported the kriging technique is better than any other technique (Robinson and Metternicht 2006). Three geostatistical techniques such as kriging, IDW, and radial basis function (RBF Spline) were adopted to examine the spatial distribution of the soil pH and organic content but the results of kriging was the most suitable among other techniques. The kriging is a well-established geostatistical interpolation model which is based on a logic of weighted moving average (Theodossiou and Latinopoulos 2006).

Spatial variability of soil physical properties such as penetration resistance, bulk density (BD), and aggregate stability is highly affected by soil management practices such as tillage (Gómez et al. 2005; Pramanik et al. 2013). Analysis of the maps indicates that the spatial distribution of SOC and CEC was more diverse in the topsoil than subsoil layer with generally lower values of both variables in the latter (Usowicz and Lipiec 2017). There is a general similarity in the distributions of the inherent sand, silt, and clay contents between the topsoil and subsoil layers.

A study by Maity (2006) showed that the prediction map of PR (measured at field capacity water content) was prepared for 9 contour classes (kPa) namely 1200–1400, 1400–1600, 1600–1800, 1800–2000, 2000–2200, 2200–2400, 2400–2600, 2600–2800, and 2800–3000 (Fig. 14.1). The areas with PR classes having value >2000 kPa were treated as compacted areas. Results revealed that before plowing the 0–15 cm soil layer showed contours with values <2000 kPa but subsurface had contours with values ranging between 2400 and 3000 kPa. The prediction maps of BD for surface layer before plowing showed the presence of 1.2–1.3, 13–1.4, 1.4–1.5 contour classes (Fig. 14.1) whereas the prediction maps of BD for subsurface 15–30 cm layer showed the 1.5–1.6, 1.6–1.7, and 1.7–1.8 contour

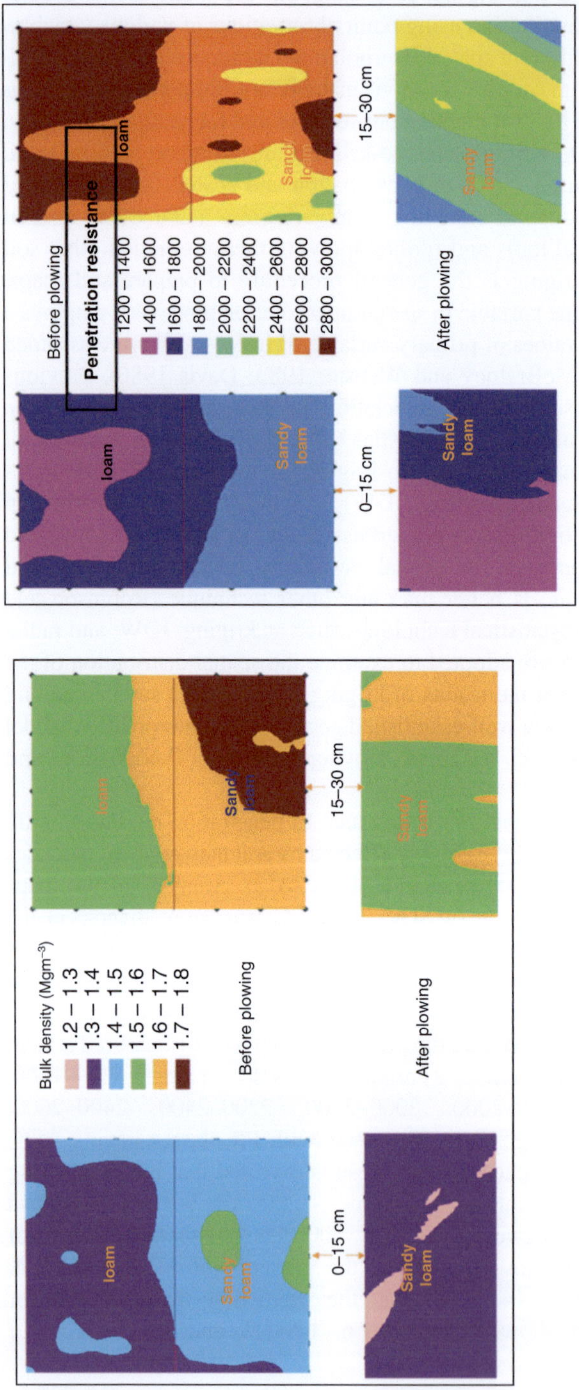

Fig. 14.1 Prediction map of penetration resistance (kPa) and bulk density (Mg m^{-3}) of IARI farm before and after plowing (source: Maity 2006)

classes. The map thus clearly showed the presence of plow pan area with BD >1.6 Mg/m^3 in subsurface in the entire area of study.

14.7 Future Research Priorities and Conclusion

1. Spatial interpretation (by GIS mapping) and modeling study (more specific to soil C) is going to be researchable priorities for monitoring soil health and quality in the coming years.
2. Secondly, standardization of analytical techniques (say for microbial biomass C, total organic C) for evaluating a specific soil health indicator in a similar agro-ecosystem/region/landscape should be another newly emerging researchable area, since it will provide more accuracy in estimation.
3. Thirdly, comparison of indexing methods for assessing the soil health/quality would be another researchable issue.
4. Fourthly, moving towards the concept of "soil security," a wider, integrative concept than "soil quality," "soil health," or "soil protection." Moreover, soil plays a vital role in the global environmental sustainability challenges of food security, water security, energy security, climate stability, biodiversity, and ecosystem services (McBratney et al. 2014).
5. Meta-analysis of the soil quality data with variable weather parameters of a specific agro-climatic zone over a gap of certain period may reveal the effect of climate change towards soil quality.
6. Soil quality research needs a holistic approach by involving all the stakeholders for creating awareness for optimum use of the soil taking into full consideration its health.

Soil quality index is, thus, potentially unique and diverse; lately emerged but created interests among scientists of natural resource management. Not only for agricultural system, but the scope of SQI has also been extended for contaminated soils, mine soils, polluted soils, and others, where the holistic and quantitative judgment of soil has become more crucial for a longer sustenance. With the support of new-age tools and logistics, GIS based spatial approaches of measuring soil quality has become more demanding. So is the web-based application or decision support system (DSS) of quantifying soil quality for greater reach to mass and faster approachability. Days are not far when indices of "soil quality" and "soil health" can interchangeably be calculated by simple multiplication/division factors, so that angle of theoretical knowledge can be converted to application point of view. Certain challenges are there but with help of evolving science and scientific understandings, there is an enormous possibility, too.

References

Acosta-Martínez V, Rowland D, Sorensen RB, Yeater KM (2008) Microbial community structure and functionality under peanut-based cropping systems in a sandy soil. Biol Fertil Soils 44:681–692

Acton DF, Gregorich LJ (1995) Understanding soil health. In: Acton DF, Gregorich LJ (eds) The health of our soils—towards sustainable agriculture in Canada, vol 11. Center for Land and Biological Resources Research, Research Branch, Agriculture and Agri-Food Canada, Ottawa, pp 5–10

Aderonke DO, Gbadegesin GA (2013) Spatial variability in soil properties of a continuously cultivated land. Afr J Agric Res 8(5):475–483

Altieri MA, Nicholls CI (2012) Agroecological diversification strategies to enhance biological pest regulation in horticultural systems. In: XXVIII International Horticultural Congress on Science and Horticulture for People (IHC2010): International Symposium on 933, pp 35–42. https://doi.org/10.17660/ActaHortic.2012.933.2

Andrews SS, Carroll CR (2001) Designing a soil quality assessment tool for sustainable agroecosystem management. Ecol Appl 11(6):1573–1585

Andrews SS, Mitchell JP, Mancinelli R, Karlen DL, Hartz TK, Horwath WR (2001) On-farm assessment of soil quality in California's central valley. Agron J 94:12–23

Andrews SS, Karlen DL, Cambardella CA (2004) The soil management assessment framework: a quantitative soil quality evaluation method. Soil Sci Soc Am J 68:1945–1962

Andrews SS, Karlen DL, Mitchell JP (2002a) A comparison of soil quality indexing methods for vegetable production systems in Northern California. Agric Ecosyst Environ 90:25–45

Andrews SS, Mitchell JP, Mancinelli R (2002b) On-farm assessment of soil quality in California's central valley. Agron J 94:12–23

Aranyos J, Tomócsik A, Makádi M, Mészáros J, Blaskó L (2016) Changes in physical properties of sandy soil after long-term compost treatment. Int Agrophys 30:269–274

Araujo Q, Ahnert D, Loureiro G, Faria J, Fernandes C, Baligar V (2018) Soil quality index for cacao cropping systems. Arch Agron Soil Sci 64(13):1892–1909. https://doi.org/10.1080/03650340.2018.1467005

Armenise E, Redmile-Gordon MA, Stellacci AM, Ciccarese A, Rubino P (2013) Developing a soil quality index to compare soil fitness for agricultural use under different managements in the Mediterranean environment. Soil Tillage Res 130:91–98

Askari MS, Holden NM (2014) Indices for quantitative evaluation of soil quality under grassland management. Geoderma 230-231:131–142

Awe GO, Reichert JM, Timm LC, Wendroth OO (2015) Temporal processes of soil water status in a sugarcane field under residue management. Plant Soil 387:395–411

Baghel BS, Gupta M, Tiwari R (2003) Potential and prospects of protective cultivation of horticultural crops. In: Proceedings of All India on Seminar Potential and Prospects for Protected Cultivation, organized by the Institute of Engineers, Ahmednagar, 12–13 December 2003

Baruah N, Medhi BK, Borah N, Baruah AM, Saikia GK (2017) Effect of long term tea cultivation in soil quality in deep, fine loamy, well drained soil of Jorhat district, Assam. J Soil Water Conserv 16(4):347–355. https://doi.org/10.5958/2455-7145.2017.00051.0

Baveye PC, Laba M (2015) Moving away from the geostatistical lamppost: why, where, and how does the spatial heterogeneity of soils matter? Ecol Model 298:24–38

Behera SK, Singh MV, Singh KN, Todwal S (2011) Distribution variability of total and extractable zinc in cultivated acid soils of India and their relationship with some selected soil properties. Geoderma 162:242–250

Behera SK, Shukla AK (2015) Spatial distribution of surface soil acidity, electrical conductivity, soil organic carbon content and exchangeable potassium, calcium and magnesium in some cropped acid soils of India. Land Degrad Dev 26:71–79

Bhaduri D, Purakayastha TJ (2014) Long-term tillage, water and nutrient management in rice-wheat cropping system: assessment and response of soil quality. Soil Tillage Res 144:83–95

Bhaduri D, Purakayastha TJ, Patra AK, Chakraborty D (2014) Evaluating soil quality under a long-term integrated tillage–water–nutrient experiment with intensive rice–wheat rotation in a semi-arid Inceptisol, India. Environ Monit Assess 186(4):2535–2547

Biswas D, Raina JN, Das KK (2017) Assessment of soil health under protected cultivation by soil quality indexing and variability analysis. Int J Curr Microbiol App Sci 6(4):2546–2556. https://doi.org/10.20546/ijcmas.2017.604.297

Bo S, Zhang TL, Zhao Q (1995) Comprehensive evaluation of soil fertility in the hilly and mountainous region of southeastern China. Acta Pedol Sin 32:369–375

Bouma J (2002) Land quality indicators of sustainable land management across scales. Agric Ecosyst Environ 88:129–136

Bouma J, Van Ittersum MK, Stoorvogel JJ, Batjes NH, Droogers P, Pulleman MM (2017) Soil capability: exploring the functional potentials of soils. In: Field D, Morgan CL, McBratney AB (eds) Global soil security. Springer International, Cham, pp 27–44

Bünemanna KK, Bongiorno G, Bai Z, Creamer RE, Deyn GD, Goede RD, Fleskens L, Geissen V, Kuyper TW, Mädera P, Pulleman M, Sukkel W, Groenigen JWV, Lijbert Brussaard L (2018) Soil quality – a critical review. Soil Biol Biochem 120:105–125

Carter MR, Gregorich EG, Anderson DW, Doran JW, Janzen HH, Pierce FJ (1997) Concepts of soil quality and their significance. In: Gregorich EG, Carter MR (eds) Developments in soil science. Elsevier, Amsterdam, pp 1–19

Cecillon L, Barthes BG, Gomez C, Ertlen D, Genot V, Hedde M, Stevens A, Brun JJ (2009) Assessment and monitoring of soil quality using near-infrared reflectance spectroscopy (NIRS). Eur J Soil Sci 60:770–784

Chandel S, Tripathi D, Kakar R (2017) Soil health assessment under protected cultivation of vegetable crops in North West Himalayas. J Environ Biol 38(1):97–103

Chaneton EJ, Avado RSL (1996) Soil nutrients and salinity after long-term grazing exclusion in flooding pampa grassland. J Range Manag 49:182–187

Chaudhury J, Mandal UK, Sharma KL, Ghosh H, Mandal B (2005) Assessing soil quality under long-term rice-based cropping system. Commun Soil Sci Plant Anal 36(9–10):1141–1161

Cherubin MR, Karlen DL, Cerri CE, Franco AL, Tormena CA, Davies CA, Cerri CC (2016) Soil quality indexing strategies for evaluating sugarcane expansion in Brazil. PLoS One 11(3): e0150860

Christensen NL, Bartuska AM, Brown JH et al (1996) The report of the Ecological Society of America committee on the scientific basis for ecosystem management. Ecol Appl 6:665–691

Davidson DA (2000) Soil quality assessment: recent advances and controversies. Prog Environ Sci 2:342–350

Davis JC (1986) Statistics and data analysis in geology. John Wiley and Sons, New York, NY

Diack M, Stott D (2001) Development of a soil quality index for the Chalmers Silty Clay Loam from the Midwest USA. In: Stott DE, Mohtar RH, Steinhardt GC (eds) Sustaining the global farm. Selected papers from the 10th International Soil conservation Organization Meeting (ISCO99). Purdue University: USDA-ARS National Soil Erosion Research Laboratory, West Lafayette, pp 550–555

Diacono M, Rubino P, Montemurro F (2013) Precision nitrogen management of wheat. Rev Agron Sustain Dev 33:219–241

Diacono M, Fiore A, Farina R, Canali S, Di Bene C, Testani E, Montemurro F (2016) Combined agro-ecological strategies for adaptation of organic horticultural systems to climate change in Mediterranean environment. Ital J Agron 11(2):85–91. https://doi.org/10.4081/ija.2016.730

Doran JW, Parkin TR (1994) Defining and assessing soil quality. In: Doran JW et al (eds) Defining soil quality for a sustainable environment, SSSA special publication no. 35. Soil Science Society of America, Madison, pp 1–22

Doran JW, Parkin TB (1996) Quantitative indicators of soil quality: a minimum data set. In: Doran JW, Jones AJ (eds) Methods for assessing soil quality, SSSA special publication no. 49. Soil Science Society of America, Madison, pp 25–37

Doran JW, Zeiss MR (2000) Soil health and sustainability: managing the biotic component of soil quality. Appl Soil Ecol 15:3–11

Dumanski J, Pieri C (2000) Land quality indicators: research plan. Agric Ecosyst Environ 81:93–102

FAO (1997) Land quality indicators and their use in sustainable agriculture and rural development. Land Water Bull 5:208

Gadja AE, Czyż A (2016) Dexter effects of long-term use of different farming systems on some physical, chemical and microbiological parameters of soil quality. Int Agrophys 30:165–172

Gandariasbeitia M, Besga G, Albizu I, Larregla S, Mendarte S (2017) Prediction of chemical and biological variables of soil in grazing areas with visible- and near-infrared spectroscopy. Geoderma 305:228–235

Garriguesa E, Corsona MS, Angersc DA, van der Werf HM, Walterb C (2012) Soil quality in life cycle assessment: towards development of an indicator. Ecol Indic 18:434–442

Glover JD, Andrews PK, Reganold JP (1998) Applying a soil quality index to conventional, integrated, and organic apple production systems. In: International Conference on Integrated Fruit Production 525, pp 217–228

Gómez JA, Vanderlinden K, Nearing MA (2005) Spatial variability of surface roughness and hydraulic conductivity after disk tillage: implications for runoff variability. J Hydrol 311(1–4):143–156

Gong L, Ran Q, He G, Tiyip T (2015) A soil quality assessment under different land use types in Keriya River basin, Southern Xinjiang, China. Soil Tillage Res 146:223–229

Goovaerts P (1998) Geostatistical tools for characterizing the spatial variability of microbiological and physico-chemical soil properties. Biol Fertil Soils 27:315–334

Govindarajan K, Thangaraju M (2001) Azospirillum—a potential inoculant for horticultural crops. South Indian Hortic 49:223–235

Gregorich EG, Carter MR, Angers DA et al (1994) Towards a minimum data set to assess soil organic matter quality in agricultural soils. Can J Soil Sci 74:367–385

Gui DW, Lei JQ, Zeng FJ, Mu GJ, Yang FX, Zhu JT (2010) Changes and analysis of soil quality under different land use types in oasis rim. Huan Jing Ke Xue 31(9):2248–2253

Haberern J (1992) Viewpoint: a soil health index. J Soil Water Conserv 47:6

Hanudin E, Setiawan M, Ikawati Z (2015) Relationship of soil quality and Vitexicarpine content in the leaves of Vitex trifolia L. Afr J Agric Res 10(28):2680–2686. https://doi.org/10.5897/AJAR2015.10015

Harris RF, Bezdicek DF (1994) Descriptive aspects of soil quality/health. In: Doran JW, Coleman DC, Bezdicek DF, Stewart BA (eds) Defining soil quality for a sustainable environment, SSSA special publication no. 35. Soil Science Society of America, Madison, pp 23–35

Harris RF, Karlen DL, Mulla DJ (1996) A conceptual framework for assessment and management of soil quality and health. In: Doran JW, Jones AL (eds) Methods for assessing soil quality, SSSA special publication no. 49. American Society of Agronomy and Soil Science Society of America, Madison, pp 61–82

Hazarika S, Thakuria D, Ganeshamurthy AN, Sakthivel T (2014) Soil quality as influenced by land use history of orchards in humid subtropics. Catena 123:37–44. https://doi.org/10.1016/j.catena.2014.07.006

Hussain I, Olson KR, Wander MM, Karlen DL (1999) Adaptation of soil quality indices and application to three tillage systems in southern Illinois. Soil Tillage Res 50:237–249

IPCC (2006) IPCC guidelines for national greenhouse gas inventories. In: Eggleston HS, Buendia L, Miwa K, Ngara T, Tanabe K (eds) Prepared by The National Greenhouse Gas Inventories Programme. IGES, Hayama

Iqbal J, Thomasson JA, Jenkins JN, Owens PR, Whisler FD (2005) Spatial variability analysis of soil physical properties of alluvial soils. Soil Sci Soc Am J 69(4):1338–1350

Journel AG, Huijbregts CJ (1987) Mining geostatistics. Academic Press, London

Kang GS, Beri V, Sidhu BS, Rupela OP (2005) A new index to assess soil quality and sustainability of wheat-based cropping systems. Biol Fertil Soils 41(6):389–398

Karlen DL, Andrews SS, Doran JW (2001) Soil quality: current concepts and applications. In: Sparks DL (ed) Advances in agronomy, vol 74. Academic Press, San Diego, pp 1–40

Karlen DL, Eash NS, Unger PW (1992) Soil and crop management effects on soil quality indicators. Am J Altern Agric 7:48–55

Karlen DL, Mausbach MJ, Doran JW, Cline RG, Harris RF, Schuman GE (1997) Soil quality: a concept, definition, and framework for evaluation. Soil Sci Soc Am J 61:4–10

Karlen DL, Wollenhaupt NC, Erbach DC, Berry EC, Swan JB, Eash NS, Jordahl JL (1994) Crop residue effects on soil quality following 10-years of no-till corn. Soil Tillage Res 31:149–167

Katyal JC, Rao NH, Reddy MN (2001) Critical aspects of organic matter management in the tropics: the example of India. Nutr Cycl Agroecosyst 61:77–88

Kinoshita R, Moebius-Clune BN, van Es HM, Hively WD, Bilgili AV (2012) Strategies for soil quality assessment using visible and near-infrared reflectance spectroscopy in a western Kenya chronosequence. Soil Sci Soc Am J 76:1776–1788

Lal R (1993) Tillage effects on soil degradation, soil resilience, soil quality, and sustainability—introduction. Soil Tillage Res 27:1–8

Lal R (1998) Soil quality and sustainability. In: Lal R et al (eds) Methods for assessment of soil degradation. CRC Press, Boca Raton, pp 17–30

Larson WE, Pierce FJ (1994) The dynamics of soil quality as a measure of sustainable management. In: Doran JW, Coleman DC, Bezdicek DF, Stewart BA (eds) Defining soil quality for a sustainable environment, SSSA special publication no. 35, vol 5. American Society of Agronomy and Soil Science Society of America, Madison, pp 37–51

Larson WE, Pierce FJ (1991) Conservation and enhancement of soil quality. In: Evaluation for sustainable land management in the developing world. Vol. 2. IBSRAM proceedings no. 12(2): technical papers. International Board for Soil Research and Management, Bangkok, pp 175–203

Lin GF, Chen LH (2004) A spatial interpolation method based on radial basis function networks incorporating a semivariogram model. J Hydrol 288(3–4):288–298

Liu J, Yang H, Zhao M, Zhang XH (2014) Spatial distribution patterns of benthic microbial communities along the Pearl Estuary, China. Syst Appl Microbiol 37:578. https://doi.org/10.1016/j.syapm.2014.10.005

Lombini A, Dinelli E, Ferrari C, Simoni A (1999) Plant-soil relationships in the serpentine screes of Mt. Prinzera (Northern Apennines, Italy). J Geochem Explor 64:19–33. https://doi.org/10.1016/S0375-6742(98)00017-X

Maity P (2006) Spatial variability analysis of soil penetration resistance of IARI farm to delineate compact zones (unpublished master's thesis). Indian Agricultural Research Institute, New Delhi, India

Mandal UK, Sharma KL, Venkanna K et al (2017) Sustaining soil quality, resilience and critical carbon level under different cropping systems in semi-arid tropical Alfisol soils. Curr Sci 112:1882–1895

Masto RE, Chhonkar PK, Singh D, Patra AK (2007) Soil quality response to long-term nutrient and crop management on a semi-arid Inceptisol. Agric Ecosyst Environ 118(1–4):130–142

Masto RE, Chhonkar PK, Purakayastha TJ, Patra AK, Singh D (2008a) Soil quality indices for evaluation of long-term land use and soil management practices in semi-arid sub-tropical India. Land Degrad Dev 19(5):516–529

Masto RE, Chhonkar PK, Singh D, Patra AK (2008b) Alternative soil quality indices for evaluating the effect of intensive cropping, fertilisation and manuring for 31 years in the semi-arid soils of India. Environ Monit Assess 136:419–435

McBratney AB, Webster R (1983) Optimal interpolation and isarithmic mapping of soil properties: V. Coregionalization and multiple sampling strategy. J Soil Sci 34(1):137–162

McBratney A, Field DJ, Koch A (2014) The dimensions of soil security. Geoderma 213:203–213

McKenzie DC, Shaw AJ, Rochester IJ, Hulugalle NR, Wright PR (2003) Soil and nutrient management for irrigated cotton. NSW Agriculture AGDEX 151/510, no. P5.3.6. NSW Agriculture, Orange, NSW

Meyer JL, Swank WT (1996) Ecosystem management challenges ecologists. Ecol Appl 6:738–740

Moebius-Clune BN, Moebius-Clune DJ, Gugino BK, Idowu OJ, Schindelbeck RR, Ristow AJ, Van Es HM, Thies JE, Shayler HA, McBride MB, Wolfe DW, Abawi GS (2016) Comprehensive assessment of soil health. The Cornell framework manual, 3rd edn. Cornell University, Geneva
Moeskops B, Buchan D, De Neve S, De Gusseme B, Widowati LR, Setyorini D, leutel S (2012) Soil quality indicators for intensive vegetable production systems in Java, Indonesia. Ecol Indic 18:218–226. https://doi.org/10.1016/j.ecolind.2011.11.011
Mohanty M, Painuli DK, Misra AK, Ghosh PK (2007) Soil quality effects of tillage and residue under rice–wheat cropping on a Vertisol in India. Soil Tillage Res 92(1):243–250
Montanaro G, Tuzio AC, Xylogiannis E, Kolimenakis A, Dichio B (2017) Carbon budget in a Mediterranean peach orchard under different management practices. Agric Ecosyst Environ 238:104–113. https://doi.org/10.1016/j.agee.2016.05.031
Muhammad AM, Shahid H, Tariq A, Muhammad A (2013) Sustainable agriculture through integrated soil fertility management on degraded lands. In: Developments in soil salinity assessment and reclamation. Springer, Dordrecht, pp 759–768
Mukherjee A, Lal R (2014) Comparison of soil quality index using three methods. PLoS One 9(8): e105981
Mukhopadhyay S, Masto RE, Yadav A, George J, Ram LC, Shukla SP (2016) Soil quality index for evaluation of reclaimed coal mine spoil. Sci Total Environ 542:540–550
Munoz P, Anton A (2008) Comparing the environment impact of greenhouse versus open field tomato production in Mediterranean region. Acta Hortic 801(2):1591–1596. https://doi.org/10.17660/ActaHortic.2008.801.197
Nabiollahi K, Taghizadeh-Mehrjardi R, Kerry R, Moradian S (2017) Assessment of soil quality indices for salt-affected agricultural land in Kurdistan Province, Iran. Ecol Indic 83:482–494
Nannipieri P, Grego S, Ceccanti B (1990) Ecological significance of the biological activity in soils. In: Bollag JM, Stotzky G (eds) Soil biochemistry. Marcel Dekker, New York, pp 293–355
Nannipieri P, Kandeler E, Ruggiero P (2002) Enzyme activities and microbiological and biochemical processes in soil. In: Enzymes in the environment. Marcel Dekker, New York, pp 1–33
Nielsen DR, Wendroth O (2003) Spatial and temporal statistics: sampling field soils and their vegetation. Catena Verlag, Reiskirchen
Obratov-Petković D, Popović I, Belanović S, Kadović R (2006) Ecobiological study of medicinal plants in some regions of Serbia. Plant Soil Environ 52(10):459–467
Pankhurst CE, Doube BM, Gupta VVSR (1997) Biological indicators of soil health: synthesis. In: Pankhurst CE, Doube BM, Gupta VVSR (eds) Biological indicators of soil health. CAB International, Wallingford, pp 419–435
Panwar P, Pal S, Reza SK, Sharma B (2011) Soil fertility index, soil evaluation, factor and microbial indices under different land uses in acidic soils of humid subtropical India. Commun Soil Sci Plant Anal 42:2724–2737. https://doi.org/10.1080/00103624.2011.622820
Papendick RI, Parr JF (1992) Soil quality—the key to sustainable agriculture. Am J Altern Agric 7:2–3
Parr JF, Papendick RI, Hornick SB, Singh RP (1992) Soil quality: attributes and relationship to alternative and sustainable agriculture. Am J Altern Agric 7:5–11
Patra DD, Anwar M, Chand S (2000) Integrated nutrient management and waste recycling for restoring soil fertility and productivity in Japanese mint and mustard sequence in Uttar Pradesh, India. Agric Ecosyst. Environ. 80(3):267–275
Patzel N, Sticher H, Karlen DL (2000) Soil fertility - phenomenon and concept. J Plant Nutr Soil Sci 163:129–142
Paz-Ferreiro J, Fu S (2013) Biological indices for soil quality evaluation: perspectives and limitations. Land Degrad Dev 27:14. https://doi.org/10.1002/ldr.2262
Paz-Kagan T, Shachak M, Zaady E, Karnieli A (2014) A spectral soil quality index (SSQI) for characterizing soil function in areas of changed land use. Geoderma 230-231:171–184
Pierce FJ, Larson WE (1993) Developing criteria to evaluate sustainable land management. In: Kimble JM (ed) Proceedings of the eighth international soil management workshop: utilization

of soil survey information for sustainable land use. USDA, Soil Conservation Service, National Soil Survey Center, Washington, pp 7–14
Pramanik P, Aggarwal P, Dey P (2013) Model for calculation of penetration resistance from easily measurable soil physical properties. Indian J Agric Sci 83(3):294–299
Purakayastha TJ, Kamble K, Pathak H, Kumari S, Singh A (2016) Decision support system for online assessment of soil health. http://ssacdss.iari.res.in/home.php
Purakayastha TJ, Pathak H, Kumari S, Biswas S, Chakrabarti B, Padaria N, Kamble K, Pandey M, Sasmal S, Singh A (2019) Soil health card development for efficient soil management in Haryana, India. Soil Tillage Res 191:294. https://doi.org/10.1016/j.still.2018.12.024
Rezaei SA, Gilkes RJ, Andrews SS (2006) A minimum data set for assessing soil quality in rangelands. Geoderma 136(1–2):229–234
Robinson TP, Metternicht GM (2006) Testing the performance of spatial interpolation techniques for mapping soil properties. Comput Electeron 50:97–108
Rodale Institute (1991) In: Conference Report and Abstracts. International Conference on the Assessment and Monitoring of Soil Quality. Rodale Press, Emmaus, PA
Romig DE, Garlynd MJ, Harris RF (1996) Farmer-based assessment of soil quality: a soil health scorecard. In: Doran JW, Jones AJ (eds) Methods for assessing soil quality. Soil Science Society of America, Madison, pp 39–60
Rossiter DG (1996) A theoretical framework for land evaluation. Geoderma 72:165–190
Saito H, McKenna A, Zimmerman DA, Coburn TC (2005) Geostatistical interpolation of object counts collected from multiple strip transects: ordinary kriging versus finite domain kriging. Stoch Env Res Risk A 19:71–85
Serrano JM, Peça JO, da Silva JRM, Shaidian S (2010) Mapping soil and pasture variability with an electromagnetic induction sensor. Comput Electron Agric 73(1):7–16
Shahid M, Nayak AK, Shukla AK, Tripathi R, Kumar A, Mohanty S, Bhattacharyya P, Raja R, Panda BB (2013) Long-term effects of fertilizer and manure applications on soil quality and yields in a sub-humid tropical rice-rice system. Soil Use Manag 29(3):322–332
Sharma KL, Mandal UK, Srinivas K, Vittal KPR, Mandal B, Grace JK, Ramesh V (2005) Long-term soil management effects on crop yields and soil quality in a dryland Alfisol. Soil Tillage Res 83:246–259
Sharma KL, Grace JK, Mishra PK, Venkateswarlu B, Nagdeve MB, Gabhane VV, Maruthi SG, Korwar GR, Chary G, Rao R, Srinivasa C, Gajbhiye PN, Madhavi M, Mandal UK, Srinivas K, Kausalya R (2011) Effect of soil and nutrient-management treatments on soil quality indices under cotton-based production system in rainfed semi-arid tropical vertisol. Commun Soil Sci Plant Anal 42(11):1298–1315
Sharma KLN, Raju KR, Das SK (2009) Soil fertility and quality assessment under tree, crop and pasture based land use systems in a rainfed environment. Commun Soil Sci Plant Anal 40:1436–1461. https://doi.org/10.1080/00103620902818096
Sikora LJ, Stott DE (1996) Soil organic carbon and nitrogen. In: Doran JW, Jones AJ (eds) Methods for assessing soil quality, SSSA special publication no. 49. Soil Science Society of America, Madison, pp 157–167
Singh SK, Pandey CB, Sidhu GS, Sarkar D, Sagar R (2011) Concentration and stock of carbon in soils affected by land uses and climate in the western Himalaya, India. Catena 87:78–89. https://doi.org/10.1016/j.catena.2011.05.008
Sinha NK, Chopra UK, Singh AK, Mohanty M, Somasundaram J, Chaudhary RS (2014a) Soil physical quality as affected by management practices under maize–wheat system. Natl Acad Sci Lett 37(1):13–18
Sinha NK, Chopra UK, Singh AK (2014b) Cropping system effects on soil quality for three agro-ecosystems in India. Exp Agric 50(03):321–342
Smith JL, Elliott LF (1990) Tillage and residue management effects on soil organic matter dynamics in semiarid regions. In: Advances in soil science. Springer, New York, NY, pp 69–88

Sofi JA, Bhat AG, Kirmai NA, Wani JA, Lone AH, Ganie MA Dar GIH (2016) Soil quality index as affected by different cropping systems in northwestern Himalayas. Environ Monit Assess 188 (3):161. https://doi.org/10.1007/s10661-016-5154-1

Stefanic G, Eliade G, Chirnogeanu I (1984) Researches concerning a biological index of fertility. In: Fifth symposium on soil biology. Bucharest, Romania, pp 35–45

Stenberg B, Rossel RAV, Mouazen AM, Wetterlind J (2010) Visible and near infrared spectroscopy in soil science. In: Sparks DL (ed) Advances in agronomy, vol 107. Academic Press, Amsterdam, pp 163–215

Stewart BA (1995) Soil management: experimental basis for sustainability and environmental quality, vol 4. CRC Press, Boca Raton, FL

Stocking MA (2003) Tropical soils and food security: the next 50 years. Science 302 (5649):1356–1359

Surchand-Singh K, Hazarika S, Thakuria D, Bordoloi LJ, Deka BC (2017) Long-term management impact on soil quality under peach (Prunus persica L.) orchard in humid subtropics of eastern Himalayas. J Indian Soc Soil Sci 65(4):401–409. https://doi.org/10.5958/0974-0228.2017.00047.0

Theodossiou N, Latinopoulos P (2006) Evaluation and optimisation of groundwater observation networks using the Kriging methodology. Environ Model Softw 21(7):991–1000

Usowicz B, Lipiec J (2017) Spatial variability of soil properties and cereal yield in a cultivated field on sandy soil. Soil Tillage Res 174:241–250

Vasu D, Singh SK, Ray SK et al (2016) Soil quality index (SQI) as a tool to evaluate crop productivity in semi-arid Deccan plateau, India. Geoderma 282:70–79

Wang X, Gong Z (1998) Assessment and analysis of soil quality changes after eleven years of reclamation in subtropical China. Geoderma 81:339–355

Wanshnong RK, Thakuria D, Sangma CB, Ram V, Bora PK (2013) Influence of hill slope on biological pools of carbon, nitrogen, and phosphorus in acidic alfisols of citrus orchard. Catena 111:1–8

Webster R, Oliver MA (2001) Geostatistics for environmental scientist. John Wiley and Sons, Ltd., Chichester, pp 47–131

Wei JB, Xiao DN, Zeng H, Fu YK (2008) Spatial variability of soil properties in relation to land use and topography in a typical small watershed of the black soil region, northeastern China. Environ Geol 53:1663–1672

Weller U, Castell WZ, Sommer M, Wehrhan M (2002) Kriging and interpolation with radial base functions a case study. http://citeseerx.ist.psu.edu/viewdoc/versions?doi=10.1.1.73.1305

Yao R-J, Yang J-S, Gao P, Zhang J-B, Jin W-H, Yu S-P (2013) Soil-quality-index model for assessing the impact of groundwater on soil in an intensively farmed coastal area of E China. J Plant Nutri Soil Sci 177:330. https://doi.org/10.1002/jpln.201200383

Zare-mehrjardi M, Taghizadeh-Mehrjardi R, Akbarzadeh A (2010) Evaluation of geostatistical techniques for mapping spatial distribution of soil PH, salinity and plant cover affected by environmental factors in southern Iran. Not Sci Biol 2(4):92–103

Zhang SW, Huang YF, Shen CY, Ye HH, Du YC (2012) Spatial prediction of soil organic matter using terrain indices and categorical variables as auxiliary, information. Geoderma 171-172:35–43

Zornoza R, Acosta JA, Bastida F, Domínguez SG, Toledo DM, Faz A (2015) Identification of sensitive indicators to assess the interrelationship between soil quality, management practices and human health. Soil 1(1):173–185. https://doi.org/10.5194/soil-1-173-2015

Nanobiosensors: Recent Developments in Soil Health Assessment

15

Nintu Mandal, Samrat Adhikary, and Rajiv Rakshit

Abstract

Soil health is the dynamic equilibrium among physical, chemical, and biological properties of soil. Real-time assessment of soil health is of utmost importance for managing soil effectively. Sensors based precision nutrient and moisture management are keys of precision agriculture. Nanomaterials by virtue of their higher specific surface along with increased chemical reactivity seem to play an important role in biosensors development. Due to high specificity, accuracy, stability, and quick reactivity and specially different physico-chemical properties, catalytic activity makes nanobiosensors more usable. Multimodal nanosensors have been developed for detection and removal of mercury. Quantum dots and carbon nanodots based sensor system for detection of heavy metals have been developed. Nanobiosensors' use for urea detection and monitoring of enzymatic activity in soil have been developed under laboratory condition and have been applied under controlled condition. Proper and controlled use of nanobiosensor can support sustainable agriculture by using optimum resource and ultimately enhancing the crop productivity.

Keywords

Nanobiosensors · Multimodal nanosensor · Quantum dots · Carbon nanodots

N. Mandal (✉) · R. Rakshit
Department of Soil Science and Agricultural Chemistry, Bihar Agricultural University, Bhagalpur, Bihar, India

S. Adhikary
Department of Agricultural Chemistry and Soil Science, Bidhan Chandra Krishi Viswavidyalaya, Nadia, West Bengal, India

A. Rakshit et al. (eds.), *Soil Analysis: Recent Trends and Applications*,
https://doi.org/10.1007/978-981-15-2039-6_15

15.1 Introduction

Nanotechnology is the manipulation of atoms or molecules at nanoscale. Nanomaterials are defined as materials having at least one dimension within 1–100 nm range (USEPA 2007). Higher specific surface areas coupled with increased chemical reactivity are unique characteristic at nanoscale. Nanomaterials are being applied in various disciplines of agricultural sciences viz. smart agrochemicals formulations, purification of metals and metalloids, increasing vase life of cut flowers, and reducing post-harvest losses.

Real-time management of agricultural production system is of utmost importance. Application of sensor technology in precision management of fertilizers and pesticides in the field need to be prioritized. Nanobiosensor structure is miniature in nature as compared to conventional biosensors. Nanomaterials can be integrated into other technologies such as lab-on-a-chip to facilitate molecular analysis. Several applications of nanobiosensors such as detection of analytes like urea, pesticides, monitoring of metabolites glucose, and detection of various pathogens have been documented. Nanobiosensors may be of great value for effective monitoring of soil quality in terms of its constituents, pH, humidity, microbial load, etc.

15.2 Development of Biosensors: Chronological

First biosensor was invented in the year 1967 which eventually led to the development of other sensors. Biosensors development and its application were restricted to laboratory conditions only up to early twentieth century. Technological development in biosensors led to field scale application of sensors. Three generations of biosensors have been classified as described hereunder:

1. First generation biosensors: Mode of operation is based on electrical response.
2. Second generation biosensors: Specific "mediators" between the reaction and the transducer generates improved response.
3. Third generation biosensors: Reaction itself causes the response. Here no product or mediator diffusion is directly involved.

Developments in nanobiosensors have been indicated in Table 15.1. Nanomaterials based biosensors came into existence during 2000 (Table 15.1).

15.3 Nanobiosensors: Definition

Nanobiosensor is a modified version of a biosensor which may be defined as a compact analytical device or biologically derived sensitized element linked to a physico-chemical transducer. This is usually built on the nanoscale to obtain, process, and analyze the data at the level of atomic scale (Rai et al. 2012).

Table 15.1 Historical development in biosensing technology

Year	Developments
1962	Amperometric enzyme electrode (glucose sensor) was described by Clark
1969	Potentiometric biosensor: Urease immobilized on an ammonia electrode detected urea by Guilbault and Montalva
1975	First commercial biosensor (Yellow Spring Instrument biosensor)
1975	First microbe based biosensor (first immunosensors)
1982	First fiber-optic based biosensor for glucose
1983	First surface plasmon resonance (SPR) immunosensors
1987	Blood glucose biosensor launched by Medisense Exactech
2000	Nano Technology biosensor, chip, quantum dots, etc.

Technological advances in twenty-first century made it possible for development of miniature form of biosensors. Integration of nanotechnology, instrumentation, electronics, and biology made it possible for development of biosensors having higher temporal and spatial resolution with precise detection limits. Nanosensors with immobilized baroreceptor probes that are selective for target analyte molecules are called nanobiosensors. Precision of nanobiosensors are up to atomic or molecular level.

15.4 Soil Health: Characteristics and Its Methods of Evaluations

15.4.1 Characters of Healthy Soils (As adopted from www.css.cornell.edu/extension/soil-health/manual.pdf)

- *Tilth*: Soil tilth refers to physical condition of the soils with respect to its productive capacity. Soils with poor tilth appears lifeless or being cloddy. Soil with great tilth is brittle.
- *Sufficient depth*: For proper root growth and drainage, a soil with sufficient depth is needed. Soil depth can influence the plant that grows in them. Deeper soils provide more water and nutrients to plants than more shallow soils.
- *Good water storage and good drainage*: Soil should have a good water storage capacity and good drainage for effective plant growth. An ideal soil should have stable pore to take in water.
- *Sufficient supply, but not excess of nutrients*: Nutrient supply is the central idea for ideal plant development. An abundance of nutrient may also lead to potential contamination of ground water besides leading to aberrations to the climate in reference to greenhouse gases.

- *Population of beneficial organisms*: Soil living organisms are important for the effective working of the soils. Soil micro-organisms are involved in nutrient cycling, maintains soil structure, and so on. Biological reactions in soils are catalyzed by this beneficial organism in soils.
- *Low weed pressure*: Weeds are competitors of crops for every component there in soils which are important for plant growth and development. Low weed pressure will make the soil functional to achieve the goals of sustainable plant production.
- *Free of chemicals and toxins*: Presence of toxic substances and synthetic compounds may lead to restrict the functions of soils by affecting the microbes therein.
- *Resilience when unfavorable conditions occur*: A sound soil will bounce back more rapidly after a negative occasion, for example, collecting under wet soil conditions, or if land limitations confine or change arranged turns.

15.4.2 What Indicates Soil Health?

Indicators viz. physical, chemical, and biological are there to determine the status of soil health. An indicator of soil health should be (1) sensitive to soil management practices, (2) correlated with soil functions or variables which are difficult to measure, (3) relates to ecosystem processes, (4) comprehensive for decision-making, and (5) cheap and easy to measure (Parisi et al. 2005).

15.4.3 Assessment of Soil Health and Its Approach

Assessment of soil health is needed to increase the consciousness among all about the importance of soil health. Assessment of soil health is to understand beyond problems related with nutrient deficiencies and excesses. Assessment of soil health comprises three main steps, as presented in (Fig. 15.1). It includes the selection of dataset of relevant attributes of soils; secondly to quantify these soil attributes through lab analysis; and finally to integrate these attributes to construct the final index by criteria for defining weight to each attributes (Rinot et al. 2019; Rakshit et al. 2018). The basic approach for soil health management starts from determining the background of the farm or location with proper understanding of the management history. The focus should be on the targeted function of the soils considering the constraints, possibilities, and priorities. The final idea is to create short and long term management plans with integration of agronomic science with the real situation needed for growers. This could only be achieved once we monitor the changes, repeated testing of the attributes and its evaluation over the time.

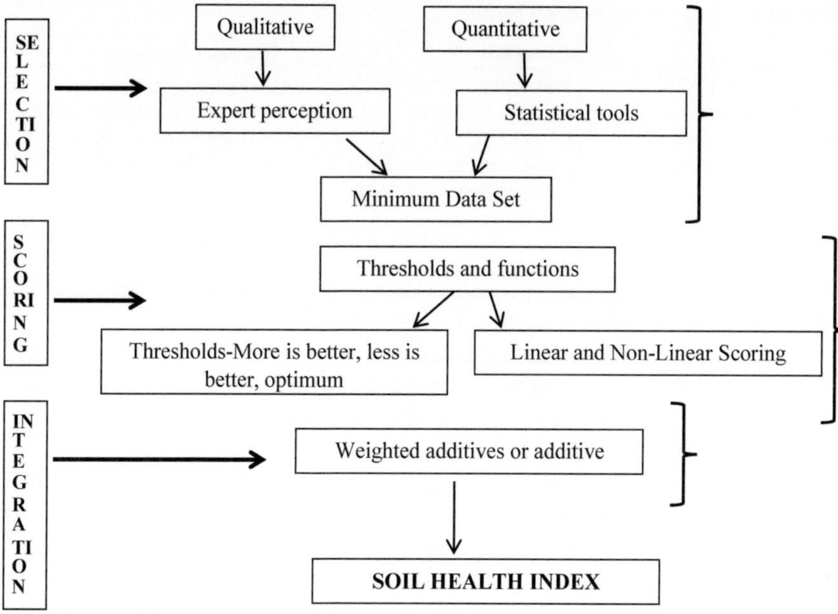

Fig. 15.1 Steps used for assessment of soil health (modified from Rinot et al. 2019)

Fig. 15.2 Integration of the physical, biological, and chemical components of the soil to achieve soil health index (adapted from the Rodale Institute)

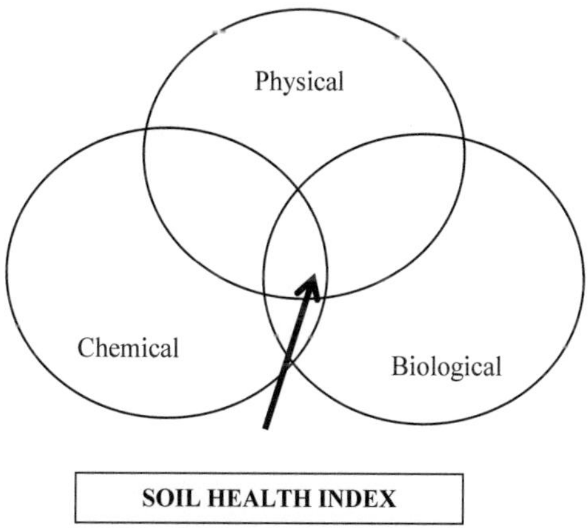

15.4.4 Soil Quality Assessment Procedures

A number of procedures were proposed by various groups or institutions. The Cornell Comprehensive Assessment of Soil Health (CASH) protocol highlights the amalgamation of biological, physical, and chemical properties of soils (Fig. 15.2). These measurements include the measurement of physical, chemical, and biological parameters of soils. There is a scope to add supplementary indicators as add-ons specific for the area under study. To validate the attributes, a team of farmers,

Table 15.2 Markers to monitor soil health

Physical	Biological	Chemical
Texture, bulk density, porosity, hydraulic conductivity, aggregate size, available water	Potentially mineralizable nitrogen, cellulose decomposition rate, respiration rate of microbes, active carbon, nematode population, protein	Phosphorus, nitrogen, potassium, DTPA extractable micronutrients in soil, heavy metals, salinity, sodicity, exchangeable acidity

teachers, specialist, and experts used to re-examine throughout the years, and few markers were identified as presented in Table 15.2.

Although the indicators are proposed and established well, it is of utmost importance to appraise the soil health in the farm itself. Preliminary evaluation in farm is done by visual observations. On farm assessment of soil health is recommended to support farmers to estimate the effects of their management on productivity. This approach permits interaction among beneficiaries and experts with an idea to interpret the farm knowledge to soil health information. The main challenge still remains to develop soil quality and soil health standards to assess changes which are useful to farmers. This methodology is purely emotional and reflects the expert's inclination. The scorecards developed for on farm assessment emphasizes the qualitative observations of soil. These cards may be developed to evaluate soil health through farmer observation of soil physical, chemical, and biological properties. Examples includes the observation on earthworm numbers which can indicate the biological activity in the soil. Besides, assessment kits are available through which farmers can appraise the attributes qualitatively and these kits are important education tool to raise the awareness among the cultivators.

Recently, Rinot et al. (2019) proposed a soil health management structure based on ecosystem services in which the first step is to take samples from a wide range of soils and to measure the physical, chemical, and biological traits of soils for minimizing the data set. The second step converts the data into normalized scores termed as scoring functions (expressed as percent 0–100%). The scale and magnitude are mainly determined on soil provisioning services of a specific area. When the entire soil ecosystem service is considered, the functions may be more than the conventional linear and non-linear and interactions between different attributes. The third step focuses on integration through least squares models. With this, a coefficient will be derived for each attribute which expresses its contribution to ecosystem service.

15.4.5 Limitations of Indices of Soil Quality

Disciplinary biases sometimes evolve in expert opinion based soil quality assessment approaches. There are reports to add sub-indicators reflecting various faces of soil quality to overcome localized problems. A high degree of spatio-temporal variation in microbial property within a given land interferes in the calculation of these indices. Another limitation is the ignorance of the attributes across the soil profile as only top layers are generally considered for the soil quality outcomes (Sparling et al. 2004). There is still a scope to propose a globally acceptable method as there is intrinsic intricacy within soils. Several soil indicators have been proposed to define health (Morrow et al. 2016), but their usefulness of merging these indicators into a comprehensive one remains incomprehensible. Thus, soil quality assessment has changed through time in terms of objectives, tools and methods, and overall approach (Bünemann et al. 2018).

15.5 Nanobiosensors: Characteristics

Specific characteristic of nanobiosensors have been outlined as hereunder:

- *Specificity*: Biosensors are specific, i.e., it can distinguish between analyte and other than analyte.
- *Stability*: Sensors are stable at normal storage conditions.
- *Interaction*: Interaction of biosensors with analyte is independent of any parameters such as stirring, pH, and temperature.
- *Response time*: Reaction time should be minimal.
- *Accuracy*: Accurate, reproducible, precise, and linear response over the useful analytical range and also be free from electrical noise should be there.
- *Feature*: The nanobiosensor must be nontoxic, biocompatible, tiny, and non-antigenic.
- *Flexibility*: Should be portable, cheap, and capable of being used by semi-skilled operators.

15.5.1 Constituents of Nanobiosensors

There are basically three component parts of nanobiosensors (Fig. 15.3).

- *Probe*: This is the biologically sensitized element. The biologically sensitized elements (probe) including nucleic acids, antibodies, enzymes, receptors, molecular imprints, tissue, lectins, organelles, micro-organisms, etc., are either a biologically derived material or bio-mimic component that receives signals from the analytes (sample) of interest and transmits it to transducer.
- *Transducer*: Its acts as an interface between probe and detector. It measures the physical change that occurs due to interaction of probe and detector and transforms the energy change into electrical signal.
- *Detector*: Signals received from the transducer are passed to a microprocessor where they are amplified and analyzed. Data generated is then transferred to user friendly output and displayed/stored (Hassani 2016).

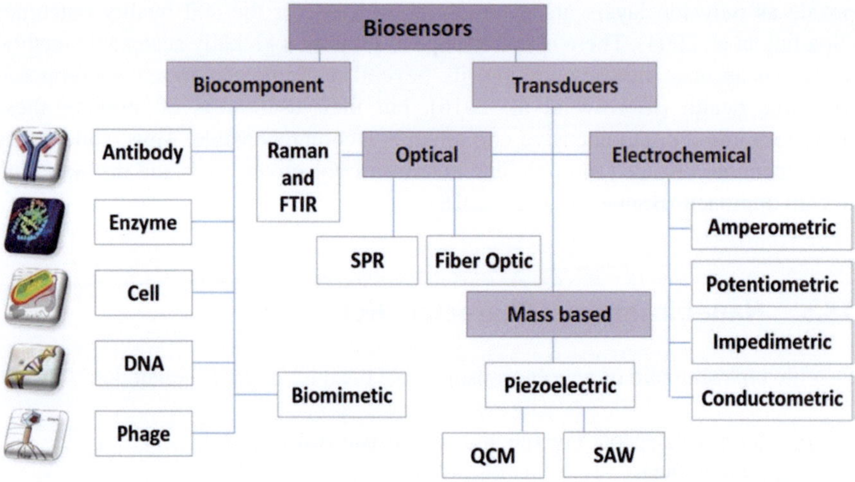

Fig. 15.3 Classification of biosensors (adapted from Hassani 2016)

15.5.2 Classifications of Biosensors

Some most important biosensors are briefly elaborated according to their mode of function:

- *Optical biosensors*

  An optical biosensor is a compact analytical device that is able to detect a desired analyte by assembly of a light source, optical transducer system, sensing element, and a detector. The optical transducer transmits the light and acts as the substrate for the sensing material that has been immobilized on a suitable optical surface. Thus, the interactions of sensing material with analyte initiate some physico-chemical changes that are produced in response to light signals, and finally, a detector measures the transduced light as the output.
- *Electrochemical biosensors*

  In the simplest definition, an electrochemical biosensor directly converts a biological event to an electronic signal. Despite the diversity in bio-recognition elements, enzymes are the predominant substrate in electrochemical detection techniques due to their specific binding capabilities and bio-catalytic activity (Grieshaber et al. 2008). Researchers have introduced a different classification for electrochemical biosensors based on signal property, which measures the variation of biological changes in solution by means of potential, charge accumulation, current, conductance, or impedance. Typically, amperometric, potentiometric, impedimetric, and conductometric biosensors have been organized. The precision and sensitivity of these systems mainly correlate with the bio-recognition element. The sensitivity also depends on the conductivity of the materials.

15.5.3 Comparison of Conventional Analytical and Biosensing Techniques

Conventional analytical techniques		Biosensors	
HPLC, LC, GC-MS, fluorimetry, and SPR		Electrochemical; optical; mass based; bio-component	
Advantages	Disadvantage	Advantages	Disadvantage
Sensitive	Time consuming	Rapid real-time detection	Limited commercial application
Specific	Expensive	Cost-effective	
	Laboratory monitoring	Portable	
	Trained laboratory personnel	Simple use	
	High tech equipment	Highly sensitive	
	Extensive sample preparation	Limited sample preparation	
	Not reusable	Reusable	
	More organic solvent consumption	Less organic solvent consumption specific	

15.6 Implications of Nanobiosensors in Agriculture and Allied Sector

Presently, nanomaterial-based biosensors exhibit fascinating prospects over traditional biosensors. Nanobiosensors have marked advantages such as enhanced detection sensitivity/specificity and possess great potential for its applications in different fields including environmental and bioprocess control, nutrient monitoring in soils (nutrient status checking through different technique: Nutrient detection based on sufficiency or deficiency aspcct). But here we are concerned with the role of nanobiosensor in agriculture.

15.6.1 Nanobiosensors Use for Urea Detection

Presence of urea can be detected by monitoring urease (Ur) and glutamate dehydrogenase (GLDH). Ur catalyzes decomposition of urea into hydrogen bicarbonate and ammonium ions (NH_4^+). NH_4^+ ions are unstable and easily disperse in environment. Metal oxide nanoparticles-chitosan (CH) based hybrid composites have attracted much interest for the development of a desired biosensor (Kaushik et al. 2009). The proposed biochemical reaction during the urea detection in two steps are: First, a Schiff base intermediate being formed between NH_3 and α-KG and the Schiff base intermediate are protonated due to the transfer of the hydride ion from NADH resulting in L-glutamate (Fig. 15.4).

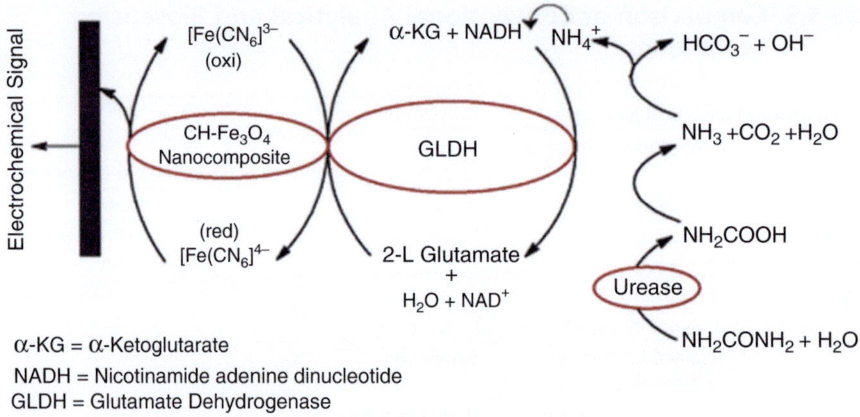

Fig. 15.4 Biochemical reaction during electrochemical detection of urea using Ur-GLDH/CH-Fe_3O_4 nanobiocomposite (Kaushik et al. 2009)

15.6.2 Nanobiosensor Use for Heavy Metal Detection

Heavy metal ions are regarded as one of the most toxic substance affecting the environment. In recent years, biosensors are gaining importance as suitable detectors for heavy metal ions. They prove very promising for environmental monitoring, since the system is simple, rapid, and selective. Several techniques based on spectroscopy, ion-selective electrodes, polarography, and voltammetry have been described in the past. Zhylyak et al. (1995) *developed a urease based* conductometric biosensor for the determination of heavy metal ions in wastewater (Fig. 15.5).

15.6.3 Diagnostic Tool for Soil Quality and Disease Assessment

Nanosensors may be used to diagnose soil disease (caused by infecting soil microorganisms, such as viruses, bacteria, and fungi) via the quantitative measurement of differential oxygen consumption in the respiration (relative activity) of "good microbes" and "bad microbes" in the soil. The measurement proceeds through the following steps: two sensors impregnated with "good microbes" and "bad microbes," respectively, are immersed in a suspension of soil sample in buffer solution and the oxygen consumption data by two microbes were detected. By comparing two data, we can easily decide which microbe favors the soil. Apart from that, we can also predict whether or not soil disease is ready to break out in the tested soil. So, it is to be emphasized that the biosensor offers an innovative technique of diagnosing soil condition based on semi-quantitative approach.

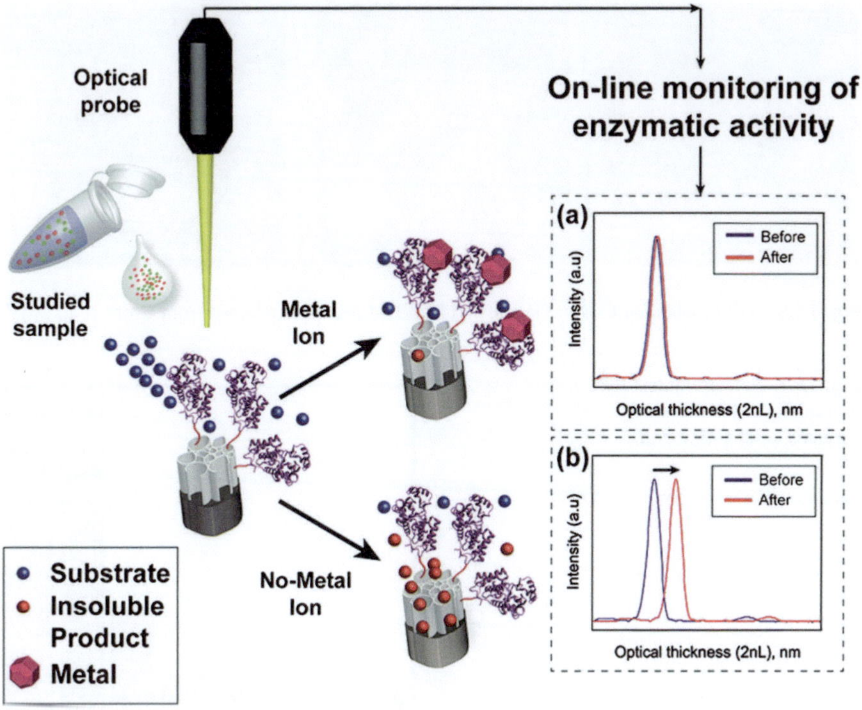

Fig. 15.5 Mechanisms of conductometric biosensor for heavy metal determination

15.6.4 Quantum Dots and Carbon Nanodots Based Sensor System for Detection of Heavy Metals

Förster resonance energy transfer (FRET) based novel optical system is an alternate approach for detection of heavy metals pollutants (Chini et al. 2019). The sensing system comprises of graphene quantum dots (GQDs) as donor and carbon nanodots (C-Dots) as an acceptor component (Fig. 15.6). When these fluorescent nanodots are within the FRET distance, fluorescence of the donor GQDs is quenched by the non-radiative energy transfer to acceptor C-Dots. Fluorescence lifetime is measured by time resolved photoluminescence spectroscopic study to validate the FRET efficacy of the mix-dot based sensor system. Upon gradual addition of heavy metals like arsenic (As^{5+}) and mercury (Hg^{2+}) into this sensor system, there is a significant amount of reduction in the investigated FRET signal.

Noteworthy, As^{5+} has shown a faster decay profile owing to its higher positive charge and strong electron affinity in comparison to other metal ions. The strong affinity of As^{5+} towards the carboxylic group present on the surface of GQD and C-Dot also leads to enhanced quenching. The selectivity of the FRET sensor to other metal ions are also tested for 150 μM concentration of different metal ions and

Fig. 15.6 AFM images of C-Dot, GQDs, and Mix-Dot particles (adapted from Chini et al. 2019)

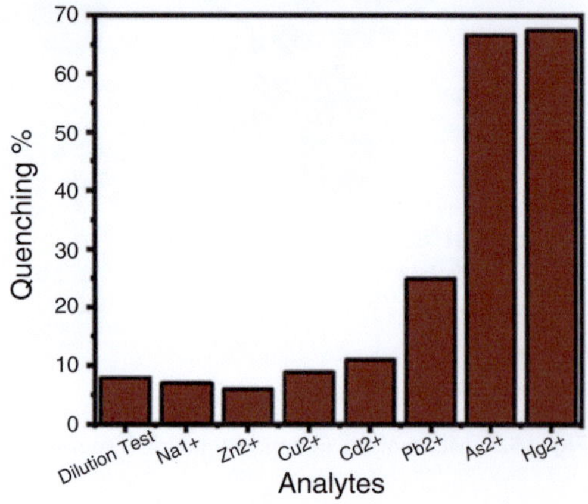

Fig. 15.7 Selectivity test for investigated FRET signals (adapted from Chini et al. 2019)

showed in Fig. 15.7. Finally, subsequent experiments analysis of real samples was performed for As^{5+}, Hg^{2+}, and Pb^{2+}.

The GQD: C-Dot FRET pair based optical sensors are used for detecting different heavy metals like As^{5+}, Hg^{2+}, and Pb^{2+} in aqueous solution. When the increasing concentrations of the analytes are added to the FRET sensor, the fluorescence intensity decreases rapidly (Fig. 15.8). The sensitivity for these three metal ions towards FRET sensor followed order: $As^{5+} > Hg^{2+} > Pb^{2+}$.

15.6.5 Multimodal Nanosensor: Detection and Removal of Mercury

A highly sensitive and environment friendly multimodal nanosensor encompassing magnetic and fluorescent functionality was designed for the simultaneous detection and removal of mercury ion in water (Satapathi et al. 2018). A significant fluorescence quenching is observed with the increasing concentration of Hg^{2+} with surprisingly low limit of detection. The detected analyte is successfully removed with the help of a bar magnet leaving no residual secondary pollution. The superparamagnetic Fe_2O_3 nanoparticles were prepared by the chemical co-precipitation method (Fig. 15.9). Initially, the pristine Fe_2O_3NPs tend to

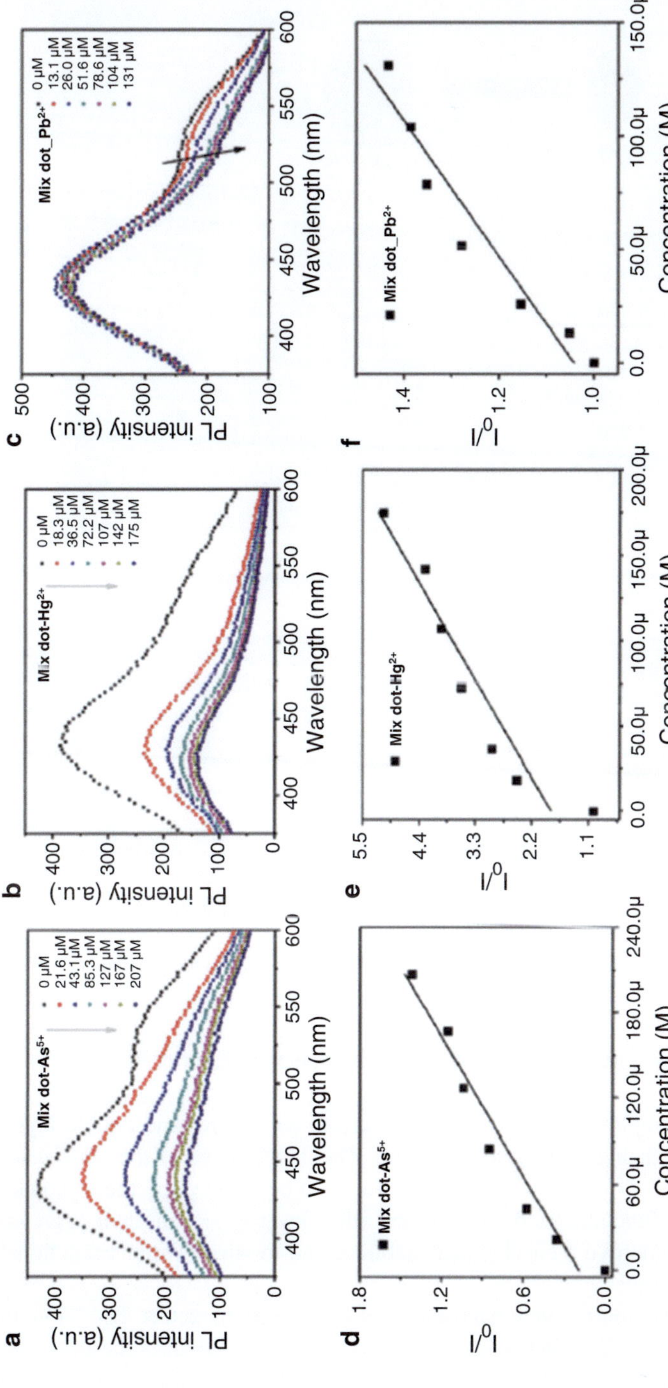

Fig. 15.8 Fluorescence intensity quenching with the presence of **(a)** As^{5+}, **(b)** Hg^{2+}, and **(c)** Pb^{2+}. **(d–f)** Stern–Volmer constant of As^{5+}, Hg^{2+}, and Pb^{2+}. (Adapted from Chini et al. 2019)

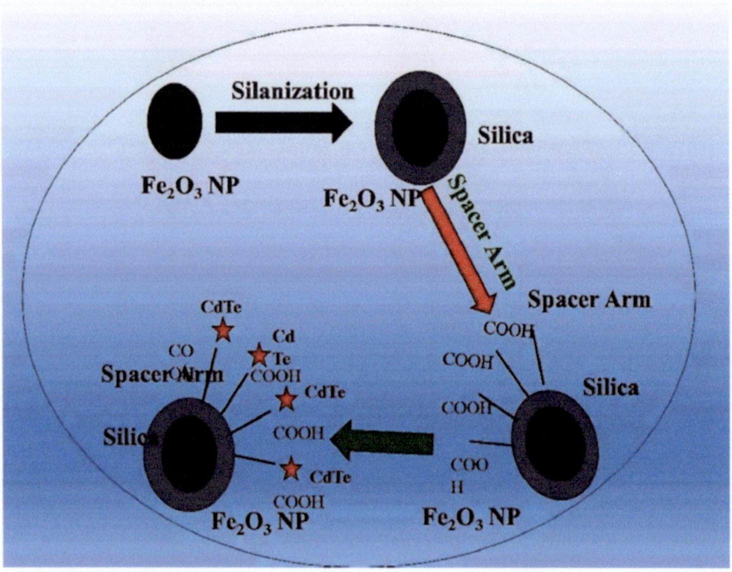

Fig. 15.9 Design of multimodal biosensor (adapted from Satapathi et al. 2018)

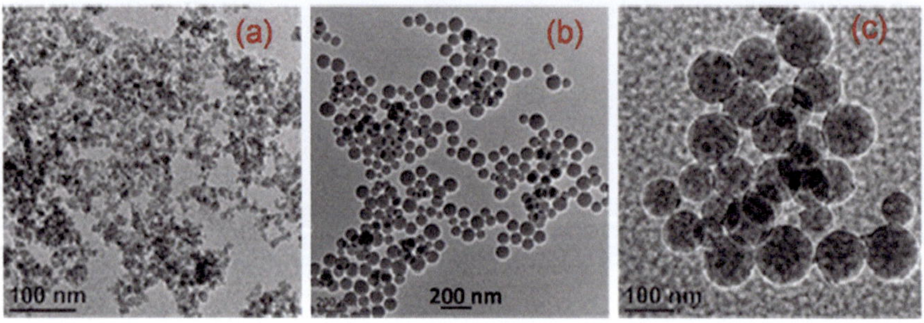

Fig. 15.10 TEM image of Fe_2O_3 NP (**a**), silica coated Fe_2O_3 NP (**b**), and multimodal nanosensor (**c**). (Adapted from Satapathi et al. 2018)

aggregate into large clusters, and thus lost the special magnetic properties associated with single domain.

Citric acid is employed as a surfactant to create an electrostatic double layer, thereby, reducing their tendency to aggregate. The TEM image of the citric acid-stabilized Fe_2O_3 NPs was shown in Fig. 15.10. The surface of the magnetic nanoparticles was functionalized with an inert silica layer to prevent their aggregation in liquid, to improve their chemical stability, and to attach various functional groups capable of sensing analytes.

The synthesized multimodal nanosensor is employed to detect Hg^{2+} ion in aqueous environment. The increasing concentration of Hg^{2+} reduces the PL emission intensity of nanosensor. A significant signal off of the nanosensor is observed in

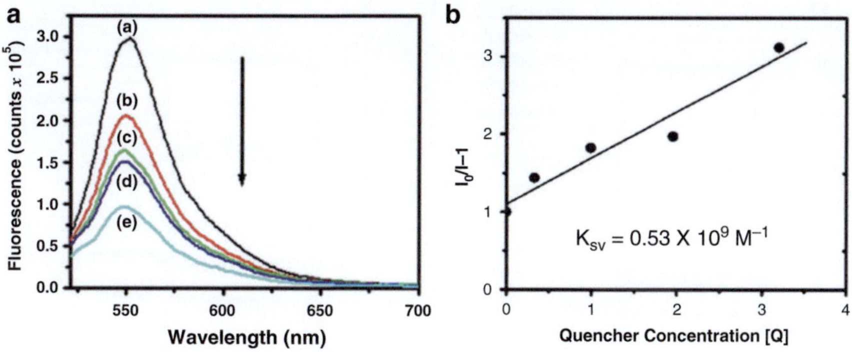

Fig. 15.11 PL intensity of nanosensor with changing concentration of Hg^{2+} (**a**); calculation of Stern–Volmer constant (**b**) (adapted from Satapathi et al. 2018)

the presence of increasing concentrations of Hg^{2+}. Figure 15.11a depicted the PL intensity of nanosensor with changing the concentration of Hg^{2+}. Quantitatively, the PL quenching sensitivity could be correlated to the Stern–Volmer constant (KSV), determined by the equation,

$$I_0/I = 1 + \mathrm{KSV}$$

where, I_0 is the initial PL intensity, I is the resulting PL intensity upon addition of the analyte, and Q is the analyte concentration. A larger KSV value represented a higher sensitivity of the fluorophore toward the analyte. The Stern–Volmer constant is calculated to be $0.14 \times 109\ M^{-1}$ (Fig. 15.11b). Sensing experiments are performed with the different Hg^{2+} concentration and the lowest detection limit is found to be 0.49 nm which is within the range set by USEPA. This also reveals that this multimodal nanosensor probe is more sensitive than the previously reported sensing probes in the literature.

15.6.6 Surface Plasmon Resonance (SPR) Nanosensor for Detection of Zn(II) Ions

A novel Zn(II) ions imprinted poly (2-hydroxyethyl Methacrylate-N-methacryloyl-(L)-histidine methyl ester) surface plasmon resonance (SPR) nanosensor is designed for detection of Zn(II) ions in aqueous solution and artificial plasma providing a low cost, rapid and reliable results compared to other techniques such as atomic absorption spectroscopy, inductively coupled plasma-mass spectrometer, X-ray fluorescence with synchrotron radiation (Jalilzadeh et al. 2018).

The selectivity of Zn(II) ions imprinted nanosensor have been examined through the adsorption of Cd(II), Cu(II), Fe(II), Pb(II), and Zn(II) ions. Selective recognition of Zn(II) ions with Zn(II) ions imprinted nanosensor was investigated using 0.25 mg/mL nitrate solution of Cd(II), Cu(II), Fe(II), Pb(II), and Zn(II) (Fig. 15.12).

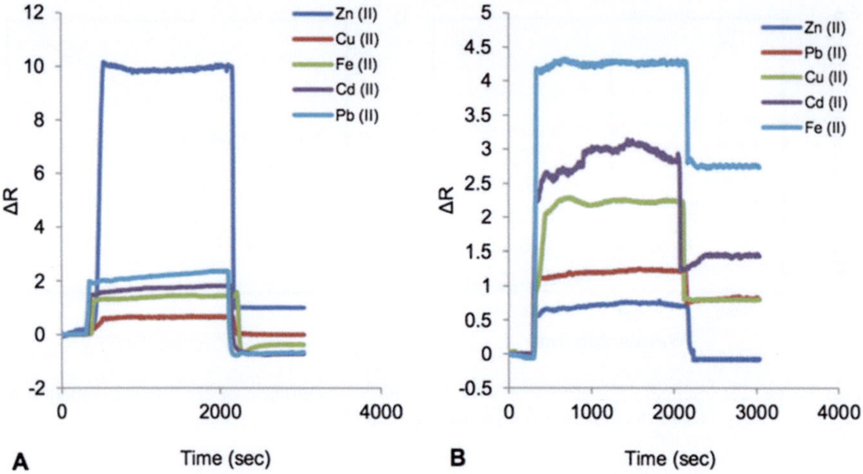

Fig. 15.12 The SPR responses of (**a**) Zn(II) ions imprinted nanosensor and (**b**) non-imprinted nanosensors (concentration of all metal ions is 0.25 μg/mL). (Adapted from Jalilzadeh et al. 2018)

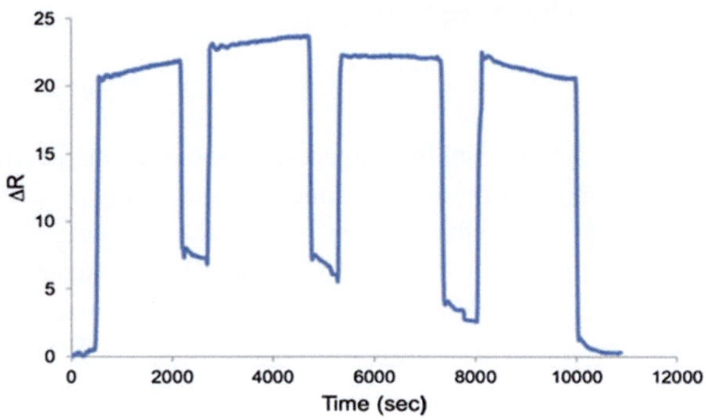

Fig. 15.13 Reproducibility of Zn(II) ions imprinted SPR nanosensors (Zn(II) ion concentration was 0.75 lg/mL in all measurements, 25 mM acetate buffer, pH 5.0 at a flow rate 150 mL/min). (Adapted from Jalilzadeh et al. 2018)

According to the experimental data, the selectivity coefficients (k) and relative selectivity coefficients (k') of Cu(II), Fe(II), Cd(II), and Pb(II) ions with respect to Zn(II) ions are calculated for Zn(II) ions imprinted and non-imprinted nanosensors.

Reproducibility of Zn(II) ions imprinted SPR nanosensor response was examined by repeating of equilibration–adsorption–regeneration for four times using aqueous zinc nitrate solution with the concentration of 0.75 mg/mL (Fig. 15.13). As seen in the figure, reproducibility of reflective response during four cycles was demonstrated by Zn(II) ions imprinted nanosensor. Stability results for Zn(II) ions imprinted PHEMAH based SPR nanosensor shown in Fig. 15.14 indicate that Zn(II) ions imprinted PHEMAH based SPR nanosensor are stable under long-term storage conditions. After 3 weeks of storage of the SPR based nanosensor, 87% of the initial activity of the SPR nanosensor remained.

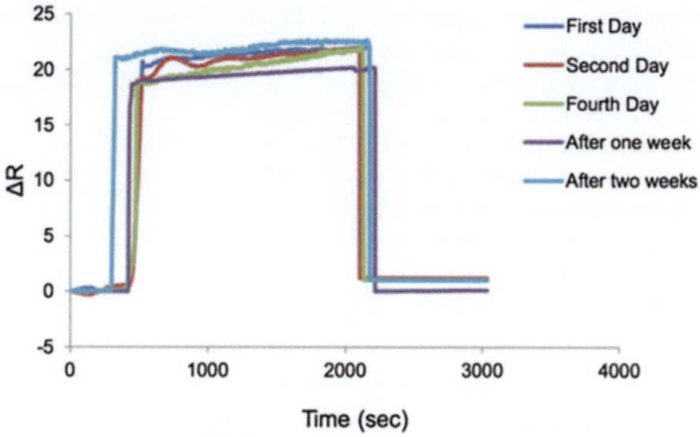

Fig. 15.14 Reusability on different days of Zn(II) ions imprinted SPR nanosensors (Zn(II) ion concentration was 0.75 lg/mL in all measurements, 25 mM acetate buffer, pH 5.0 at a flow rate 150 mL/min). (Adapted from Jalilzadeh et al. 2018)

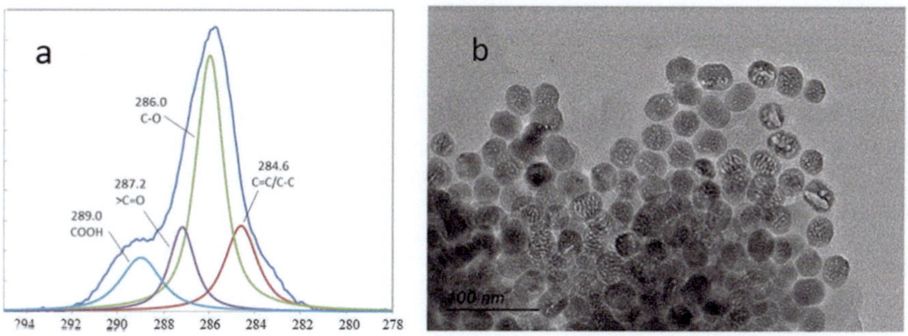

Fig. 15.15 (**a**) Deconvolution of the C1s band of the XPS spectra. (**b**) TEM image of the CDs. (Adapted from Hernández-Rodríguez et al. 2018)

Designed nanosensor was applied for selective detection of Zn(II) ions in aqueous solution within the range of 0.5–1.0 μg/mL. The limit of detection (LOD) and limit of quantification (LOQ) are calculated as 0.19 and 0.64 ng/mL, respectively.

15.6.7 Carbon Dots as Temperature Nanosensors

Carbon quantum dots (CQDs, C-dots, or CDs) are small carbon nanoparticles (less than 10 nm in size) with some form of surface passivation. Different types of carbon dots were reported in literature viz. carbon quantum dots (CQDs), carbon nanodots (CNDs), and polymer dots (PDs) each one of them with different structure and photoluminescence properties (PL) (Hernández-Rodríguez et al. 2018).

The results from the XPS indicated a composition in the surface mainly of carbon, oxygen, nitrogen, and small amounts of other elements such as sodium and boron. Analysis of the C1s (Fig. 15.15a) band indicated a high degree of oxidation on the surface of the CDs. TEM analysis (Fig. 15.15b) showed quasi-spherical particles with an average diameter of 31 nm and low size dispersion (from 25 to 38 nm).

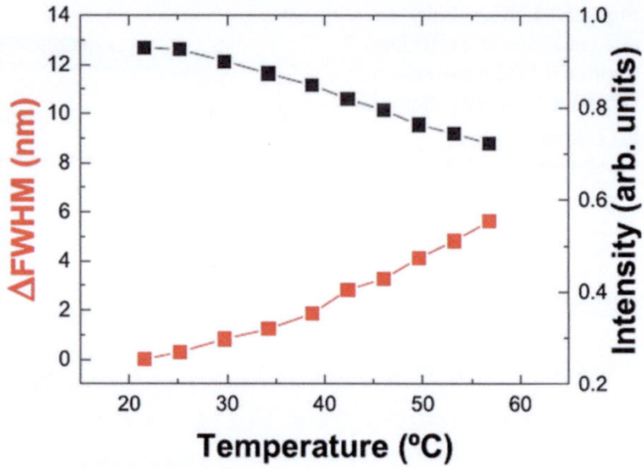

Fig. 15.16 Temperature dependence of the intensity of the emission spectrum of the carbon dots upon 405 nm excitation and dependence of the increment of the FWHM of the emission band with respect to the value obtained at 20 °C (139 nm). (Adapted from Jalilzadeh et al. 2018)

The temperature dependence of the increment of the FWHM (full width at half maximum) of the emission band can be seen in Fig. 15.16. On the other hand, in this temperature range the emission band intensity decreases with the temperature (Fig. 15.16). Carbon quantum dots systems are promising temperature nanosensors for in vivo physiological applications.

15.7 Nanobiosensor Promotes Sustainable Agriculture

- The nanofertilizers and nanobiosensor should show sustained release and sense of nutrients on demand while preventing them from prematurely converting into chemical/gaseous forms that cannot be absorbed by plants.
- Nanofertilizer allows selective nitrogen release linked to time, environmental and soil nutrient condition.
- Zeolites are naturally occurring crystalline aluminosilicates that can (a) enable better plant growth; (b) improve the efficiency and value of fertilizer; (c) improve water infiltration and retention; (d) improve yield; (e) retain nutrients for use by plants; (f) improve long-term soil quality; and (g) reduce loss of nutrients in soil. Zeolite holds nutrients in the root zone for plants to use when required. An added benefit of zeolite application is that unlike other soil amendments (gypsum and lime), it does not break down over time but remains in the soil to help improve nutrient and water retention permanently. With subsequent applications, the zeolite will further improve the soil's ability to retain nutrients and produce improved yields. Zeolites linked to a nanobiosensor can modernize agriculture in the sense that the biosensor can sense the deficiency in either plant or soil and control the release of water/nutrients retained in the zeolite. Pesticides inside nanoparticles are being developed that can be timed-release or have release linked to an environmental trigger. Also, combined with a smart delivery system, herbicide could be applied only when necessary, resulting in greater production of crops and less injury to agricultural workers.

15.8 Conclusions

Nanotechnology specially invented nanobiosensor is revolutionizing the development of biosensors in recent years. Due to high specificity, accuracy, stability, and quick reactivity and specially different physico-chemical properties, catalytic activity makes nanobiosensor more usable. Enhanced detection according to sensitivity has great potential for its applications in different fields including environmental and bioprocess control and agriculture (diagnostic tool for soil quality and disease assessment, agent to promote sustainable agriculture, device to detect contaminants and other molecules). Proper and controlled use of nanobiosensor can support sustainable agriculture by using optimum resource and ultimately enhancing the crop productivity.

References

Bünemann EK, Bongiorno G, Bai Z, Creamer RE, de Deyn G, de Goede R, Fleskens L, Geissen V, Kuyper TW, Mäder P, Pulleman M, Sukkel W, Van Groenigen JW, Brussaard L (2018) Soil quality – a critical review. Soil Biol Biochem 120:105–125

Chini MK, Kumar V, Javed A, Satapathi S (2019) Graphene quantum dots and carbon nano dots for the FRET based detection of heavy metal ions. Nano-Structures Nano-Objects 19:100347

Content adapted from resources developed by the Cornell Waste Management Institute (2007) Soil quality and testing and the healthy soils. Healthy Communities Project. www.css.cornell.edu/extension/soil-health/manual.pdf

Grieshaber D, MacKenzie R, Vörös J, Reimhult E (2008) Electrochemical biosensors – sensor principles and architectures. Sensors 8:1400–1458

Hassani A (2016) Novel approach to enhance TCP throughput in wireless sensor networks. Int J Adv Technol 7(3):159

Hernández-Rodríguez MA, Afonsoa MM, Palenzuel JA, Martín IR, Soler-Carracedo K (2018) Carbon dots as temperature nanosensors in the physiological range. J Lumin 196:313–315

Jalilzadeh M, Çimen D, Özgür E, Esen C, Denizli A (2018) Design and preparation of imprinted surface plasmon resonance (SPR) nanosensor for detection of Zn(II) ions. J Macromol Sci Part A Pure Appl Chem 56:877–886

Kaushik A, Solanki PR, Ansari AA, Sumana G, Ahmad S (2009) Iron oxide-chitosan nanobiocomposite for urea sensor. Sensors Actuators B Chem 138:572–580

Morrow JG, Huggins DR, Carpenter-Boggs LA, Reganold JP (2016) Evaluating measures to assess soil health in long-term agroecosystem trials. Soil Sci Soc Am J 80:450–462

Natural Resources Conservation Services (2012) Soil health. http://www.nrcs.usda.gov/wps/portal/nrcs/main/soils/health/

Parisi V, Menta C, Gardi C, Jacomini C, Mozzanica E (2005) Microarthropod communities as a tool to assess soil quality and biodiversity: a new approach in Italy. Agric Ecosyst Environ 105:323–333

Rai V, Acharya S, Dey N (2012) Implications of nanobiosensors in agriculture. J Biomaterials Nanobiotechnol 3:315–324

Rakshit R, Das A, Padbhushan R, Sharma RP, Kumar S (2018) Assessment of soil quality and identification of parameters influencing system yield under long-term fertilizer trial. J. Indian Soc Soil Sci 66:166–171

Rinot O, Levy GJ, Steinberger Y, Svoray T, Eshel G (2019) Soil health assessment: a critical review of current methodologies and a proposed new approach. Sci Total Environ 648:1484–1491

Satapathi S, Kumar V, Chini MK, Bera R, Halder KK, Patra A (2018) Highly sensitive detection and removal of mercury ion using a multimodal nanosensor. Nano-Structures Nano-Objects 16:120–126

Sparling GP, Schipper LA, Bettjeman W, Hill R (2004) Soil quality monitoring in New Zealand: practical lessons from a 6-year trial. Agric Ecosyst Environ 104:523–534

U.S. Environmental Protection Agency (2007) U.S. Environmental Protection Agency Nanotechnology White Paper. EPA 100/B-07/001. U.S. Environmental Protection Agency, Washington

Zhylyak GA, Dzyadevich SV, Korpan YI, Soldatkin AP, El'skaya AV (1995) Application of urease conductometric biosensor for heavy-metal ion determination. Sensors Actuators 24:145–148

Forensic Pedology: From Soil Trace Evidence to Courtroom

16

Tan Wei Ling Stella, Sanjay Swarup, Mandi See Suet Ning, Nur Qasrina Bte Iskandar Lim, Samantha Phua Mun Lin, Terry Tan Boon Jay, and Subhadip Ghosh

Abstract

Forensic pedology is the study of soil to answer legally related questions and problems. Soil evidence could play an important role to answer legal questions in court and to solve crimes. In this chapter, it is explained how soil is as a type of trace evidence in terms of its properties. There are various techniques and methods to study soil. In the National University of Singapore, the Forensic Science Research team uses the scanning electron microscopy coupled with energy-dispersive X-ray spectroscopy, which is a relatively new method to profile soil. Here, the team's methodology is explained. Finally, case studies in Singapore show the importance of soil science expertise in solving rape and murder cases.

Keywords

Forensic pedology · Forensic science · Soil forensics · Soil properties · Scanning electron microscopy/microscope · Energy dispersive X-ray

T. W. L. Stella · M. See Suet Ning (✉) · N. Q. B. I. Lim · S. Phua Mun Lin · T. Tan Boon Jay
Forensic Science Research Laboratory, Department of Biological Sciences, Faculty of Science, National University of Singapore (NUS), Singapore, Singapore
e-mail: A0114551@u.nus.edu

S. Swarup
Small Molecule Biology Laboratory, Department of Biological Sciences, Faculty of Science, National University of Singapore (NUS), Singapore, Singapore

S. Ghosh
Centre for Urban Greenery and Ecology, National Parks Board, Singapore, Singapore

A. Rakshit et al. (eds.), *Soil Analysis: Recent Trends and Applications*,
https://doi.org/10.1007/978-981-15-2039-6_16

16.1 What Is Trace Evidence in Forensic Science?

In 1910, French criminologist and pioneer in forensic science Edmond Locard founded the fundamental principle to associative evidence that states: "Every contact leaves a trace," which is now commonly termed as Locard's exchange principle (Fig. 16.1).

The contact between an individual with another individual or a location will result in the introduction or removal of something from the environment. An example of a one-way transfer will be a criminal leaving their fingerprints in a crime scene. However, two-way transfers are more likely to occur when trace evidence is contributable, whereby the criminal will also take away something from the scene like hair or glass particles. Such trace evidence is crucial in establishing the elements in the crime and linking the players in a rational, objective, and well-supported manner in the court.

> Wherever he steps, whatever he touches, whatever he leaves, even unconsciously, will serve as a silent witness against him. Not only his fingerprints or his footprints, but his hair, the fibers from his clothes, the glass he breaks, the tool mark he leaves, the paint he scratches, the blood or semen he deposits or collects. All of these and more bear mute witness against him. This is evidence that does not forget. It is not confused by the excitement of the moment. It is not absent because human witnesses are. It is factual evidence. Physical evidence cannot be wrong, it cannot perjure itself, it cannot be wholly absent. Only human failure to find it, study and understand it, can diminish its value.
>
> Paul L. Kirk

Fig. 16.1 Edmond Locard with his microscope. Adapted from London multimedia news, 2015. Retrieved from https://londonmultimedianews.com/2015/03/01/forensics-the-anatomy-of-crime-opens-at-the-wellcome-collection/10077873-edmond-locard-at-a-microscope-with-his-so/

Fig. 16.2 Four-way linkage theory

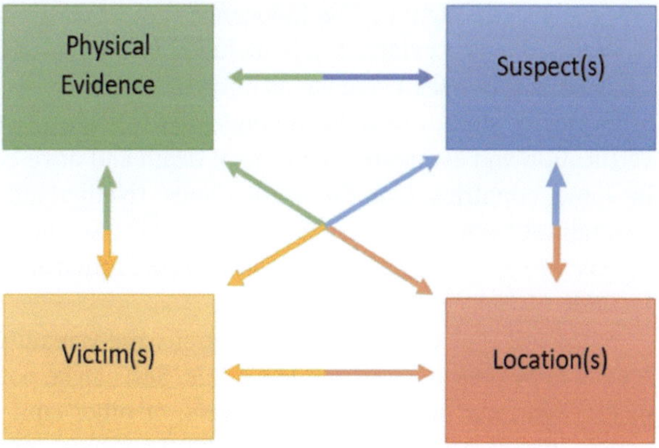

Section 5 of the Evidence Act (EA) states that "Evidence may be given in any suit or proceeding of the existence or non-existence of every fact in issue and of such other facts as are hereinafter declared to be relevant, and of no others" (Singapore Legal Advice 2019) (Fig. 16.2).

Evidence admissible can be in the form of documentary evidence, oral testimonies, digital and physical evidence, which can be recorded and retrieved during the investigation process. Trace evidence, a subset of physical evidence, is essential in providing source- (association between suspect, victim, and location) and activity-level (how, what, and where) information needed for a logical crime scene reconstruction. The value of trace evidence is due to its microscopic size and individualistic nature (Blackledge and Jones 2007). Examples of trace evidence include glass fragments, hair, fibers, gunshot residues, paint particles, and the primary objective of this chapter: soil.

As you will see in this chapter, soil has helped to solve crime and has its place in a courtroom. Methods used to study soil would also be briefly described, especially in the context of National University of Singapore's Forensic Science Research lab. The comprehensive details of methodologies will not be provided, but this sneak peek might have you find soil an interesting topic in academia henceforth. This chapter will end with legal case studies in Singapore, where soil had contributed to the prosecution of a cold-blooded murderer and a rapist—truly showing you how soil evidence makes its way to the courtroom.

16.2 Soil as Trace Evidence

The potential of soil as highly valuable trace evidence has been gaining traction over the years. The National Academy of Sciences (NAS) report by the National Institute of Justice (NIJ) in 2009 reported on the need to recognize and respond to the immediate need for significant improvements in many aspects of forensic science, especially in the light of The Innocence Project (National Institute of Justice 2009). A section of trace evidence was included in the report but made no mention on the study of soil in trace evidence as a forensic science discipline. However, in current times, many studies have been conducted to learn the use of soil as a tool for site verification and estimating the time of death and have been applied in real life cases in some countries like the Netherlands (Netherlands Forensic Institution) and Australia (Centre for Australian Forensic Soil Sciences). As of 2016, a total of US $1,059,776 have been awarded by the NIJ to fund forensic soil research (National Institute of Justice 2016).

Since soil is abundant in many locations and is readily available, it is highly likely for soil to be present or related to a crime. Soil can be potentially collected from soles of footwear, car tires, clothing, shovels, or other equipment and related to a wide variety of crimes like sexual assault, homicide, and kidnapping (Stam 2004; Uitdehaag et al. 2016). Forensic examiners will carefully sieve out soil that has been likely disturbed from human activity for comparison against a control to calculate the degree of similarity. The weight (or evidential value) of the comparison will be assessed accordingly in a case trial (e.g., Bayesian method) (Finkelstein and Fairley 1970).

Blackledge and Jones (2007) lined out six properties of ideal trace evidence which applies to soil: nearly invisible, highly individualistic, high probability of transfer and retention, easy collection, separation and concentration, and easy characterization against a database, of which some will be further explained below. These properties, therefore, make soil a highly valuable piece of trace evidence in the criminal investigation process. Figure 16.3 below shows the various properties of soil that could influence soil forensics. Each property is further explained thereafter.

16.2.1 Class and Individual Characteristics of Soil

Soil is a cornucopia of information as it plays host to a wide diversity of organisms and chemical compounds. Microorganisms, pollen, and chemical compounds present in soil are commonly used to assess the similarity and differences between a seized and control soil sample. While water creates a homogenous environment, the coarse nature of soil particles promotes heterogeneity due to the promotion of niche differentiation and the creation of highly diverse microhabitats.

The difference in abundance of bacteria generalists (e.g., phyla *Proteobacteria, Verrucomicrobia, and Acidobacteria*, which are class characteristics) and presence of individualistic and novel species set the basis for comparison between different locations (β-diversity). Biogeography of microorganisms has proved useful in providing invaluable pieces of evidence in forensic science through the process of elimination and association (Schauser et al. 2016). Similarly, fungi have also been proved to be useful in identifying sites at a broader scale (Shinde et al. 2003).

Other class characteristics include color, soil class, presence of chemical elements, compounds, and molecules, which can be determined through analysis

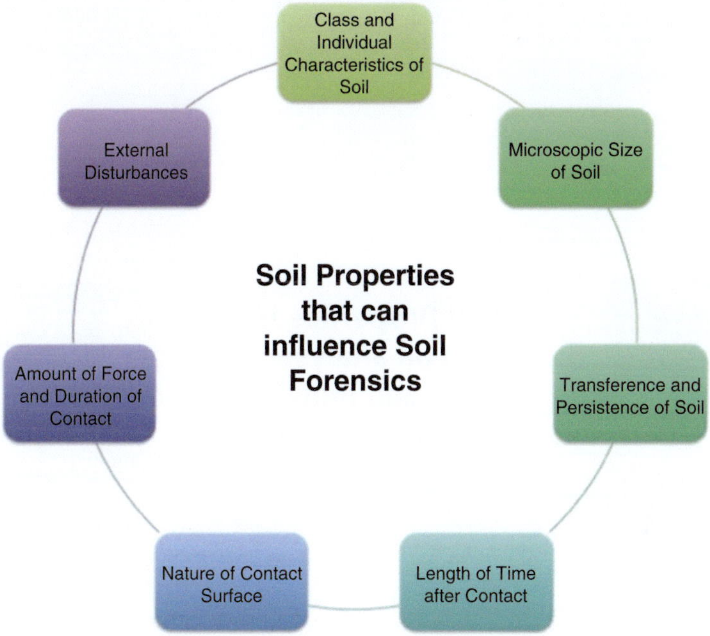

Fig. 16.3 Figure of various soil properties that can influence forensic pedology, which is the study of soil in the service of law

that is further explained later in the chapter. Figure 16.4 above shows the various types of aforementioned class and individual characteristics of soil in graphical form.

16.2.2 Microscopic Size of Soil

The microscopic size of soil makes it near invisible to the naked eye. It is expected for suspects fleeing the crime scene to dispose of more obvious and noticeable evidence like physical objects (e.g., murder weapons), blood stains, or even lipstick. However, the minute size of soil makes it harder to detect and hence, disposed of.

A case study outlined in Fitzpatrick et al. (2009) observed that the yellow-brown color of the fine clay and silt (<50 μm fraction) was hard to visually detect under the larger and coarser gravel soil. Therefore, even if the suspect were to brush off the visible soil, it is difficult to remove all traces of soil particulates completely.

16.2.3 Transference and Persistence of Soil

The key characteristic that makes soil the ideal form of trace evidence is its strong capacity to transfer (primary transfer) from one object to another, forming a direct link to associate otherwise discrete elements in the crime. However, the loss of soil

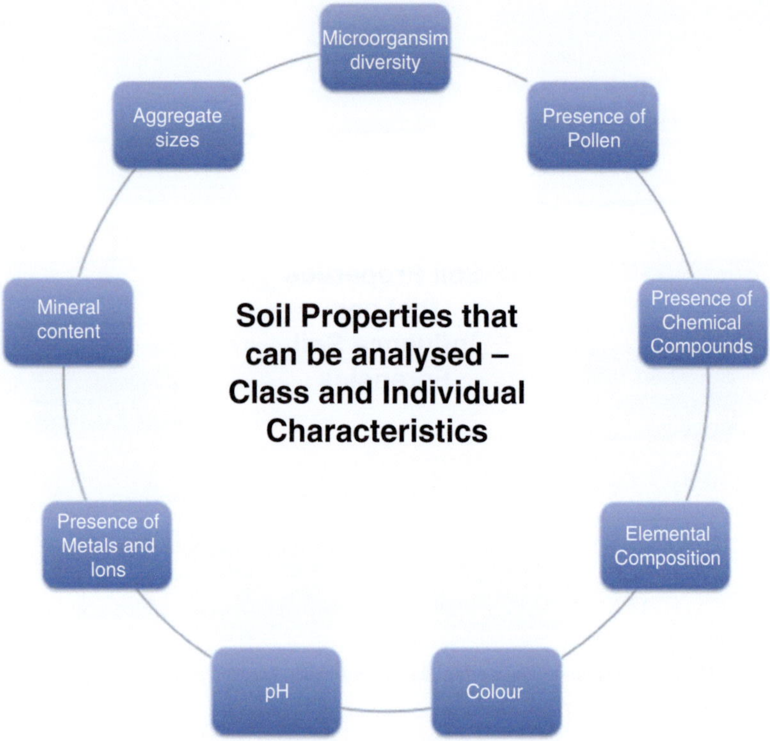

Fig. 16.4 Class and individual characteristics of soil. These properties can be analyzed in a laboratory

trace evidence through secondary transfer, which is defined to be the exchange of evidence that is not entirely associated with the crime itself, emphasizes the imminent need for investigators to retrieve the soil particles as early as possible.

The factors that determine the transfer and persistence of soil are as follows: (1) length of time after the contact, (2) nature of contact surface, (3) amount of force and duration of contact, and (4) external disturbances following contact.

16.2.4 Length of Time After Contact

The decay of soil has been determined to be similar to other forms of trace evidence (Bull et al. 2006). In general, there is an initial exponential decrease in original trace evidence with the highest proportion remaining after 4 h to be 18%. This is followed by a transient increase and then a slow decrease of trace evidence over time.

The mechanism suggested by Pounds and Smalldon (1975a, b, c) described the process to be determined by the strength of binding of the soil particulates onto the surface. A strongly bound soil particle will be less susceptible to the force of nature

Fig. 16.5 The Health Sciences Authority headquarters at Outram Road, Singapore. By Phoebedechin—Own work, CC BY-SA 3.0, https://commons.wikimedia.org/w/index.php?curid=4851411

or human activity than a weakly bound particle. In the first few hours, weakly bound particles are easily brushed off, resulting in a steep decline in the original fiber proportion. The short-term increase can be explained by the re-introduction of the trace evidence that fell from another area of the object (e.g., shirt) to another. As time since contact increases, the remaining fibers are mostly strongly bound and are increasingly harder to remove (Pounds and Smalldon 1975a, b, c). Nevertheless, particles can still be retrieved even after 7 days.

Larger particles were proposed to be more inclined to decay as compared to smaller particles (Brocard and Peyrot 2004). Grain size distributions on the sole of footwear were uni-modal for intermediate sized particles and bimodal for extreme sized particles (<0.02 mm and >4.00 mm), making the assessment of the latter useful in determining time since contact.

A potential issue highlighted in Morgan et al. (2009) is that the longer the time since the initial contact, the more problematic the interpretation of the evidence will be. In soils collected on the soles of footwear, the subsequent movement will introduce soil contaminants from irrelevant locations, producing a complex mixture of samples that rarely retain layers following a chronological order (Fig. 16.5).

Running, which is to be expected of a suspect fleeing the crime scene, further exacerbates the mixing of layers. Therefore, forensic examiners and lab technicians in the Health Sciences Authority (HSA) should exercise caution when retrieving samples for further experiments.

16.2.5 Nature of Contact Surface

Bull et al. (2006) determined that the material type majorly affects the extent of transference and persistence, even more than the particulate type. Coarser materials with a more open matrix (e.g., wool) are more likely to promote the transfer of trace evidence as compared to smoother surfaces (e.g., wood) (Lepot et al. 2015; Pounds and Smalldon 1975a, b, c). Soil particles bind strongly and weakly to rough and smooth surfaces, respectively.

A coarse donor surface is less likely to transfer soil particles, but a coarse recipient surface is more efficient in receiving the particles. The opposite is true for smoother surfaces. Therefore, in a secondary transfer, the contact between a smooth donor surface and a rough recipient surface causes a higher proportion of soil particles transferred as the soil particles are more readily dislodged.

16.2.6 Amount of Force and Duration of Contact

Transfer of soil particles may not occur through direct contact alone. Some amount of force is needed to guarantee soil transference (McDermott 2009). An increase in force results in a more significant proportion of trace particles transferred. For example, colliding into another individual will result in a higher force that shakes off the bound soil particles which increases particle transfer, as compared to merely brushing by.

Likewise, more violent crimes will likely result in a larger amount of soil deposited on the clothes. A maximum pressure of 250 kg/m^2 is the threshold whereby increasing pressure did not result in an increase in transferred fiber, though such amount of pressure is unlikely to occur in an individual's daily activities (Kidd et al. 1981). Likewise, an increase in duration of contact causes a higher proportion of trace evidence transferred.

16.2.7 External Disturbances

Secondary transfer of soil occurs naturally in their daily activities. Activities that require more rigorous contact (i.e., washing clothes and taking public transport) increase the decay of trace evidence. Across all material types, the number of soil particles found remaining decreases drastically, though a small amount of soil particles continues to persist. As only a small amount of soil sample is required for analysis, soil as a form of trace evidence is still highly valuable (Uitdehaag et al. 2016).

16.3 Soil and Its Components

To the naked eye, soil looks nothing more than a mere bunch of mess with neither structure nor organization. The image that is often ingrained in people's minds when the word "soil" is being mentioned is something that is abundant yet seemingly has nothing much to it. However, with soil, there is actually more than meets the eye.

For one, the most important pharmaceuticals like antibiotics are derived from the microbes that are found in soil (Shamarez and Manvi 2010). Without antibiotics, many of the diseases that are commonly caused by bacteria can do devastating damages to the human population, effectively sending us back to the dark ages. Apart from humans, soil benefits plants, too. Plants grow best in various types of soil. Certain species grow better in more acidic soil, and others are healthier in more neutral soil. This is crucial in parts of the world where production of food is key and yet is not as blessed with good environmental conditions. Achieving the best crop yield with the least amount of resources used is one of the sustainable goals of these countries (Oshunsanya and Nwosu 2018).

With such areas of research that yielded results as mentioned, how did researchers actually know what to look for in the seemingly pile of chaos? Well, soil actually has structure to it and can be broken down into the various components to be analyzed. Components include pH, microbial communities, metals and ions, elemental composition, and texturing, just to name a few. There are papers published that indicate these parameters do have a link with each other, meaning that getting a certain set of data could possibly predict what lies inside that area of soil. Take the example of the studies done in an urban park in Manhattan and an urban ring road in Beijing (Reese et al. 2015; Yan et al. 2016). The urban park soil samples were found to be more acidic in nature, which could mean a greater microbial growth. The urban ring road soil samples were found to be more neutral in nature, which could mean a greater diversity of microbial species. Both were later confirmed to be in line with the hypotheses made.

16.4 Forensic Soil Science in National University of Singapore

The examples mentioned above are global in scale. Now, let us take a look at Singapore, which has a much smaller area. Singapore's land area hovers around 721.5 km^2, about 23 times smaller than that of Beijing's land area of 16,808 km^2. Yet, differences in the soil community were noted. Whether it is within a park or along a road, the composition shifts. To guide you along the whole process of obtaining the soil samples, till ending up with the data charts, real life works on soil analysis will be shared. These works were carried out by the Forensic Science Research Lab at the National University of Singapore, in collaboration with other labs that have expertise in the area of soil.

Fig. 16.6 Photo of site after digging was done to get soil samples. The five holes can be seen in this picture

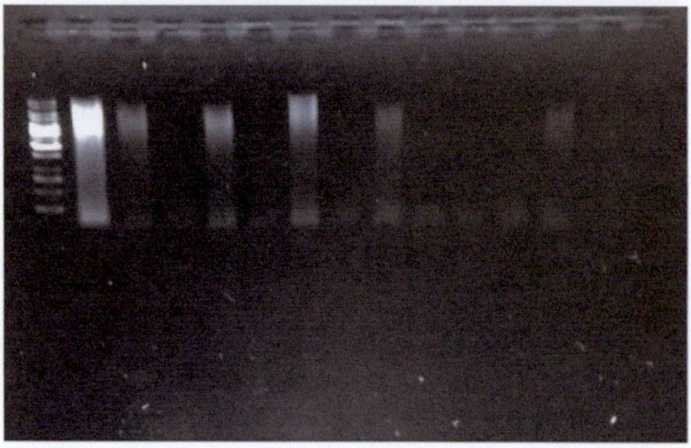

Fig. 16.7 Gel electrophoresis of some soil samples that were dug

First, a grid (1 m × 1 m) was used to locate and identify the area to be dug. Soil samples were dug from the 4 corners of the grid, as well as 1 more from the center of the grid (Fig. 16.6). During excavation, cores of different colors were already noted within the grid, highlighting the possible presence of various types of metals or elements.

The samples were then brought back to the lab and processed accordingly, either by air-drying or freeze-drying. One method of classifying soil by its color can be done through the use of the Munsell color system.

Back at the lab, soil is tested for its microbial and non-microbial aspects. To get a preliminary understanding of the soil microbial community, DNA had to be extracted from the soil samples. Here, we could tell the amount of DNA in the soil, which could translate to the amount of microbes in the soil. Below is a gel electrophoresis picture of some of the samples that were dug (Fig. 16.7).

Table 16.1 quantifies the actual amount of DNA within the sample. As shown, a brighter band corresponds to a higher amount of DNA that can be found in the soil, indicating a more abundant microbial community. Subsequently, the soil samples

Table 16.1 Quantification of the actual DNA amount in each sample

Lane	Quantity
A	35.8 ng/μL
B	10.7 ng/μL
C	Low
D	19.3 ng/μL
E	3.48 ng/μL
F	31.8 ng/μL
G	Low
H	12.3 ng/μL
I	Low
J	2.50 ng/μL
K	5.36 ng/μL
L	10.7 ng/μL

The letters correspond to the lanes shown in Fig. 16.4

Fig. 16.8 SEX-EDX machine in National University of Singapore

were sent for sequencing, to finally determine the microbial species that the soil samples contains.

In terms of non-microbial aspect, elemental composition analysis could be performed. One such method that has been used by the NUS Forensic Science lab to study the elemental composition of soil is through the use of SEM-EDX (Scanning Electron Microscopy with Energy Dispersive X-ray) (Fig. 16.8).

Fig. 16.9 SEX-EDX machine SE mode

The SEM-EDX machine enables you to take pictures of the soil sample using 2 modes, SE (secondary electrons) mode, which gives you a 3D-image of the soil samples, and BSE (back scatter electrons) mode, which shows the distribution of heavier and lighter elements.

Quantification and distribution of the elements within the soil samples are of interest to those in the field of pedology. Such information is useful to assist in the soil–microbial relationship research, as microbes do rely on elements in the soil to support their basic metabolic functions. The information also serves to give an indicator about the potential richness of the soil (Figs. 16.9 and 16.10).

The images above were taken from one of the soil samples that were obtained by the lab. From the pictures, there is a distinguishable difference between the 3D-image from the SE mode, and the distribution of heavier and lighter elements from the BSE mode (Figs. 16.11 and 16.12).

After seeing the data in Fig. 16.9, the surprising element that might have caught your eye might be that of titanium. In fact, there exists a variety of rare elements and metals that can be found in soil which you might not expect, but definitely not as abundant as the more common types like iron (Taylor 2006).

With characteristics that differ from each other, soil is able to present a rather unique profile, depending on which part of the country it is found in. In Singapore, soil is being utilized as part of forensic investigations. The uniqueness of the profile of soil has enabled investigators to narrow down their search to an area where the crime might have occurred. Drawing some similarities from the CSI crime shows, it is indeed as exciting and adrenaline pumping when the work to link soil to its location is a race against time as well.

Fig. 16.10 SEM-EDX machine BSE mode

Fig. 16.11 Elemental composition from the soil sample displayed in Figs. 16.7 and 16.8

Spectrum: Point

Element	AN	Series	Net	norm. C [wt.%]	Atom. C [at.%]	Error [%]
Oxygen	8	K-series	64613	55.76	70.17	4.9
Silicon	14	K-series	63682	21.30	15.27	0.7
Aluminium	13	K-series	52049	16.23	12.11	0.6
Iron	26	K-series	5961	6.24	2.25	0.2
Titanium	22	K-series	814	0.47	0.20	0.0
			Total:	100.00	100.00	

16.5 Case Studies in Singapore

Now that we have covered what soil is about and how it is studied, we will elaborate on two cases in Singapore where soil was used as trace evidence to help solve the crime.

16.5.1 Kallang Body Parts Case: PP v. Leong Siew Chor

16.5.1.1 Background

Liu Hong Mei, a 22-year-old Chinese national working in Singapore, had an affair with Leong Siew Chor in mid-2004 (Lum 2006; Tay 2011). On 13 June 2005, the

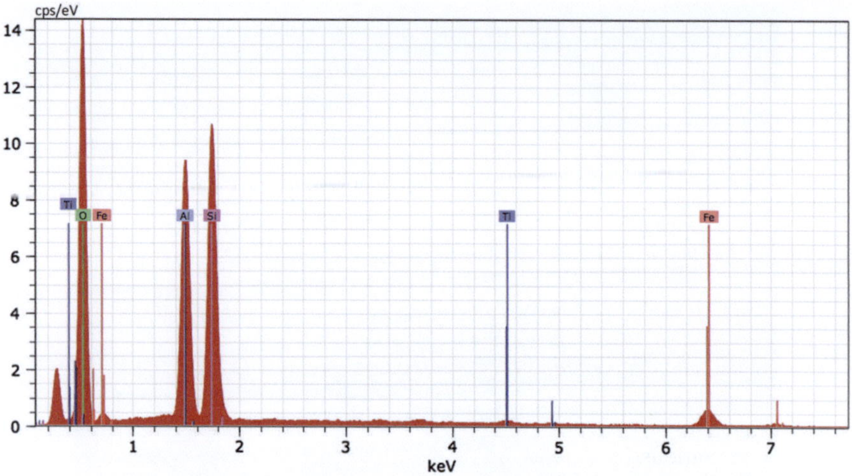

Fig. 16.12 Graph showing counts per second of each element

Fig. 16.13 Ms Liu Hong Mei from China was 22 years of age, who worked under Leong as a production operator. Adapted from Singapore Police Force, 2016. Retrieved from https://www.straitstimes.com/singapore/courts-crime/guilty-as-charged-leong-siew-chor-killed

couple checked into Hotel 81 Gold in Geylang, and while Liu Hong Mei was pre-occupied in the shower, Leong Siew Chor stole her automatic teller machine (ATM) card. On the very same day, Leong attempted to withdraw money from Liu's bank account at various ATMs (Tanjong Katong Complex, Joo Chiat Complex, Haig Road, and Beach Road). He went to Haig Road's ATM again on 14 June 2005, and out of these attempts, three were successful, leading to over $2000 being withdrawn (Tay 2011; Leong Siew Chor v. PP [2006] SGCA 38) (Fig. 16.13).

On 14 June 2005, Liu realized her ATM card was missing and discovered unauthorized withdrawals made from her account. She contacted Leong and lodged a police report about it. The next morning, Leong requested Liu to visit him and strangled Liu to death with a towel for fear that she will find out that he was the

Fig. 16.14 Leong Siew Chor, who killed his lover and chopped her body into seven parts. Adapted from Singapore Police Force, 2016. Retrieved from https://www.straitstimes.com/singapore/courts-crime/guilty-as-charged-leong-siew-chor-killed-lover-and-cut-up-her-body

perpetrator. He subsequently took her body to kitchen and dismembered her. He then wrapped the various body parts with newspaper and placed them in either plastic bags or cardboard boxes, which were disposed of at various locations via various modes of transportation (Tay 2011; Leong Siew Chor v. PP [2006] SGCA 38) (Fig. 16.14).

Leong cycled to Ubi Road, where he disposed her clothes, shoes, and feet into separate rubbish bins. He disposed her lower legs and head, and lower and upper torso into Singapore River and Kallang River, respectively. Liu's handbag and its contents were dumped into the rubbish bin outside Ang Mo Kio MRT station (Fong 2006, Tay 2011).

On 17 June 2005, the body parts which were dumped into Kallang River surfaced and drifted to the bank, where the lower torso were then discovered by a cleaner. The police then discovered the upper torso later that day. The head and legs in bags from Singapore River were recovered en route to Tuas incineration plant. Liu's feet, clothes, and belongings were never found (Tay 2011).

16.5.1.2 Investigative Process

Leong Siew Chor was arrested on 17 June 2005 and charged with murder the following day. Due to forensic investigations conducted by the police, Liu's funeral was delayed. It was a complicated and difficult process due to the decomposed nature of the victim's body parts (Tay 2011). Leong had his trial in May 2006.

Trace evidence (e.g., soil), as well as DNA evidence, were admitted into court. They conclusively linked Leong Siew Chor to the murder of Liu Hong Mei. Leong was then convicted of murder and sentenced to death by Justice Tay Yong Kwang on 19 May 2006 (The Straits Times 2016).

Soil particles were recovered from Leong Siew Chor's sandal. Upon further inspection, the origin of the soil was traced back to that from Kallang River Bank. There were bougainvillea thorns, and small seashells found in the sandy soil particles, consistent with vegetation and soil at the Kallang River Bank (Singapore Academy of Law 2009) (Fig. 16.15).

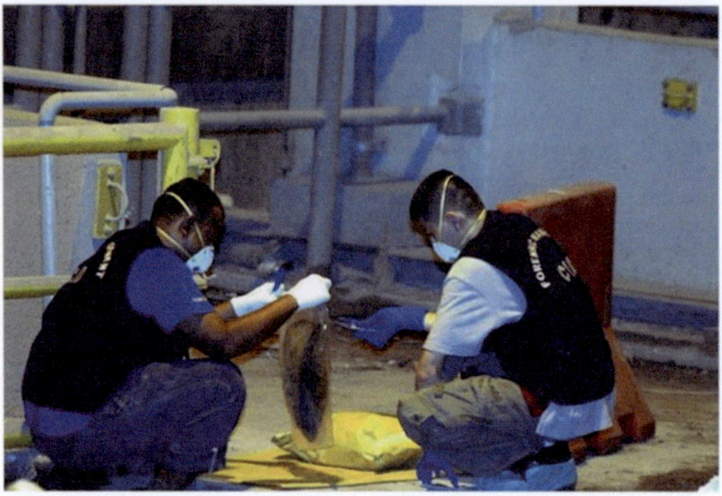

Fig. 16.15 The police found the murdered woman's head. Photo adapted from the Straits Times, Singapore, 2016. Retrieved from https://www.straitstimes.com/singapore/courts-crime-guilty-as-charged-leong-siew-chor-killed-lover-and-cut-up-her-body-to-cover

16.5.1.3 Importance of Soil Evidence

Soil recovered from Leong's sandals provided undisputed evidence that Leong was indeed at the site of disposal. This shows that soil evidence indeed can play an important role in linking suspects to the location of interest, by looking at the various properties or materials trapped in the soil particles. The sandy soil, bougainvillea thorns and small seashells found could provide enough clues that it belongs to Kallang River Bank, and hence investigators can form linkage between the suspect and the location at which the body was disposed of.

16.5.2 Rape Case: PP v. Lim Choon Beng

16.5.2.1 Background

On the morning of 9 February 2013, Lim Choon Beng raped and sexually assaulted a 24-year-old woman thrice at three different locations along Martin Road, all in a span of 20 min (Lum 2016).

The victim, a Chinese national, had been working in Singapore as a performing artiste (Lum 2016). She was walking home in the early morning on 9 February 2013 from Havelock Road alone. In order to reach her house, she would have to cross a bridge at Saiboo Street, walk along Martin Road, and then turn onto River Valley Close. At this time, Lim was near Saiboo Street and had been drinking at a bar. As the victim was walking along Martin Road, she noticed the accused crossing the road and appeared to be approaching her. The accused engaged the victim when she slowed her pace so that he could walk ahead of her (PP v. Lim Choon Beng [2016]

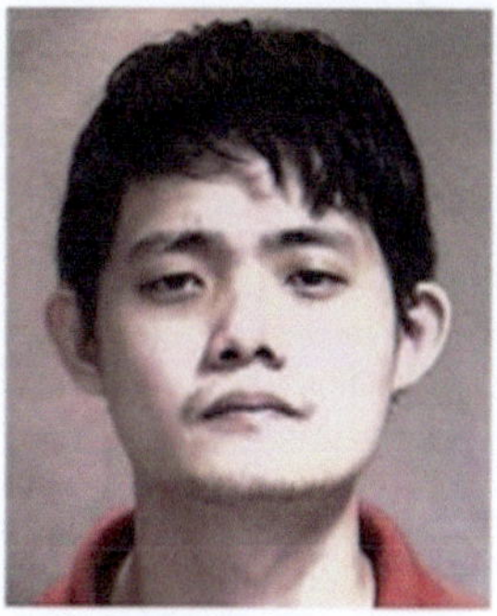

Fig. 16.16 Lim Choon Beng raped the victim three times in 20 min. Photo from Singapore Police Force, 2016. Retrieved from https://www.straitstimes.com/singapore/courts-crime/man-who-raped-woman-3-times-gets-close-to-17-years-jail-and-22-strokes

SGHC 169). Lim then sexually assaulted and raped the victim thrice at three different locations along Martin Road (Fig. 16.16).

The victim managed to escape as Lim stood up to put on his pants. She stopped a car and requested the female driver to drive her to the police; however, the female driver could not locate the police station. Hence, the victim phoned a friend and her friend called the police on behalf of the victim. The victim was then instructed to return to the area of River Valley Close where the last sexual intercourse happened. She found police officers with the accused and identified the accused to the police as the perpetrator (PP v. Lim Choon Beng [2016] SGHC 169).

16.5.2.2 Investigative Process

Upon investigation, the woman's torn panties were found among the vegetation at the first spot and the zipper of her bloodstained dress was dislodged from the right side. Soil was found on a few areas of the victim's dress and her shawl. There were also soil-like stains that were found on the lower half of the sleeves of the accused as well as the front right region of his shirt. Similarly, soil-like stains were found on the front right and left knee regions of the jeans of the accused (PP v. Lim Choon Beng [2016] SGHC 169).

Lim Choon Beng was subsequently found guilty and was sentenced to 17 years' jail and 22 strokes of cane by Judicial Commissioner Foo Chee Kock (Chelvan 2016).

16.5.2.3 Importance of Soil Evidence

Soil found on the victim's dress and shawl can link the victim back to the crime scene, which were the various locations she was raped and sexually assaulted. In addition, the soil-like stains on the accused clothes can link the accused to the crime scene and establish his alibi. Similarly, the location of the soil-like stains on clothes of the accused can show how the accused got his clothes stained, and in what position could he have been in when sexually assaulting and raping the victim.

References

Blackledge RD, Jones EL (2007) All that glitters is gold! In: Blackledge RD (ed) Forensic analysis on the cutting edge. John Wiley & Sons, Inc, Hoboken, NJ, pp 1–32. https://doi.org/10.1002/9780470166932.ch1

Brocard C, Peyrot B (2004) Particle size analysis of soils under simulated scene of crime conditions: the interest of multivariate analyses. Forensic Sci Int 140:159–166. https://doi.org/10.1016/j.forsciint.2003.11.032

Bull PA, Morgan RM, Sagovsky A, Hughes GJA (2006) The transfer and persistence of trace particulates: experimental studies using clothing fabrics. Sci Justice J Forensic Sci Soc 46 (3):185–195. https://doi.org/10.1016/S1355-0306(06)71592-1

Chelvan VP (2016) Nearly 17 years' jail, 22 strokes of cane for man who raped woman thrice in public. Channel NewsAsia. Retrieved from https://www.channelnewsasia.com/news/singapore/nearly-17-years%2D%2Djail-22-strokes-of-cane-for-man-who-raped-woman-8051290. Accessed 25 Feb 2019

Finkelstein MO, Fairley WB (1970) A Bayesian approach to identification evidence. Harv Law Rev 83(3):489. https://doi.org/10.2307/1339656

Fitzpatrick RW, Raven MD, Forrester ST (2009) A systematic approach to soil forensics: criminal case studies involving transference from crime scene to forensic evidence. Crim Environ Soil Forensics 2:105–127. https://doi.org/10.1007/978-1-4020-9204-6_8

Kidd CBM, Robertson J, College R, Street G, W, U. K. G. I. I. X (1981) The transfer of textile fibres during simulated contacts. J Forensic Sci Soc 22(3):301–308

Leong Siew Chor v. PP [2006] SGCA 38

Lepot L, Vanden Driessche T, Lunstroot K, Gason F, De Wael K (2015) Science and justice fibre persistence on immersed garment — Influence of knitted recipient fabrics. Sci Justice 55 (4):248–253. https://doi.org/10.1016/j.scijus.2015.02.006

Lum S (2006) Report of stolen ATM card led to killing, court told. The Straits Times

Lum S (2016) Trial starts for man who allegedly raped woman 3 times in 3 places in 20 minutes. The Straits Times. Retrieved from https://www.straitstimes.com/singapore/courts-crime/trial-starts-for-man-who-allegedly-raped-woman-3-times-in-3-places-in-20. Accessed 26 Feb 2019

McDermott SD (2009) Trace evidence: transfer, persistence, and value. In: Wiley encyclopedia of forensic science. John Wiley & Sons, Ltd, Chichester. https://doi.org/10.1002/9780470061589.fsa200

Morgan RM, Freudiger-Bonzon J, Nichols KH, Jellis T, Dunkerley S, Zelazowski P, Bull PA (2009) The forensic analysis of sediments recovered from footwear. In: Ritz K, Dawson L (eds) Criminal and environmental soil forensics. Springer, Cham, pp 278–294

National Institute of Justice (2009) Strengthening forensic science in the United States: a path forward

National Institute of Justice (2016) Trace evidence. Retrieved 24 February 2019 from https://www.nij.gov/topics/forensics/evidence/trace/pages/welcome.aspx#

Oshunsanya S, Nwosu N (2018) Soil-water-crop relationship: a case study of cassava in the tropics. https://doi.org/10.5772/intechopen.71968

Pounds CA, Smalldon KW (1975a) The transfer of fibres between clothing materials during simulated contacts and their persistence during wear part I – fibre transference. J Forensic Sci Soc 15(1):17–27

Pounds CA, Smalldon KW (1975b) The transfer of fibres between clothing materials during simulated contacts and their persistence during wear part II – fibre persistence. J Forensic Sci Soc 15(1):29–37

Pounds CA, Smalldon KW (1975c) The transfer of fibres between clothing materials during simulated contacts and their persistence during wear part III – a preliminary investigation of the mechanisms involved. J Forensic Sci Soc 15(3):197–207

PP v. Lim Choon Beng [2016] SGHC 169

Reese AT, Savage A, Youngsteadt E, Mcguire KL, Koling A, Watkins O et al (2015) Urban stress is associated with variation in microbial species composition—but not richness —in Manhattan. ISME J 10:751–760
Schauser L, Dawson L, Franqueville L, Simonet P (2016) Microbial soil community analyses for forensic science: application to a blind test. Forensic Sci Int 270:153–158. https://doi.org/10.1016/j.forsciint.2016.12.004
Shamarez AM, Manvi FV (2010) Production of antibiotics from soil-isolated actinomycetes and evaluation of their antimicrobial activities. Trop J Pharm Res 9:373–377
Shinde D, Lai Y, Sun F, Arnheim N (2003) Taq DNA polymerase slippage mutation rates measured by PCR and quasi-likelihood analysis: (CA/GT)n and (A/T)n microsatellites. Nucleic Acids Res 31(3):974–980. Retrieved from http://www.ncbi.nlm.nih.gov/pubmed/12560493
Singapore Academy of Law (2009) Inter Se. Mediactive, Singapore
Singapore Legal Advice (2019) Admissibility of Evidence in the Singapore Courts. Retrieved 24 February 2019 from https://singaporelegaladvice.com/lawarticles/admissibility-of-evidence-in-the- singapore-courts/
Stam M (2004) Soil as significant evidence in a sexual assault/attempted homicide case. Geol Soc Lond, Spec Publ 232(1):295–299. https://doi.org/10.1144/GSL.SP.2004.232.01.28
Tay YK (2011) Kallang body parts murder. National Library Board. Retrieved from http://eresources.nlb.gov.sg/infopedia/articles/SIP_1793_2011-03-17.html. Accessed 23 February 2019
Taylor P (2006) Communications in soil science and plant analysis. Access Details: [subscription number 794614672]
The Straits Times (2016) Guilty as charged: Leong Siew Chor killed lover and cut up her body to cover up theft of ATM card. Retrieved from https://www.straitstimes.com/singapore/courts-crime/guilty-as-charged-leong-siew-chor-killed-lover-and-cut-up-her-body-to-cover. Accessed 23 Feb 2019
Uitdehaag S, Quaak F, Kuiper I (2016) Soil comparisons using small soil traces, a case report. In: Soil in criminal and environmental forensics. Springer, Cham, pp 61–69. https://doi.org/10.1007/978-3-319-33115-7_5
Yan B, Li J, Xiao N, Qi Y, Fu G, Liu G et al (2016) Urban-development-induced changes in the diversity and composition of the soil bacterial community in Beijing. Nat Publ Gr 2–10

Harnessing Soil Microbiomes for Creating Healthy and Functional Urban Landscapes

17

Aditya Bandla, Shruti Pavagadhi, and Sanjay Swarup

Abstract

Urban soil microbiomes are attractive interventional targets for creating healthy and functional urban landscapes. In this chapter, we introduce molecular meta-omics techniques that can be used to study the composition and functioning of such microbiomes in a high-throughput and culture-independent manner. We highlight studies in which such approaches have been applied to soil microbiomes in both natural and managed ecosystems. We then discuss how data from such approaches can be interpreted using ecological frameworks and discuss how such information can in turn be used to develop sustainable solutions for managing urban landscapes and increasing the productivity of urban agroecosystems.

Keywords

Soil · Soil microbiome · Urban landscapes · Sustainability · Multi-omics

A. Bandla
Singapore Center for Environmental Life Sciences Engineering, National University of Singapore, Singapore, Singapore

S. Pavagadhi
Singapore Center for Environmental Life Sciences Engineering, National University of Singapore, Singapore, Singapore

Department of Biological Sciences, National University of Singapore, Singapore, Singapore

S. Swarup (✉)
Singapore Center for Environmental Life Sciences Engineering, National University of Singapore, Singapore, Singapore

Department of Biological Sciences, National University of Singapore, Singapore, Singapore

NUS Environmental Research Institute, National University of Singapore, Singapore, Singapore

Synthetic Biology for Clinical and Technological Innovation, National University of Singapore, Singapore, Singapore
e-mail: sanjay@nus.edu.sg

A. Rakshit et al. (eds.), *Soil Analysis: Recent Trends and Applications*,
https://doi.org/10.1007/978-981-15-2039-6_17

17.1 Introduction

By 2050, more than two-thirds of the world's population will reside in urban areas (United Nations 2014). Creating healthy and functional urban landscapes capable of supporting such a population is therefore a global priority. A vast majority of urban landscapes have soils as their foundational basis which in turn are extensively modified and actively managed to provide a multitude of services for urbanites. They support infrastructure, green cover, cycle nutrients, regulate runoff and act as a sink for pollutants (Wall et al. 2012). In recent years, efforts have renewed to utilize urban soils for developing sustainable urban agroecosystems to produce fresh and nutritious vegetables for local consumption (Toju et al. 2018). Yet, we understand little about the biological diversity that underpin the provision of most such services. This is in part due to the widely held assumption that urban soils have been modified to such an extent that they lack the ability to support species-rich biological communities. However, studies in recent years have convincingly shown that urban soils support complex communities of macroorganisms (such as plants and insects) and microorganisms (such as bacteria, archaea, and fungi) that actively interact to drive ecological processes (McGuire et al. 2013; Ramirez et al. 2014). This observation has prompted scientists and policy makers alike to analyze how such communities function and in turn how such information can be leveraged to manage and manipulate entire biological communities for applicable benefits.

Typically, microorganisms vastly outnumber macroorganisms in most ecosystems and exist as complex communities termed as microbiomes (Flemming and Wuertz 2019). In natural ecosystems, soil microbiomes have been shown to possess the requisite genetic machinery to regulate soil carbon stocks, cycle essential plant-nutrients, confer resistance to plants from invasive pathogenic microorganisms, and degrade pollutants (Bell et al. 2016; Fierer 2017; Schimel and Schaeffer 2012; Van Der Heijden et al. 2008). Pioneering efforts have revealed that urban soil microbiomes can be highly diverse and can comprise several novel microorganisms (Ramirez et al. 2014). In addition to being species-rich, molecular surveys have shown that urban soils harbor genetic novelty of medical and biotechnological relevance not found in other ecosystems (Charlop-Powers et al. 2016). Although these observations highlight the tremendous potential of soil microbiomes as interventional targets, they remain to be systematically explored across different urban landscapes.

Here, we outline strategies for studying different facets of species-rich urban soils and discuss scientific challenges specific to the molecular investigation of entire soil microbiomes. We then highlight how such information can be leveraged to manage microbiomes in urban landscapes with a view to optimize the provision of microbiome-mediated ecosystem services and for developing novel microbial solutions that can increase the productivity of urban agroecosystems in a sustainable manner.

17.2 Disentangling Soil Microbiomes Using Molecular Meta-Omics

The highly diverse nature of soil microbiomes in general and our current inability to culture the vast majority of its members mean that we require approaches that enable us to study them at a high resolution and in a culture-independent manner. Molecular meta-omics approaches allow us to do this and encompass a wide range of techniques that can be used to study different facets of entire microbiomes. Typically, such techniques involve the extraction of biomolecular fractions (DNA, RNA, proteins, and metabolites) directly from environmental samples and profiling them using either high-throughput sequencers or mass spectrometers (Franzosa et al. 2015).

While, such techniques have been successfully applied to study diverse environmental and host-associated microbiomes, their efficacy is often limited when it comes to examining soil microbiomes. For example, the activity of ribonucleases—enzymes that degrade nucleic acids—are often elevated in soils in comparison to other systems (Keown and Greenfield 2004) which in turn prohibit a representative fraction of DNA or RNA from being obtained under standard sampling and laboratory working conditions. Similarly, the co-extraction of such enzymes and a milieu of other chemicals present in the soil matrix can limit downstream steps such as PCR amplification (Schrader et al. 2012) or analyte separation using liquid chromatography (Bundy et al. 2009). In addition to such analytical challenges, the amount of data required to capture the diversity of soil microbiomes is often large which in turn necessitates high-end computational infrastructure (Kyrpides et al. 2016) that is not commonly available to the vast majority of scientists. Addressing these issues requires new considerations when using individual techniques as well as designing studies that integrate multiple techniques. In the following sections, we outline the utility of such techniques for studying different aspects of the urban soil microbiome and also highlight challenges specific to each technique.

17.3 Quantifying Microbiome Composition and Function Using Metagenomics and Metatranscriptomics

Characterizing microbiome composition (identities of resident microorganisms and their relative abundance within the community) is typically the first step in any study that seeks to either understand the role of microbiomes under a given context or how they can be managed for applicable benefits. This can be accomplished using a technique termed metagenomics which involves the direct recovery of DNA from samples, fragmenting them into short pieces and profiling millions of such fragments in a random fashion using high-throughput sequencers (Fig. 17.1). The identity of different microorganisms is inferred by matching the nucleotide sequences of these fragments to sequences in reference databases. The relative abundance of the different microorganisms can then be deduced by calculating the number of times

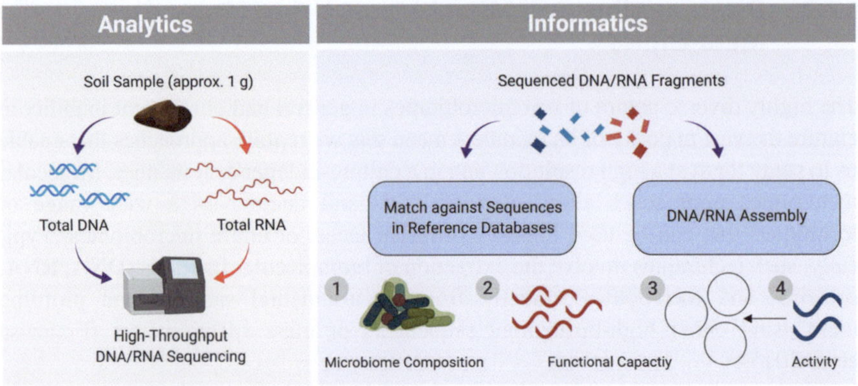

Fig. 17.1 Analytical and informatics workflow for metagenomics and metatranscriptomics. Total DNA/RNA is recovered from soil samples, fragmented, and characterized using high-throughput sequencing. Sequenced fragments are processed in two ways. By matching DNA fragments to sequences in reference databases, information of microbiome composition (*1*) and collection of genes (*2*) can be obtained. Collection of transcripts (*2*) can be obtained in a similar fashion. By assembling DNA fragments and binning them, one can obtain a collection of genomes (*3*), often termed metagenome assembled genomes (MAGs). In a similar fashion, entire transcripts can be assembled (*4*) and mapped back to MAGs to infer organismal origin

each fragment assigned to a particular microorganism occurs within the dataset. In addition to providing information on the identities of the different microorganisms, nucleotide sequences also provide information about the entire collection of genes that exist within the microbiome under study.

Examining collections of genes is not necessarily the objective of studies which seek to examine only the composition of microbiomes, in which case, sequencing single genes or regions of such genes which can act as a reliable molecular marker for different microorganisms can be pursued. In the case of microbiomes, this marker is typically the gene encoding for the small sub-unit of the 16S/18S rRNA. Entire genes or regions of such genes are selectively amplified from a sample's DNA pool (in its entirety) prior to sequencing. Identities of different microorganisms and their abundances are inferred in a manner similar to the one outlined earlier. Since this approach only involves the sequencing of amplified targets, it should not be confused with metagenomics. However, it is frequently included under the umbrella of metagenomic techniques as a way to examine microbiome composition. Using this approach, the earth microbiome project, one of the largest scientific collaborative efforts in recent times to catalogue the earth's microbial diversity revealed that soils around the world are highly diverse and are largely composed of oligotrophic microorganisms. By correlating microbial community diversity with environmental factors such as pH and temperature, this work also highlighted that soil diversity is highest in regions with a relatively low mean temperature (about 10 °C) and at near neutral pH (around 7) (Thompson et al. 2017). Therefore, in addition to revealing microbial community composition, the utility of this approach can be expanded by coupling it with associative statistical modeling to reveal environmental markers that structure such communities.

As noted previously, metagenomic techniques also provide information about the entire collection of genes which exist within the microbiome under examination. Gene identities and abundances can provide insights into a microbiome's collective functional capacity. For example, soil microbiomes in New York's central park were shown to harbor several novel gene clusters encoding for natural products of biotechnological relevance through a targeted metagenomics approach (Charlop-Powers et al. 2016). However, a more powerful approach comprises the assembly of short DNA fragments into longer ones termed contigs which can subsequently be separated into metagenomic bins each of which provide a strong working hypothesis about the genomes of individual microorganisms. This approach thus allows one to link the identity of microorganisms to their functional capacity which otherwise cannot be obtained just by examining collections of genes. Using this approach, the capacity to degrade lignocellulose in forest soils was found predominantly within the members of the family *Caulobacteraceae,* highlighting their potential importance in contributing to decomposition processes in such ecosystems (Wilhelm et al. 2019).

Although metagenomics is a powerful technique for characterizing the composition and functional capacities of microbiomes, it does not quantify their activity. A direct measure of the functional activity of microbiomes can be obtained by quantifying either RNA transcripts, proteins, or metabolites. In a technique termed metatranscriptomics, the entire set of RNA produced by a microbial community is recovered directly from samples, converted to cDNA, and profiled in a manner similar to metagenomic techniques. As such, the identity and abundance of different transcripts in itself provides rich information on microbiome functioning; however, the utility of such information is enhanced when transcripts are mapped back to genomes thereby linking activity to different groups of microorganisms. For example, this approach revealed the important role of viruses in regulating the carbon cycle within peatland soils. Specifically, viral transcripts recovered using metatranscriptomics were mapped back to viral genomes assembled from metagenomics data, thereby allowing the identification of the active subset as well as genomic features linking them to microbial populations involved in carbon turnover. Moreover, such viruses were also found to both encode and express genes involved in complex carbon degradation suggesting a direct role in cycling carbon within such ecosystems (Emerson et al. 2018).

Despite the tremendous utility of such techniques, applying them to examine more complex facets of soil microbiomes still remains a challenge. For instance, high microbial diversity that typify most soils mean that extensive sequencing data is required for obtaining a meaningful representation of the community. Shallow sequencing data precludes the assembly of genomes of most microorganisms which exist in low proportions within such microbiomes (Howe et al. 2014). Further, the high levels of genetic novelty that exist within soil microbiomes mean that only a minor fraction of metagenomes and metatranscriptomes can be annotated using current databases (Delmont et al. 2012). However, given the potential of such techniques, we expect the development of new approaches which address these challenges thereby enabling a more holistic examination of the composition and functioning of soil microbiomes.

17.4 Quantifying Functional Activity of Microbiomes Using Metaproteomics and Metabolomics

Regulatory phenomena at the community level are often mediated by sets of different proteins and metabolites. Therefore, their characterization using high-throughput methods termed metaproteomics and metabolomics, respectively, can provide complementary insights into the functioning of entire microbiomes (Fig. 17.2). Proteins and metabolites can be directly retrieved from samples in a manner similar to the recovery of nucleic acids. However, as opposed to the characterization of nucleic acids using next-generation sequencing techniques, proteins and metabolites are profiled using liquid chromatographs coupled with mass spectrometers which together provide readouts on their mass and abundance. Similar to shotgun sequencing techniques, proteins and metabolites are fragmented prior to characterization in order to provide accurate readouts. Fragmentation patterns can reveal the peptide sequence of proteins, while in the case of metabolites they provide information on their chemical composition and structure. Putative identities of proteins are then inferred by matching peptide sequences to proteins in reference databases using homology-based searches, while those of metabolites are inferred by matching fragmentation patterns of mass features to those of metabolites in curated databases. Abundances of proteins and metabolites can be subsequently inferred in a manner similar to metagenomics or metatranscriptomics.

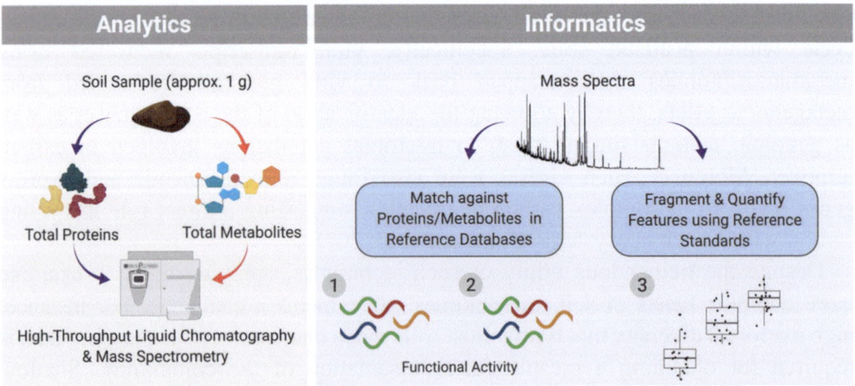

Fig. 17.2 Analytical and informatics workflow for metaproteomics and metabolomics. Total proteins/metabolites are recovered from soil samples, fragmented, and characterized using high-throughput chromatography coupled with mass spectrometry. Putative identities of mass fragments are then inferred by matching them against proteins/metabolites in reference databases (*1, 2*). Absolute quantification of metabolites (*3*) can be pursued by further fragmenting mass features of interest and quantifying such fragments using mass spectrometry (MS/MS)

Using a semi-quantitative metaproteomics approach, the vast majority of extracellular hydrolytic enzymes (such as cellulases and chitinases) involved in forest litter decomposition were shown to be of fungal origin (Schneider et al. 2012), highlighting the importance of fungi in the functioning of forest soil ecosystems. Such an approach can also reveal the physiological adaptation of microbiota to different environmental conditions. For instance, the vast majority of microorganisms in the arctic permafrost were found to express cold shock proteins presumably helping them survive under frozen conditions (Hultman et al. 2015). In addition to its individual utility, combining metaproteomics with complementary approaches such as metagenomics can offer powerful insights into microbiome functioning that cannot be obtained otherwise. For example, using a combination of metagenomics and metaproteomics, methanol-based methylotrophy in the rhizosphere of rice plants was shown to be mainly driven by the activity of bacteria linked to the genus *Methylobacterium*. In addition to helping link proteins involved in this process to specific groups of microorganisms, metagenomics substantially improved the identification of a broad range of other proteins not found in reference databases (Knief et al. 2011), further highlighting the utility of such integrative approaches.

Similarly, complementary insights into the functioning of microbiomes can also be obtained using metabolomics as well as by integrating it with other omics approaches. For example, field observations and experiments with soil isolates showed that the biological diversity of soil biocrusts were maintained in part by the capacity of resident microbial populations to utilize different classes of metabolites released by the dominant primary producer (Baran et al. 2015; Swenson et al. 2018). This was accomplished using an exometabolomics approach which characterizes the set of metabolites secreted by biological entities into their surrounding environment. Using a combination of metabolite profiling and 16S rRNA gene amplicon sequencing, benzoxazinoids, a class of defensive secondary metabolites released by plant roots were shown to significantly alter the composition and functioning of rhizosphere microbiomes which in turn impacted plant performance (Hu et al. 2018).

While the utility of metaproteomics and metabolomics for quantifying the functional activity of microbiomes is clear, their application for characterizing the functioning of soil microbiomes remains a challenge. For example, innate properties of soils (such as high salt concentrations) reduce their compatibility with standard practices in metaproteomics and metabolomics (Beale et al. 2016). They also share several challenges with techniques such as metagenomics and metatranscriptomics; for instance, reference databases currently only capture a minor fraction of the biological novelty often encountered in soils. However, given the tremendous potential of such techniques to improve our understanding of soil microbiome functioning, we expect continuation of efforts to develop new practices that address these challenges as well as the application of these techniques under new contexts.

17.5 Leveraging Molecular Meta-Omics Information for Developing Sustainable Solutions

Molecular meta-omic techniques can provide rich descriptors of microbial communities; however, the utility of such information for developing sustainable solutions depends on our ability to interpret them within a framework which can yield insights into the relationships between microbiomes and the ecosystem. A number of studies have shown that the application of ecological principles offers a powerful framework for obtaining such insights. In the following sections, we highlight how such ecological principles can be used to interpret multidimensional meta-omics data in order to develop sustainable solutions for managing soils in urban landscapes and creating highly productive urban agroecosystems.

17.6 Crafting Sustainable Urban Landscape Management Regimes

Urban landscapes with soils and vegetation as their foundational basis commonly comprise of lawns, parks, gardens (including thematic ones and roof-tops), road-side kerbs, and waterways. Management measures typically focus on maintaining soil health, establishing and sustaining target vegetation at optimal states, managing pests (including weeds and insects), and reducing greenhouse gas emissions. Ecological studies in natural soil systems have convincingly shown that soil microbiome functioning impact processes which determine such outcomes and thus make them attractive targets for planned interventions that aim to maximize desirable benefits.

To maximize desirable microbial functions, a thorough overview of the different microorganisms that exist within soil microbiomes and their functional traits is a prerequisite. This can be accomplished by surveying soils across different landscapes using amplicon sequencing and shotgun metagenomics. The measurement of environmental factors (such as pH, temperature, and landcover) is equally important as this will allow the identification of key drivers that structure such communities and in turn populate the list of modifiable factors that can be subsequently used to steer microbiomes to desirable states. In addition to such surveys, manipulative field experiments can also offer such insights. Data from such studies can be interpreted using the ecological framework on microbial community types which deals with the identification of strongly recurring patterns based on microbiome composition (Gonze et al. 2017). Such patterns have been identified, for instance, across different microbial habitats in the human body including the gut (Arumugam et al. 2011), vagina (Ravel et al. 2011), and the oral cavity (Ding and Schloss 2014). Studies have convincingly shown that communities can switch configurations and by extension functioning in response to changing environmental factors. By extension, soil microbiome datasets either from different urban landscapes or manipulative field experiments should be explored for the existence of such community types and its key drivers. This can be done by first clustering samples based on microbiome composition inferred using amplicon data; second, exploring the functional trait composition of such community configurations using metagenomics; and finally correlating this information with environmental factors or the treatments being studied. However, it should be noted that the separation of community types based solely on composition does not necessarily imply a difference in their functioning due to functional redundancy. This therefore necessitates quantifying functional activity using metatranscriptomics to reliably identify functionally different community types as well as when testing if modifiable factors (identified using associative techniques) do indeed shift community types to those that fulfil managerial targets. In case distinct community types do not exist, this strategy can be easily extended to steer sub-communities, core microbiomes (subsets of microorganisms shared across a vast majority of samples) as well as functional guilds (groups of microorganisms which perform the same function).

Several examples show that this approach is tractable for steering existing communities to desirable states. For example, pioneering studies identified diet as an important factor associated with gut enterotypes (De Filippo et al. 2010; Wu et al. 2011). Follow-up experimental studies showed that diet indeed was capable of inducing switches in community types and functioning, thus making diet a therapeutic target for regulating gut health (David et al. 2014; Hjorth et al. 2018). Experimental studies have demonstrated that the addition of pyrolyzed plant residues to soil can induce shifts in microbial diversity and biomass which in turn was shown to impact plant performance (Kolton et al. 2017; Mehari et al. 2015). Similarly, plants that grow in soil actively shape the microbiome by modifying soil properties and altering resource availability through root exudation (Hartmann et al. 2009). Thus, plants can also be used to shift soil microbiomes to states that maximize applicable benefits. A classic example of such a strategy is utilizing the capacity of legumes to enrich the soil microbiome for diazotrophs thereby establishing a self-sustaining nitrogen cycle which in turn can support the growth of plants with a high nitrogen requirement in the future (Bradshaw Anthony et al. 1982).

While managerial targets can be achieved using this strategy, it is important that future efforts also focus on understanding generative mechanisms. Only a thorough understanding of the mechanisms underlying such outcomes can help in developing management regimes that are robust in the face of unpredictable environmental change.

17.7 Developing Sustainable Urban Agroecosystems

Urban centers are redefining the ways in which farming is practiced with a view to offset a considerable proportion of the food demand they generate. Agroecosystem configurations range from patches of land earmarked for agriculture, indoor setups to vertical farms, placing soils and crops in new contexts. Obtaining optimal and nutritious outputs will depend on our ability to improve plant–soil feedbacks (PSF) under these new settings. PSFs affect plant growth, nutrition, tolerance to environmental perturbations, and susceptibility to pests and pathogens among others (van der Putten et al. 2013). Several studies have shown that this feedback is mediated to a large extent by the soil and rhizosphere—soils that lie in close vicinity to the roots—microbiomes (Fitzpatrick et al. 2018; Hu et al. 2018; Semchenko et al. 2018). Therefore, maximizing the beneficial functions of such microbiomes and engineering synthetic consortia that can confer the same are the central focus of several ongoing translative research efforts aiming to maximize agroecosystem productivity in a sustainable manner (Pavagadhi 2019).

In contrast to urban landscapes, shifting existing soil microbiomes to optimal states is not necessarily the prime objective for managing microbiome-mediated PSFs. Rather, one of the ways that this can be achieved is by facilitating the establishment of beneficial microbiomes during the early developmental stages of plants given that such stages are easily accessible to ameliorative efforts. Such efforts should be informed by studies which investigate the composition of soil and rhizosphere microbiomes at different growth stages of plants as well as efforts that seek to understand the dynamics of microbiome establishment. Amplicon sequencing and metagenomics are important tools that can be used to accomplish this as outlined in the previous section. Conceptual frameworks on microbiome assembly in turn can be used to interpret these datasets and to identify key microbial targets which influence assembly as well as timepoints for active intervention. Key microbial targets also termed core microbiomes can be inferred using network theory which delineates such subsets as those that can potentially regulate the dynamics of entire communities through a range of ecological interactions. For example, core microbiomes have been identified across a wide range of plant types (Lundberg et al. 2012; Xu et al. 2018) and how in turn they affect plant performance. Strategies for leveraging core microbiomes to enhance PSFs can range from inoculating seeds with such microbiomes to modifying factors (such as resource inputs in the form of fertilizers) which directly influence their establishment, growth, and functioning within the community. Another approach is to develop core microbiomes with different functional portfolios in vitro using ecological principles which can then be deployed in the field. This can be accomplished by culturing core microorganisms (previously identified using informatics approaches) in a high-throughput manner termed culturomics. Techniques that quantify functional activity such as metatranscriptomics, metaproteomics, and metabolomics can then be used to identify stable configurations of different core microorganisms that can confer plant-beneficial functions. Such an approach can also leverage extensive information on plant-growth promoting microorganisms from existing studies.

Although approaches outlined above remain to be tested, several studies that have examined different components of this approach show that it can be tractable. For example, a highly simplified synthetic microbial consortia could be assembled directly on Maize roots guided by ecological principles thereby enabling highly resolved examinations of community dynamics and function (Niu et al. 2017). In addition to the identification of core rhizosphere microbiomes associated with plants, a number of studies have also shown the importance of such core microbiomes in conferring plant-beneficial functions such as resistance to invasive microorganisms (Cernava et al. 2019). In terms of deploying such microbial portfolios, evidence from seed inoculation experiments with two or more microbial strains suggest that multiple strains can co-establish in a stable manner and act synergistically to enhance plant performance (Cassán et al. 2009).

Enhancing PSFs and thereby increasing urban agroecosystem productivity remains a grand challenge. Manipulating and managing soil microbiomes that are closely associated with plants under these contexts through approaches outlined above can help in achieving this goal. We expect continuation of efforts that use meta-omics approaches to gain a comprehensive understanding of different facets of such microbiomes as well as to develop solutions which have them as their focal basis.

17.8 Conclusions

Here, we have introduced and outlined the utility of meta-omics approaches for understanding the composition and functioning of urban soil microbiomes, management and manipulation of which offers an attractive way for developing healthy and functional urban landscapes. We have also discussed key challenges which limit the utility of these techniques and expect the development of methods which address these to continue. A comprehensive understanding of urban soil microbiomes can only be obtained by integrating such techniques in a manner which address the questions at hand. Finally, we highlight the importance of interpreting data obtained using such techniques within an ecological framework and discuss ways in which innovative microbiome-based solutions can be developed for managing urban landscapes and developing sustainable urban agroecosystems.

References

Arumugam M, Raes J, Pelletier E, Le Paslier D, Yamada T, Mende DR et al (2011) Enterotypes of the human gut microbiome. Nature 473:174. https://doi.org/10.1038/nature09944

Baran R, Brodie EL, Mayberry-Lewis J, Hummel E, Da Rocha UN, Chakraborty R et al (2015) Exometabolite niche partitioning among sympatric soil bacteria. Nat Commun 6:8289. https://doi.org/10.1038/ncomms9289

Beale D, Kouremenos K, Palombo E (2016) Microbial metabolomics. Springer, Cham

Bell TH, Stefani FOP, Abram K, Champagne J, Yergeau E, Hijri M, St-Arnaud M (2016) A diverse soil microbiome degrades more crude oil than specialized bacterial assemblages obtained in culture. Appl Environ Microbiol 82(18):5530. https://doi.org/10.1128/AEM.01327-16

Bradshaw Anthony D, Marrs RH, Roberts RD, Skeffington RA, Stewart William Duncan P, Rosswall T (1982) The creation of nitrogen cycles in derelict land. Philos Trans R Soc Lond B Biol Sci 296(1082):557–561. https://doi.org/10.1098/rstb.1982.0027

Bundy JG, Davey MP, Viant MR (2009) Environmental metabolomics: a critical review and future perspectives. Metabolomics 5(1):3

Cassán F, Perrig D, Sgroy V, Masciarelli O, Penna C, Luna V (2009) Azospirillum brasilense Az39 and Bradyrhizobium japonicum E109, inoculated singly or in combination, promote seed germination and early seedling growth in corn (Zea mays L.) and soybean (Glycine max L.). Eur J Soil Biol 45(1):28–35. https://doi.org/10.1016/j.ejsobi.2008.08.005

Cernava T, Erlacher A, Soh J, Sensen CW, Grube M, Berg G (2019) Enterobacteriaceae dominate the core microbiome and contribute to the resistome of arugula (Eruca sativa Mill.). Microbiome 7(1):13. https://doi.org/10.1186/s40168-019-0624-7

Charlop-Powers Z, Pregitzer CC, Lemetre C, Ternei MA, Maniko J, Hover BM et al (2016) Urban park soil microbiomes are a rich reservoir of natural product biosynthetic diversity. Proc Natl Acad Sci 113(51):14811. https://doi.org/10.1073/pnas.1615581113

David LA, Maurice CF, Carmody RN, Gootenberg DB, Button JE, Wolfe BE et al (2014) Diet rapidly and reproducibly alters the human gut microbiome. Nature 505(7484):559–563. https://doi.org/10.1038/nature12820

De Filippo C, Cavalieri D, Di Paola M, Ramazzotti M, Poullet JB, Massart S et al (2010) Impact of diet in shaping gut microbiota revealed by a comparative study in children from Europe and rural Africa. Proc Natl Acad Sci 107(33):14691. https://doi.org/10.1073/pnas.1005963107

Delmont TO, Prestat E, Keegan KP, Faubladier M, Robe P, Clark IM et al (2012) Structure, fluctuation and magnitude of a natural grassland soil metagenome. ISME J 6(9):1677–1687. https://doi.org/10.1038/ismej.2011.197

Ding T, Schloss PD (2014) Dynamics and associations of microbial community types across the human body. Nature 509:357. https://doi.org/10.1038/nature13178

Emerson JB, Roux S, Brum JR, Bolduc B, Woodcroft BJ, Jang HB et al (2018) Host-linked soil viral ecology along a permafrost thaw gradient. Nat Microbiol 3(8):870–880. https://doi.org/10.1038/s41564-018-0190-y

Fierer N (2017) Embracing the unknown: disentangling the complexities of the soil microbiome. Nat Rev Microbiol 15:579. https://doi.org/10.1038/nrmicro.2017.87

Fitzpatrick CR, Copeland J, Wang PW, Guttman DS, Kotanen PM, Johnson MTJ (2018) Assembly and ecological function of the root microbiome across angiosperm plant species. Proc Natl Acad Sci 115(6):E1157. https://doi.org/10.1073/pnas.1717617115

Flemming HC, Wuertz S (2019) Bacteria and archaea on earth and their abundance in biofilms. Nat Rev Microbiol 17(4):247–260. https://doi.org/10.1038/s41579-019-0158-9

Franzosa EA, Hsu T, Sirota-Madi A, Shafquat A, Abu-Ali G, Morgan XC, Huttenhower C (2015) Sequencing and beyond: integrating molecular omics for microbial community profiling. Nat Rev Microbiol 13:360. https://doi.org/10.1038/nrmicro3451

Gonze D, Lahti L, Raes J, Faust K (2017) Multi-stability and the origin of microbial community types. ISME J 11:2159. https://doi.org/10.1038/ismej.2017.60

Hartmann A, Schmid M, Tuinen D v, Berg G (2009) Plant-driven selection of microbes. Plant Soil 321(1):235–257. https://doi.org/10.1007/s11104-008-9814-y

Hjorth MF, Roager HM, Larsen TM, Poulsen SK, Licht TR, Bahl MI et al (2018) Pre-treatment microbial Prevotella-to-Bacteroides ratio, determines body fat loss success during a 6-month randomized controlled diet intervention. Int J Obes 42(3):580–583. https://doi.org/10.1038/ijo.2017.220

Howe AC, Jansson JK, Malfatti SA, Tringe SG, Tiedje JM, Brown CT (2014) Tackling soil diversity with the assembly of large, complex metagenomes. Proc Natl Acad Sci 111(13):4904. https://doi.org/10.1073/pnas.1402564111

Hu L, Robert CAM, Cadot S, Zhang X, Ye M, Li B et al (2018) Root exudate metabolites drive plant-soil feedbacks on growth and defense by shaping the rhizosphere microbiota. Nat Commun 9(1):2738. https://doi.org/10.1038/s41467-018-05122-7

Hultman J, Waldrop MP, Mackelprang R, David MM, McFarland J, Blazewicz SJ et al (2015) Multi-omics of permafrost, active layer and thermokarst bog soil microbiomes. Nature 521:208. https://doi.org/10.1038/nature14238

Keown H, Greenfield LC (2004) Decomposition of nucleic acids in soil. N Z Nat Sci 29:13

Knief C, Delmotte N, Chaffron S, Stark M, Innerebner G, Wassmann R et al (2011) Metaproteogenomic analysis of microbial communities in the phyllosphere and rhizosphere of rice. ISME J 6:1378. https://doi.org/10.1038/ismej.2011.192

Kolton M, Graber ER, Tsehansky L, Elad Y, Cytryn E (2017) Biochar-stimulated plant performance is strongly linked to microbial diversity and metabolic potential in the rhizosphere. New Phytol 213(3):1393–1404. https://doi.org/10.1111/nph.14253

Kyrpides NC, Eloe-Fadrosh EA, Ivanova NN (2016) Microbiome data science: understanding our microbial planet. Trends Microbiol 24(6):425–427. https://doi.org/10.1016/j.tim.2016.02.011

Lundberg DS, Lebeis SL, Paredes SH, Yourstone S, Gehring J, Malfatti S et al (2012) Defining the core Arabidopsis thaliana root microbiome. Nature 488:86. https://doi.org/10.1038/nature11237

McGuire KL, Payne SG, Palmer MI, Gillikin CM, Keefe D, Kim SJ et al (2013) Digging the New York City Skyline: soil fungal communities in green roofs and city parks. PLoS One 8(3): e58020. https://doi.org/10.1371/journal.pone.0058020

Mehari ZH, Elad Y, Rav-David D, Graber ER, Meller Harel Y (2015) Induced systemic resistance in tomato (Solanum lycopersicum) against Botrytis cinerea by biochar amendment involves jasmonic acid signaling. Plant Soil 395(1):31–44. https://doi.org/10.1007/s11104-015-2445-1

Niu B, Paulson JN, Zheng X, Kolter R (2017) Simplified and representative bacterial community of maize roots. Proc Natl Acad Sci 114(12):E2450–E2459. https://doi.org/10.1073/pnas.1616148114

Pavagadhi SS (2019) Urban farming: towards a leading role for Singapore in developing sustainable urban solutions. Adv Sci 24(1):4–5. http://www.science.nus.edu.sg/images/research/newsletter/AIS_June2019.pdf

Ramirez KS, Leff Jonathan W, Barberán A, Bates Scott T, Betley J, Crowther Thomas W et al (2014) Biogeographic patterns in below-ground diversity in New York City's Central Park are similar to those observed globally. *Proc Biol Sci 281*(1795):20141988. https://doi.org/10.1098/rspb.2014.1988

Ravel J, Gajer P, Abdo Z, Schneider GM, Koenig SSK, McCulle SL et al (2011) Vaginal microbiome of reproductive-age women. Proc Natl Acad Sci 108(Supplement 1):4680. https://doi.org/10.1073/pnas.1002611107

Schimel J, Schaeffer S (2012) Microbial control over carbon cycling in soil. Front Microbiol 3:348

Schneider T, Keiblinger KM, Schmid E, Sterflinger-Gleixner K, Ellersdorfer G, Roschitzki B et al (2012) Who is who in litter decomposition? Metaproteomics reveals major microbial players and their biogeochemical functions. ISME J 6:1749. https://doi.org/10.1038/ismej.2012.11

Schrader C, Schielke A, Ellerbroek L, Johne R (2012) PCR inhibitors – occurrence, properties and removal. J Appl Microbiol 113(5):1014–1026. https://doi.org/10.1111/j.1365-2672.2012.05384.x

Semchenko M, Leff JW, Lozano YM, Saar S, Davison J, Wilkinson A et al (2018) Fungal diversity regulates plant-soil feedbacks in temperate grassland. Sci Adv 4(11):eaau4578. https://doi.org/10.1126/sciadv.aau4578

Swenson TL, Karaoz U, Swenson JM, Bowen BP, Northen TR (2018) Linking soil biology and chemistry in biological soil crust using isolate exometabolomics. Nat Commun 9(1):19. https://doi.org/10.1038/s41467-017-02356-9

Thompson LR, Sanders JG, McDonald D, Amir A, Ladau J, Locey KJ et al (2017) A communal catalogue reveals Earth's multiscale microbial diversity. Nature 551:457. https://doi.org/10.1038/nature24621

Toju H, Peay KG, Yamamichi M, Narisawa K, Hiruma K, Naito K et al (2018) Core microbiomes for sustainable agroecosystems. Nat Plants 4(5):247–257. https://doi.org/10.1038/s41477-018-0139-4

United Nations (2014) World urbanization prospects

Van Der Heijden MGA, Bardgett RD, Van Straalen NM (2008) The unseen majority: soil microbes as drivers of plant diversity and productivity in terrestrial ecosystems. Ecol Lett 11(3):296–310. https://doi.org/10.1111/j.1461-0248.2007.01139.x

van der Putten WH, Bardgett RD, Bever JD, Bezemer TM, Casper BB, Fukami T et al (2013) Plant–soil feedbacks: the past, the present and future challenges. J Ecol 101(2):265–276. https://doi.org/10.1111/1365-2745.12054

Wall DH, Ritz K, Six J, Strong DR, van der Putten WH (2012) Soil ecology and ecosystem services. Oxford University Press, Oxford

Wilhelm RC, Singh R, Eltis LD, Mohn WW (2019) Bacterial contributions to delignification and lignocellulose degradation in forest soils with metagenomic and quantitative stable isotope probing. ISME J 13(2):413–429. https://doi.org/10.1038/s41396-018-0279-6

Wu GD, Chen J, Hoffmann C, Bittinger K, Chen Y-Y, Keilbaugh SA et al (2011) Linking long-term dietary patterns with gut microbial enterotypes. Science 334(6052):105. https://doi.org/10.1126/science.1208344

Xu J, Zhang Y, Zhang P, Trivedi P, Riera N, Wang Y et al (2018) The structure and function of the global citrus rhizosphere microbiome. Nat Commun 9(1):4894. https://doi.org/10.1038/s41467-018-07343-2

MIX
Papier aus verantwortungsvollen Quellen
Paper from responsible sources
FSC® C105338

If you have any concerns about our products,
you can contact us on
ProductSafety@springernature.com

In case Publisher is established outside the EU,
the EU authorized representative is:
Springer Nature Customer Service Center GmbH
Europaplatz 3, 69115 Heidelberg, Germany

Printed by Libri Plureos GmbH
in Hamburg, Germany